9th AIChE Southwest Process Technology Conference 2017

Galveston, Texas, USA
12-13 October 2017

ISBN: 978-1-5108-5674-5

9th AIChE Southwest Process Technology Conference 2017

Galveston, Texas, USA
12-13 October 2017

AIChE
The Global Home of Chemical Engineers

Printed from e-media with permission by:

Curran Associates, Inc.
57 Morehouse Lane
Red Hook, NY 12571

Printed by Curran Associates, Inc. (2018)

AIChE
120 Wall Street, FL 23
New York, NY 10005-4020

Phone: (800) 242-4363
Fax: (203) 775-5177

www.aiche.org

Additional copies of this publication are available from:

Curran Associates, Inc.
57 Morehouse Lane
Red Hook, NY 12571 USA
Phone: 845-758-0400
Fax: 845-758-2633
Email: curran@proceedings.com
Web: www.proceedings.com

TABLE OF CONTENTS

Chairman's Introduction

Conference: Southwest Process Technology Conference

Year: 2017

Proceeding: 9th AIChE Southwest Process Technology Conference

Group: Southwest Process Technology Conference

Session: Thursday Keynote Address

Time: Thursday, October 12, 2017 - 8:30am-8:45am

Part of the U.S. Department of Energyâs Office of Energy Efficiency and Renewable Energy, The Advanced Manufacturing Office supports R&D projects, R&D consortia, and early-stage technical partnerships with national laboratories, companies, non-profits, state and local governments, and universities to bring innovative manufacturing technologies and materials to full scale industrial use. Built upon a foundation of strong public-private partnerships, AMO is helping transition scientific innovations into new manufacturing capabilities for the security and well-being of our country and the world. AMO is also helping manufacturers reduce industrial energy consumption and drive the next generation of manufactured goods for the U.S. economy. For example, through direct investments in specific technology areas such as modular chemical process intensification and smart manufacturing, AMO assists U.S. industry in transforming traditional separations and reactions applications into modular intensified chemical processes that require less capital, operating costs and energy consumption and enables the integration of advance process control sensors to realize hierarchical levels of optimization from the processing cell to the global enterprise.

Coffee Break

Southwest Process Technology Conference

Conference: Southwest Process Technology Conference

Year: 2017

Proceeding: 9th AIChE Southwest Process Technology Conference

Group: Southwest Process Technology Conference

Session: Petroleum Refining Technology 1

Time: Friday, October 13, 2017 - 10:30am-11:00am

Propylene, originating from catalytic and thermal cracking processes like Fluid Catalytic Cracking, FCC and Delayed Coking, DCU, presents a processing challenge for petroleum refineries. Fortunately, there are many options for handling and processing refinery propylene. If the demand from the petrochemical market is strong, refinery grade propylene can be recovered in a refinery gas plant processing unsaturated or olefinic feed. A refinery can produce chemical or polymer grade propylene by adding additional treatment and distillation equipment. Derivatives like isopropyl benzene are produced in refineries that recover benzene as well as propylene. Refineries have stayed away from adding the capability for converting propylene to polypropylene.

Lacking the existence or proximity to strong propylene or derivative demand, refiners have traditionally used processes for reacting propylene to maximize production of gasoline blend stocks. These processes have historically included catalytic polymerization and acid catalyzed alkylation of isobutane. More recently, to meet environmental regulation restricting benzene in gasoline, some refiners have implemented technology to alkylate dilute benzene streams with propylene, resulting in a benzene regulation compliant, higher octane product.

This paper will define and compare key characteristics and considerations of refinery propylene processes for both fuels and petrochemicals: fractionation, solid catalyst polymerization, liquid acid catalyzed alkylation of isobutane and solid catalyst alkylation of concentrated and dilute benzene. Recent innovations in liquid acid catalyzed alkylation will also be included.

Coffee Break

Southwest Process Technology Conference

Conference: Southwest Process Technology Conference

Year: 2017

Proceeding: 9th AIChE Southwest Process Technology Conference

Group: Southwest Process Technology Conference

Session: Process Intensification

Time: Thursday, October 12, 2017 - 10:30am-11:00am

Amidst significant changes in the North American refining economic environment, octane and alkylate production have remained major profitability drivers for most refiners. Changing operating strategies and capitalizing on new catalyst technologies to maximize yields of high value products is critical to successfully navigate the current challenging environment.

In this paper, we present key aspects of the Albemarle ACTIONTM catalyst trial at Marathon Petroleum Corporationâs (MPC) Galveston Bay Refinery FCC3 unit. The main operating objective was to increase profitability by improving unit yield selectivities, especially LPG production and volume gain. Using a variety of tools, sources of data and analysis methods, including a calibrated FCC-SIM unit model, MPC and Albemarle were able to confirm that ACTION delivered the desired performance improvements. The observed in-unit yield selectivities (including coke) matched the laboratory predictions and increased unit profitability by more than $1.60/bbl.

Coffee Break

Southwest Process Technology Conference

Conference: Southwest Process Technology Conference

Year: 2017

Proceeding: 9th AIChE Southwest Process Technology Conference

Group: Southwest Process Technology Conference

Session: Process Safety

Time: Thursday, October 12, 2017 - 3:00pm-3:30pm

As construction costs have continued to rise over the last two decades, nitrogen-based fertilizer production has grown in scale in order to maintain economic feasibility. However, in some mid-continent locations in the US, fertilizer is often in demand in much smaller quantities in more localized markets. Increased transportation costs to these markets puts an economic strangle-hold on some agricultural and other smaller chemical users of ammonia (NH_{3}). Smaller âdistributedâ NH_3 plants will be required to reliably supply these markets; however, they suffer the challenge of economies of scale and high equipment counts. To achieve economic viability, the production methods for NH_3 must be simplified.

The Linde Ammonia Concept (LAC^{TM}) for the production of ammonia from natural gas or light hydrocarbons reduces process steps and equipment quantities to lower investment and operating costs. Plant start-up and operation are also simplified. LAC combines a high-purity hydrogen plant, a high-purity nitrogen plant, and an ammonia synthesis loop. Hydrogen is mixed with nitrogen from an air separation unit to give inert-free ammonia synthesis gas that is converted to NH_3 with minimum purge gas waste.

By simplifying the flow sheet, reducing equipment counts, and maximizing modularization of process units, small scale NH_3 economics can be highly improved and new projects can be competitive with sources currently trucked or railed in from coastal production.

Coffee Break

Southwest Process Technology Conference

Conference: Southwest Process Technology Conference

Year: 2017

Proceeding: 9th AIChE Southwest Process Technology Conference

Group: Southwest Process Technology Conference

Session: Chemical Process Technology 1

Time: Friday, October 13, 2017 - 10:00am-10:30am

Transport processes in packed reactors have transient behavior (e.g., adsorption, desorption, or Chemical-Looping) [1]. For reactors with a low tube-to-particle diameter ratio, particle resolved CFD simulations have been used extensively in the last years to accurately predict the steady-state conditions (see [2], [3] and [4]).

Simplified models rely on effective parameters like the effective thermal conductivity, the heat transfer coefficient, and the dispersion coefficient. A work flow based on particle resolved CFD and particle tracking is proposed to determine axial dispersion coefficients in fixed-bed reactors that can later be used in simplified models.

The effect of the bed morphology, Reynolds number, axial position, and tube-to-particle diameter ratio is thoroughly investigated.

References

[1] Adams et al. (2009) Int. J. Hydrogen Energy, 34, 8877-8891
[2] Wehinger et al. (2015) Chem. Eng. Sci., 122, 197-209
[3] Wehinger et al. (2016) AIChE J., 62(12), 4436-4452
[4] Wehinger et al. (2017) Ind. Eng. Chem. Res., 56(1), 87-99

Southwest Process Technology Conference
Year: 2017

Sessions:
12-Oct-17

Thursday Keynote Address

Chair(s):
Glasgow, I. M., IAG

Keynote Address

Papers:
8:30 AM
Chairman's Introduction
8:45 AM
Innovations in Manufacturing and Energy

Robert W. Ivester

Catalysis & Reaction Engineering

Chair(s):
Gildert, G., Vanguard Catalysts, LLC.
Co-chair(s):
Camero, A., Shell Global Solutions (US), Inc.

This session will have papers related to catalysis and reaction based engineering.

Papers:
9:30 AM
Sustaining Economic Competitiveness for Industrial Catalyzed Processes

Steve King

10:00 AM
Explosive Rate Increases Using Shaped Charges

Adrienne Van Kooperen

10:30 AM
Coffee Break
11:00 AM
A New Era for Alkylation

Elizabeth Carter

11:30 AM
Axial Dispersion in Fixed-Bed Reactors from Numerical Analysis

Nico Jurtz
Philipp Waldherr
Thomas Eppinger
Ravindra Aglave
Matthias Kraume
Chandra Tourani

Energy Efficiency

Chair(s):
Roberto, F., ExxonMobil Chemical Company
Co-chair(s):
Torres, V., University of Texas

This session will include presentations based around energy efficiency.

Papers:
9:30 AM
Integration of Divided Wall Columns in a Shale Oil Processing Unit

Sundar Narayanan
Kirtan K. Trivedi
Phillip Melancon

9:55 AM
Thermal Oxidizer Optimization Through Model Predictive Control

Paul Pathasema

10:20 AM
Coffee Break
10:45 AM
Smart Manufacturing Technologies and Data Analytics for Improving Energy Efficiency in Industrial Systems

Joe Cresko
Sachin Nimbalkar

11:10 AM

Improved Cogeneration Energy Efficiency through Gas Turbine Auto-Tuners and Optimizers

Heidi Holmes

Process Intensification

Chair(s):
Bielenberg, J., RAPID
Co-chair(s):
Johnson, A., SABIC

This session offers an opportunity to share process intensification success stories. Topics of high interest by past presenters include APC, process control strategy enhancements, effective alarm management, and wireless applications. Topics that are relatively new in the industry and should also be considered when submitting an abstract include cybersecurity, big data, Industrial Internet of Things, and predictive analytics. If you have a process intensification success story that would be interesting for others to learn about, please submit an abstract for approval.

Papers:
9:30 AM
Intensified Conversion of Natural Gas Using Proton-Conducting Ceramics

Christian Kjølseth

10:00 AM
Reverse Flow Reactors: Process Intensification for High Temperature Reactions

Dhaval Bhandari

10:30 AM
Coffee Break
11:00 AM
Intensified Applications in Centrifugation

Timothy Pomaville
Gonzalo Paz

11:30 AM
Simplifying NH3 Production for Improved Economic Feasibility

Rameshwar Hiwale
Carrie C. Lalou

Distillation & Separation

Chair(s):
Cai, T., Fractionation Research, Inc.
Co-chair(s):
Bravo, J., Fractionation Research, Inc.
Aejaz, A., Burns & McDonnell

This session will present papers on distillations or separations.

Papers:
1:30 PM
Shell ConSep™ Trays Improve Diesel Recovery and Naphtha Quality in a DHT Stripper

Gregory A. Cantley
Waldo E. De Villiers
Eric Vos
Grzegorz Kamyszek

2:00 PM
Troubleshooting Tower Carryover

Juan C. Juarez
Ed Grave
Norman K. Yeh

2:30 PM
Application of Air/Water Derived Mass Transfer Models to Distillation

Frank Seibert

3:00 PM
Coffee Break
3:30 PM
Crude Unit Stripping Sections

Andrew W. Sloley

4:00 PM
Dealing with severe service applications in refineries

Izak Nieuwoudt

Process Safety

Chair(s):
Chosnek, J., KnowledgeOne LLC

Co-chair(s):
Abrahamson, D., Abrahamson Consulting LLC
Gottimukkala, C., RISKbytes, Inc.

This session will be focused on process safety.

Papers:
1:30 PM
Lessons Learned from a Crude Oil Train Derailment

Monir Ahammad

2:00 PM
Leveraging Learnings from Incidents and Near Misses Using Process Safety Networks

John W. Champion
Sheila Van Geffen

2:30 PM
Developing Competency – a Key Ingredient in Process Safety

Ellen Ross

3:00 PM
Coffee Break
3:30 PM
Walk the Line: An Industry Model of Practice Sharing to Improve Process Safety in Industry

Jerry J. Forest

4:00 PM
The Next Generation Process Hazard Analysis

David Lewis

Thermodynamics & Process Simulation

Chair(s):
Vedapuri, D., Tridiagonal Solutions Inc.
Co-chair(s):
DeJonge, P., Honeywell Process Solutions
Rice, T., Chemstations

This session will be delivering presentations discussing thermodynamics and process simulation.

Papers:
1:30 PM
Relief System Design Breakthrough Enhances Steady-State Rigor; Seamlessly Integrates Dynamics; Makes Best-Practice Safety and Operability Analysis Cost-Efficient

Dewayne Anderson

2:00 PM
The Effect of Relative Volatility, Enthalpy, Density and Surface Tension on Distillation Column Design

Robert Wenta
Hazem Haddad

2:30 PM
Trends in Industry 4.0 and IIOT and their Impact on Process Simulation

Raul Ohaco
Paul Bonner

3:00 PM
Coffee Break
3:30 PM
Performance Enhancement & Profitability Improvements of Refinery Operations using CFD

Praveen Kumar
Damodaran Vedapuri

4:00 PM
Next Generation of Thermodynamic Models and their Impact on Process Modeling and Simulation

Dimitrios Georgis
Sathish Natarajan
Vasileios Papaioannou

Professional Ethics Workshop

Chair(s):
Turner, J., Fluor Corporation

Professional Ethics Workshop

Papers:

4:45 PM
Texas Board of Professional Engineers – Professional Practice Update / Ethics

George P. Hartmann

13-Oct-17

Chemical Process Technology 1

Chair(s):
Brigman, J., Ingenero Inc.
Co-chair(s):
Chen, I., Dow Chemical Company

This session will be presenting on chemical process technology.

Papers:
8:30 AM
Vega
9:00 AM
Chemical Process Technology: How We Got Here

Raghu S. Narayan

9:30 AM
Panel Discussion: Chemical Process Technology: Changes Expected

Raghu S. Narayan
Bill Waycaster
Gregory E. Yeo
James Bielenberg

10:00 AM
Coffee Break

Student Program & Career Fair

Chair(s):
Benson, T. J., Lamar University
Co-chair(s):
Turner, J., Fluor Corporation
Schmidt, C., Waldemar S. Nelson

This session will deliver a student program, as well as a Career Fair.

Papers:

9:30 AM
Panel Discussion: "A Day in the Life of a Chemical Engineer"
10:30 AM
Effective Interviewing

Friday Keynote Address

Chair(s):
Chin, K., AIChE

Keynote Address

Papers:
11:15 AM
Big Data Analytics Skills – A Revolution Lacking in Revolutionaries

Lloyd Colegrove

Chemical Process Technology 2

Chair(s):
Chen, I., Dow Chemical Company
Co-chair(s):
Brigman, J., Ingenero Inc.

This session will be a continuation of presentations on chemical process technology.

Papers:
1:30 PM
Building a Big Data Platform for the Process Industry

Tor Jakob Ramsøy

2:00 PM
Requirements for Successful Implementation of Big Data Analytics in the Process Industry

Pratap Nair

2:30 PM
Advances in Big Data Analytics at The Dow Chemical Company

Leo H. Chiang

Petroleum Refining Technology 2

Co-chair(s):
Lockhart, M. W., Burns & McDonnell
Amstutz, T.

This session will be a continuation of presentations discussing petroleum refining technology.

Papers:
1:30 PM
An Action Plan to Improve FCC Unit Performance at the Marathon Galveston Bay Refinery

M. Skurka
D. Zalewski
Nicholas Newlon
N. Larsen
B. Dinkel
J. Becerra
George Yaluris
Eric Rutan

2:00 PM
Desalting Fundamentals: A Process Engineer's Perspective

Andrew W. Sloley

2:30 PM
On-line Leak Detection Methods for Heat Exchangers and Valves

William Mixon

Topics:
Refining

Meet The Industry Poster Reception

Chair(s):
Benson, T. J., Lamar University
Co-chair(s):
Glasgow, I. M., IAG
Mathews, J., KBR
Schmidt, C., Waldemar S. Nelson
Chin, K., AIChE

To better serve the universities in the region, the conference is adding a graduate student "Meet the Industry" reception that will wrap-up the conference.

Papers:
3:00 PM
A New Experimental Method for a Fast and Reliable Determination of Saturates, Aromatics, Resins and Asphaltenes, SARA, in Crude Oils

Sara Rezaee
Rocio H. Doherty
Francisco Vargas

3:00 PM
A New Proactive Scheduling Methodology for Front-end Crude Oil and Refinery Operations under Uncertainty of Shipping Delay

Jialin Xu
Qiang Xu

3:00 PM
Ag-Promoted Dehydroaromatization of Ethylene over ZSM-5 Catalysts

Yunwen Zhou
Ming-Feng Hsieh
Hari Thirumalai
Lars C. Grabow
Jeffrey Rimer

3:00 PM
Anchor-Tenant Models for the Synthesis of Eco-Industrial Parks through Carbon-Hydrogen-Oxygen Symbiosis Networks

Kevin Topolski
Mohamed Noureldin
Fadwa T. Eljack
Mahmoud El-Halwagi

3:00 PM
Chemical Kinetics Mechanism Modeling of Sour Gas Combustion

Huilong Gai

3:00 PM
CO and Hydrocarbon Oxidation on Pd/Ceria-Zirconia/Alumina Three-way Catalysts

Wendy Lang
Carolyn Hubbard
Paul Laing

Yisun Cheng
Michael Harold

3:00 PM
Density Functional Theory (DFT) Based Machine & Deep Learning Strategies for Rational Screening and Directed Design of Catalytic Promoter Materials and Supported Nanoparticle Interfaces for Renewable Fuel Applications

Sashank Kasiraju
Lars C. Grabow

3:00 PM
Deposition Behavior of the Asphaltene-Stabilized Emulsion and its Interfacial Rheology at Oil-Brine Interface

Yu-Jiun Lin
Peng He
Zhuqing Zhang
Steve Kuei
Sibani Lisa Biswal

3:00 PM
Design of Diesel Oxidation Catalysts with Improved Activity at Reduced Temperature Using Computational Methods

Yuying Song
Lars C. Grabow

3:00 PM
Development and Testing of Amine-Type Scavengers for Elimination of H2S from Liquid Sour Crudes

Karishma Piler

3:00 PM
Modeling Alkane Partitioning and Phase Behavior in Non-Permeable and Permeable Slit Pores

Jinlu Liu
Walter G. Chapman

3:00 PM
Operating Procedure Analysis with Consequence Prediction by Dynamic Simulation

Yan Fang
Peyton C. Richmond

3:00 PM
Optimization of Flare Operation

Anan Wang

3:00 PM
Optimized Design of Shale Gas Processing and NGL Recovery Plant under Uncertain Feed Conditions

Rekha Reddy Asani

3:00 PM
Optimizing ZSM-11 Catalysts for Methanol-to-Hydrocarbon Reactions

Yufeng Shen
Thuy T. Le
Jeffrey Rimer

3:00 PM
Palladium Dendron Encapsulated Nanoparticles Supported on MCM-41 and SBA-15

Yueyun Lou
Daniel F. Shantz

3:00 PM
Preparation of PVMS-b-PMMA Block Copolymer with Potential for Oil-Water Separation

Baraka S Lwoya
Md Farkar Uddin
Sourav Chattejree
Julie Albert

3:00 PM
Process Intensification Using Reactive Distillation: Conversion of Crude Glycerol to Propanediol

Obakore Agbroko
Tracy J. Benson

3:00 PM
Rational Design of Block-Copolymer Additives for Polymer/Small-molecule Emulsions using Experimental and Computational Approaches

Jorge Mok

Rafael Verduzco

3:00 PM
Shale Gas Monetization: Designing Separation Units for Feedstock Variability

Eric Bohac

3:00 PM
Smart Manufacturing in Distillation

Damian Efetobore

3:00 PM
Synthesis and Characterization of Tin, Tin-Aluminum, and Tin-Boron Containing MFI Zeolites

Meysam Shahami
Ross Ransom
Daniel F. Shantz

3:00 PM
Techno-Economic Analysis of Alternative Pathways of Isopropanol Production

Warissara Panjapakkul

3:00 PM
Thermodynamic Interactions in Model Polydiene/Polyolefin Blends

Jialin Qiu
Katrina Irene S. Mongcopa
Ruixuan Han
Carlos R. López-Barrón
Megan Robertson
Ramanan Krishnamoorti

3:00 PM
Tight Oil Drilling Problems and Solutions

Lisandro Fernando

3:00 PM
Time-resolved in situ Studies of Zeolite Crystal Growth

Madhuresh K. Choudhary
Manjesh Kumar
Jeffrey Rimer

3:00 PM
Understanding Pyrolysis Chemistry to Improve Biomass-to-Chemical Process

Li Chen

3:00 PM
Viscosity Modeling of Live Oil and Viscosity Prediction of Gas Injection Blends

Jieyi Zhang
Mohammed I.L. Abutaqiya
Francisco Vargas

Topics:
Students

Innovation in Manufacturing and Energy

9th AIChE Southwest Process Technology Conference
Galveston, TX

October 12th, 2017

Robert W. Ivester, PhD
Deputy Director
Advanced Manufacturing Office
www.manufacturing.energy.gov

Energy Domination: Oil and Gas Today

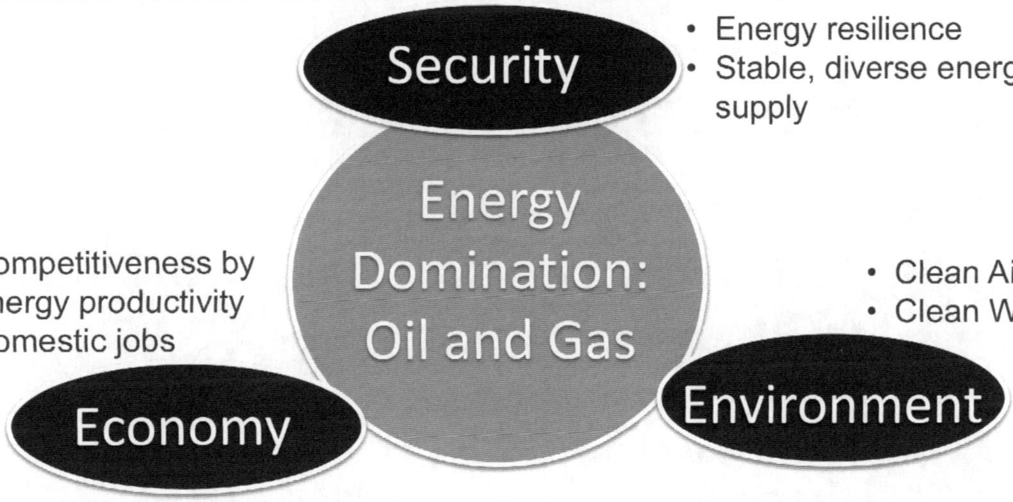

- Energy resilience
- Stable, diverse energy supply

- Competitiveness by energy productivity
- Domestic jobs

- Clean Air
- Clean Water

- Energy domination is a foundation for economic growth & jobs
- Today's low prices present opportunities to improve and innovate

U.S. DEPARTMENT OF ENERGY | Energy Efficiency & Renewable Energy

Energy Use in the US Economy

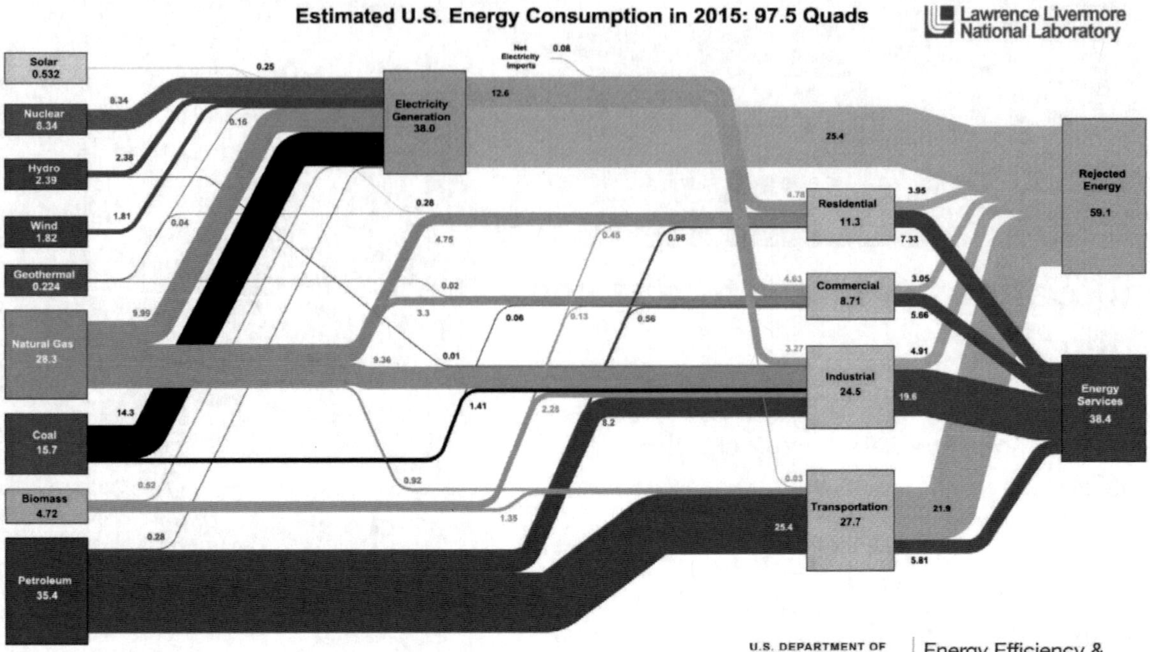

U.S. DEPARTMENT OF ENERGY | Energy Efficiency & Renewable Energy

Oil and Gas Dominance Depends on Clean Water

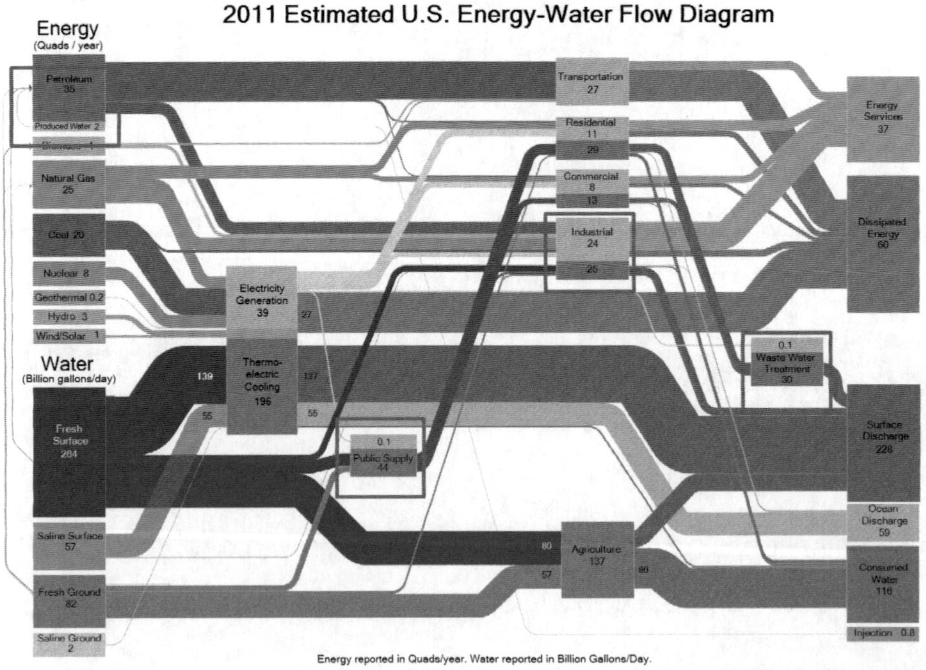

10/11/2017

U.S. DEPARTMENT OF ENERGY | Energy Efficiency & Renewable Energy

Energy Dominance = Manufacturing Dominance

Technology Innovation through Early Stage R&D in Advanced Manufacturing and Energy is a Foundation for Economic Growth and Jobs in the US

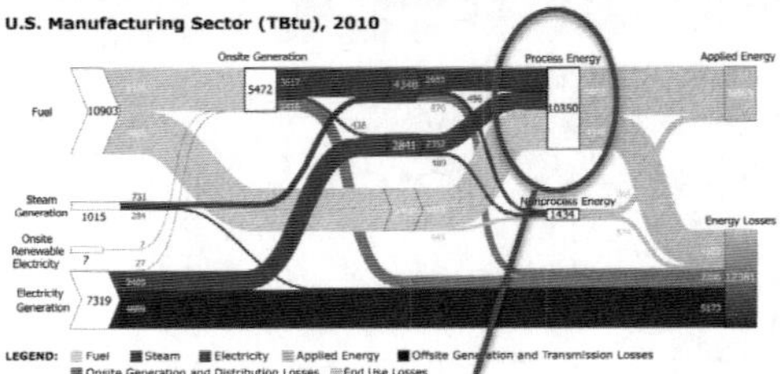

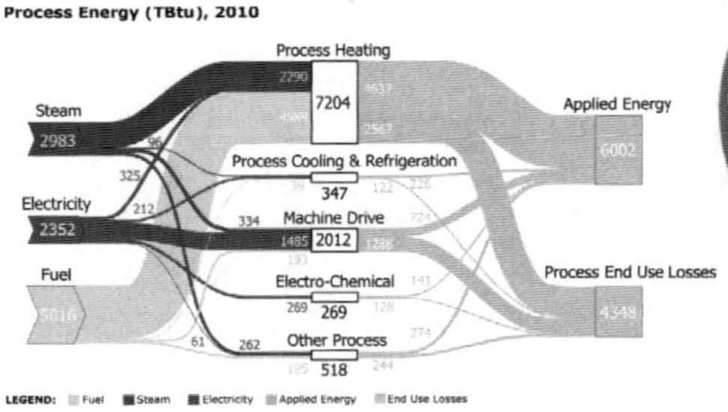

$2T Manufacturing GDP
12.4M Manufacturing Direct Employment Jobs
0.8 / 1.0 – Indirect / Direct Jobs - All Manufacturing
2.2 / 1.0 – Indirect / Direct Jobs - Advanced Sub-Sectors

24 QBTU (25% of National Total) – Manufacturing
2/3 Manufacturing Energy is in Intensive Sectors

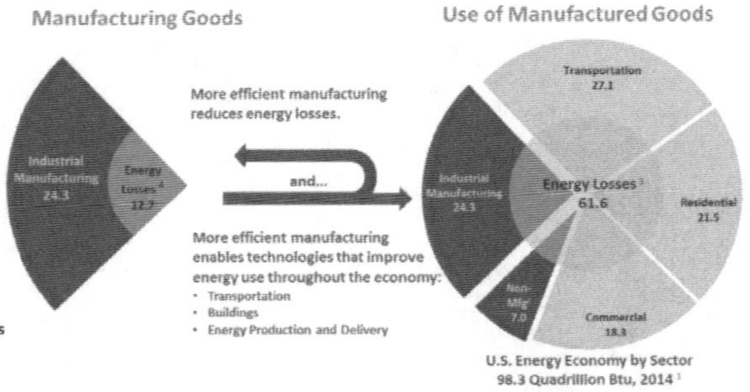

QTR, 2015

U.S. DEPARTMENT OF ENERGY | Energy Efficiency & Renewable Energy

5

Focus on manufacturing-specific R&D challenges

U.S. DEPARTMENT OF ENERGY | Office of Science
(Basic Science)

U.S. DEPARTMENT OF ENERGY | Energy Efficiency & Renewable Energy
(Applied Science)

- Only office suited to manufacturing-specific R&D challenges.
- Funded approximately $250M in R&D in FY17.
- Majority of work is through partnerships.

Solar Technologies	**Geothermal Energy**
Wind Energy	**Buildings Technologies**
Water Technologies	**Bioenergy**
Advanced Manufacturing	**Vehicles Technologies**

U.S. DEPARTMENT OF ENERGY | Energy Efficiency & Renewable Energy

6

Manufacturing Bandwidth Studies: Energy Savings Potential

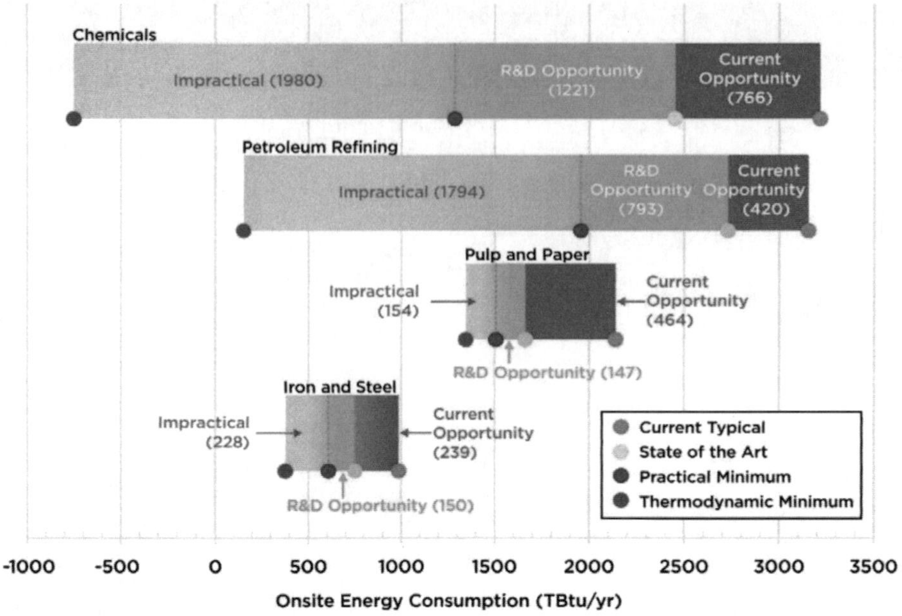

Current opportunities represent energy savings that could be achieved by deploying the most energy-efficient commercial technologies available worldwide. R&D opportunities represent potential savings that could be attained through successful deployment of applied R&D technologies under development worldwide

U.S. DEPARTMENT OF ENERGY | Energy Efficiency & Renewable Energy

7 **AMO: September 2015**

Energy Intensive Industries -Today

Primary Metals
1608 TBTU

Petroleum Refining
6137 TBTU

Chemicals
4995 TBTU

Wood Pulp & Paper
2109 TBTU

Glass & Cement
716 TBTU

Food Processing
1162 TBTU

Other Manufacturing
~1600 TBTU

U.S. DEPARTMENT OF ENERGY | Energy Efficiency & Renewable Energy

8

Third Industrial Revolution - Today

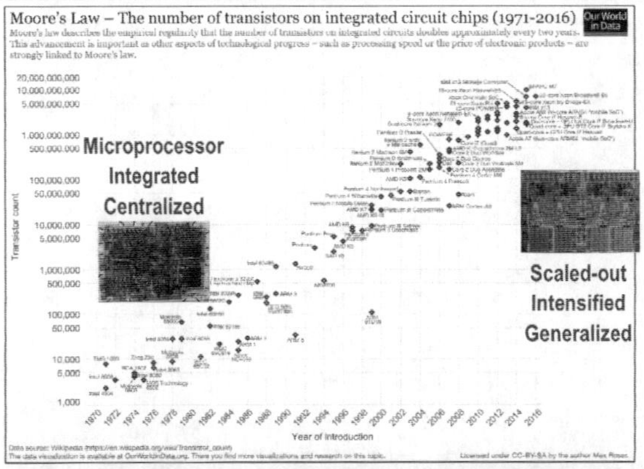

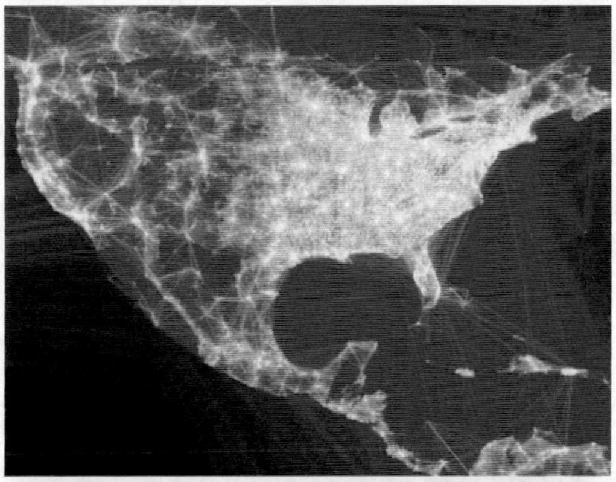

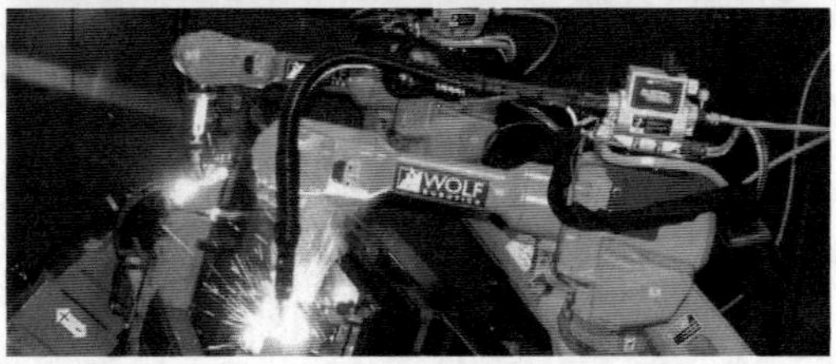

How will Manufacturing, Economy and Security of the Nation depend on Information, Computation, Actuation and Communication Technologies in the 21st Century?

U.S. DEPARTMENT OF ENERGY | Energy Efficiency & Renewable Energy

9

Processes for Clean Energy Materials & Technologies

Energy Dependence: Energy Cost Considered in Competitive Manufacturing

Solar PV Cell

Carbon Fibers

Light Emitting Diodes

Electro-Chromic Coatings

Membranes

EV Batteries

Multi-Material Joining

Water Desalination

U.S. DEPARTMENT OF ENERGY | Energy Efficiency & Renewable Energy

10

Advanced Manufacturing Office Framework

AMO Technical Focus Areas (2017 MYPP / DRAFT)

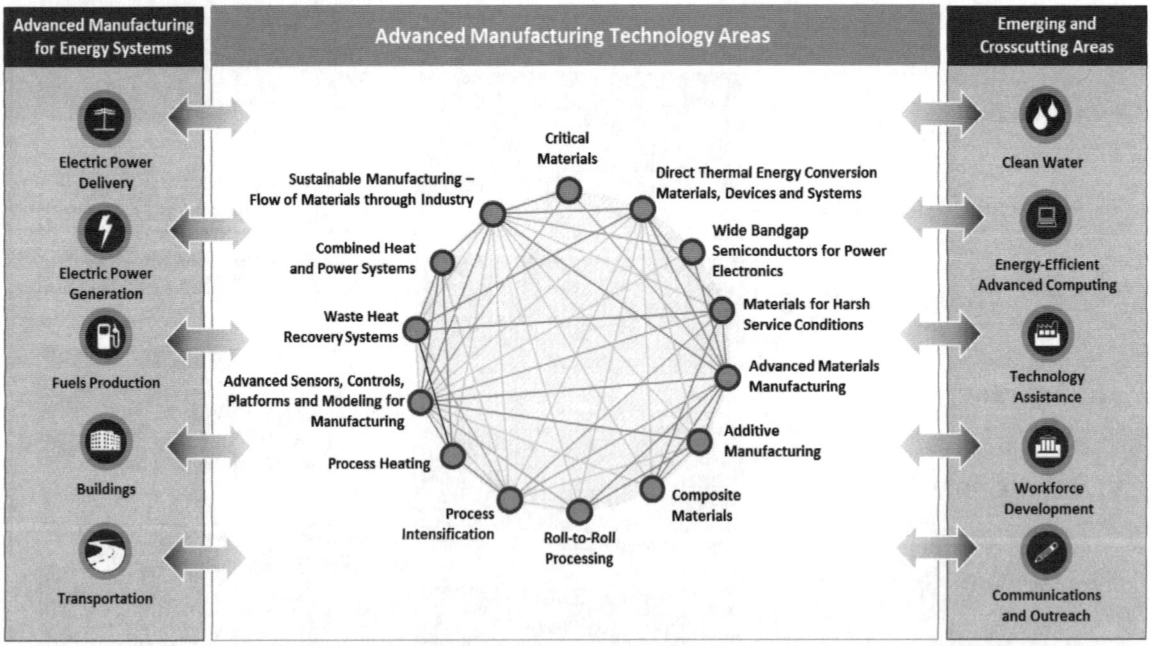

Three Pillars: Partnerships, Projects, Collaboration

- Technology Partnership Programs
- Research and Development Projects
- Research and Development Consortia
- Other Initiatives like High Performance Computing for Manufacturing (HPC4Mfg)

13

U.S. DEPARTMENT OF ENERGY | Energy Efficiency & Renewable Energy

Technology Partnership Programs

14

U.S. DEPARTMENT OF ENERGY | Energy Efficiency & Renewable Energy

Technology Partnerships Programs

Efficient On-Site Energy
CHP Technical Assistance Partnerships

Better Plants
U.S. DEPARTMENT OF ENERGY

SAINT-GOBAIN
Simplot
Bringing Earth's Resources to Life
NISSAN
legrand®
ALCOA

Energy-Saving Partnership
Better Buildings, Better Plants,
Industrial Strategic Energy Management

3M
Schneider Electric
BRIGGS & STRATTON
Cummins
Johnson Controls

Student Training & Energy Assessments
University-based Industrial Assessment Centers

U.S. DEPARTMENT OF ENERGY | Energy Efficiency & Renewable Energy

1

R&D Projects

U.S. DEPARTMENT OF ENERGY | Energy Efficiency & Renewable Energy

16

Ex: Conversion of Waste CO_2 and Shale Gas to High Value Chemicals

Project Prime: Novomer, Inc.
Project Partners: Praxair, Inc.

Technology: Develop lab-scale process that converts waste CO_2 from industrial sources and ethane-derivatives from shale gas into high value chemical intermediates (e.g., acrylic acid).

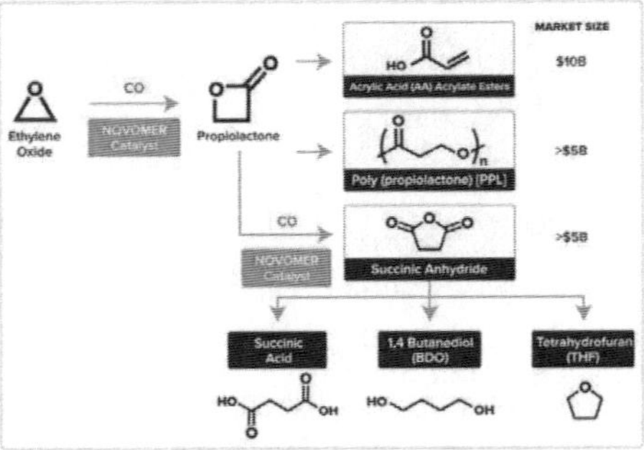

Schematic illustrating the Novomer semi-integrated process to produce high value chemical intermediates (top right) and the automated continuous system (bottom). Novomer built the automated system using private funds to allow the technology to scale more quickly. Graphic image courtesy of Novomer.

Technology Update

- Semi-integrated pilot plant scale carbonylation process has been developed and is running continuously.
- Catalyst is 99% selective and can be recovered in a continuous membrane separation process.
- SOE single cells have undergone long term cell testing.
- Novomer ended the project early in 2016 to advance the technology forward more quickly.

U.S. DEPARTMENT OF ENERGY | Energy Efficiency & Renewable Energy

Ex: New Design Methods and Algorithms for Energy Efficient Multicomponent Distillation Column Trains

Project Prime: Purdue University
Project Partners: Eastman Chemical

Technology: Develop state-of-the-art optimization algorithm to apply low-energy distillation processes that allow chemical manufacturers to reduce energy consumption between 10% and 50% at their plants.

Technology Update

- Identified new method for more efficient, faster calculations of heat duty requirement for multicomponent distillation configurations.
- Designed new dividing wall columns/distillation columns for 2 proprietary at Eastman Chemical.
- Signed agreement with ExxonMobil to use algorithms for real industrial applications at their refineries.
- Approached by SABIC regrading purchasing the software as soon as it is ready for commercialization.

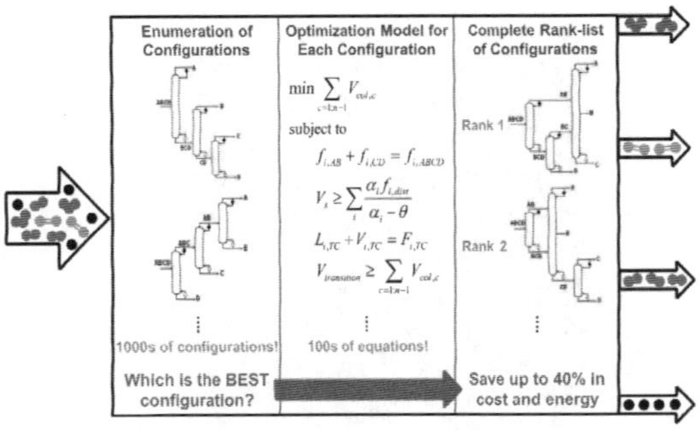

A graphic summarizing the process to calculate the globally optimal solutions for multi-component mixtures based on heat duty and total cost.
Graphic image courtesy of Purdue University

U.S. DEPARTMENT OF ENERGY | Energy Efficiency & Renewable Energy

Ex: Development & Implementation of an Automatic Continuous Online Monitoring and Control Platform for Polymerization Reactions

Project Prime: Tulane University
Project Partners: Advanced Polymer Monitoring Technologies, Inc. (APMT); Louisiana State University

Technology: Continuous, online monitoring of polymerization reactions using advanced sensor technology.

Technology Update:

- Successfully demonstrated what is believed to be the world's first manual active control of polymerization reaction at lab-scale.
- Pilot reactor designed, built, and installed ahead of schedule.
- Tulane and APMT filed a joint, non-provisional patent application based on the ACOMP/CI platform.

Academic researchers will develop the basic control and automation technology using the laboratory scale Automatic Continuous Online Monitoring of Polymerization reactions/ Control Interface (ACOMP/CI) system (left). The technology will be validated using the pilot scale ACOMP/CI (right). *Photos courtesy of Tulane University*

19

U.S. DEPARTMENT OF ENERGY | Energy Efficiency & Renewable Energy

Ex: One Step Hydrogen Generation Through Sorption Enhanced Reforming

Project Prime: Gas Technology Institute (novated from Aerojet Rocketdyne of DE, Inc.)
Project Partners: Energy & Environmental Research Center

Technology: Develop one-step hydrogen generation through a sorption enhanced reforming (SER) process that provides better energy productivity, environmental performance, product yield, and economic benefit compared to steam methane reforming.

The feasibility demonstration unit (FDU) that was recommissioned for this project. *Photo courtesy of Aerojet Rocketdyne, Inc.*

Technology Update:

- Candidate substrate materials were selected for catalyst testing.
- Test unit generating hydrogen at >80% purity.
- Demonstrated SER chemistry and process operation:
  - Over 28 hours of SER-mode operation with catalyst without degradation.
  - Achieved operation at design point (3:1 S/C ratio and 20,000 SCFD).
  - 3-fold increase in sorbent feed rate.

20

U.S. DEPARTMENT OF ENERGY | Energy Efficiency & Renewable Energy

Ex: Low-Energy, Low-Cost Production of Ethylene by Low-Temperature Oxidative Coupling of Methane

Project Prime: Siluria Technologies
Project Partners: RTI International

Technology: Develop and demonstrate a transformational catalyst technology that will be used to produce ethylene from methane.

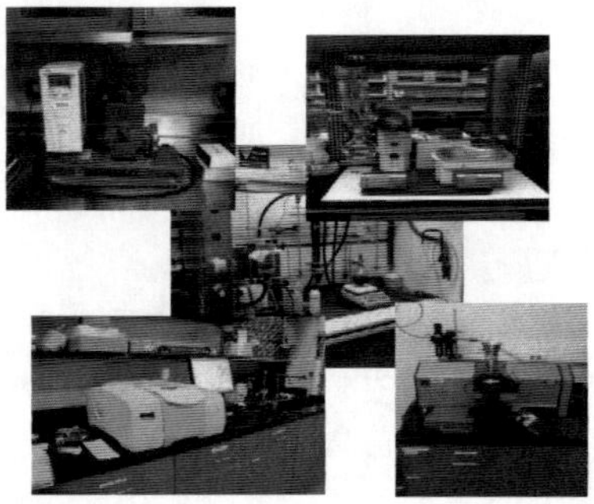

Technology Update:

- Identified novel adsorbents that can treat multicomponent stream of the OCM effluent.
- Developed process concepts that can operate at moderate pressure and temperature conditions, thus enhancing scalability.
- Developed simulation models to predict behavior of such complex multicomponent adsorbent based system.
- Fine tuning of shortlisted adsorbents to identify technically and economically most effective material(s).
- Determining the most optimum process concept(s).

21

U.S. DEPARTMENT OF ENERGY | Energy Efficiency & Renewable Energy

R&D Consortia

22

U.S. DEPARTMENT OF ENERGY | Energy Efficiency & Renewable Energy

R&D Facilities & Consortia

Address market disaggregation challenge to the industrial commons

Ford River Rouge Complex, 1920s

Photo: Library of Congress, Prints & Photographs Division, Detroit Publishing Company Collection, det 4a25915.

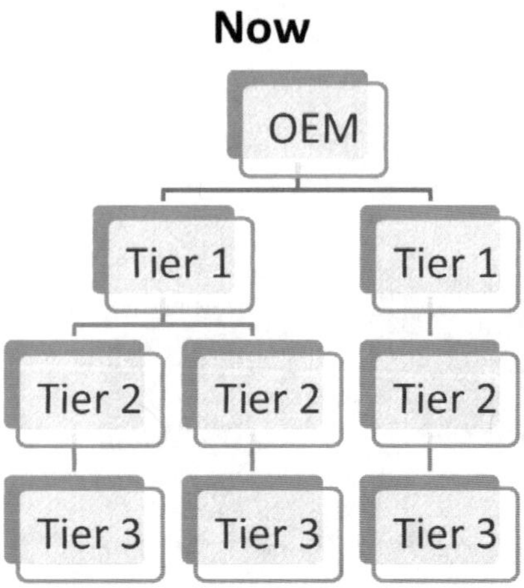

How could we get innovation into manufacturing today?

- **RD&D Consortia**
- **Workforce Development and Education**
- **Public-private Partnership to Scale**

U.S. DEPARTMENT OF ENERGY | Energy Efficiency & Renewable Energy

23

Consortia Model

<u>AMO Consortia:</u>

- Critical and Rare Earth materials
- Wide band gap semiconductors
- Carbon fiber composites
- Smart Manufacturing
- Process Intensification
- Remanufacturing and Reprocessing
- (Soon) Clean Water Production

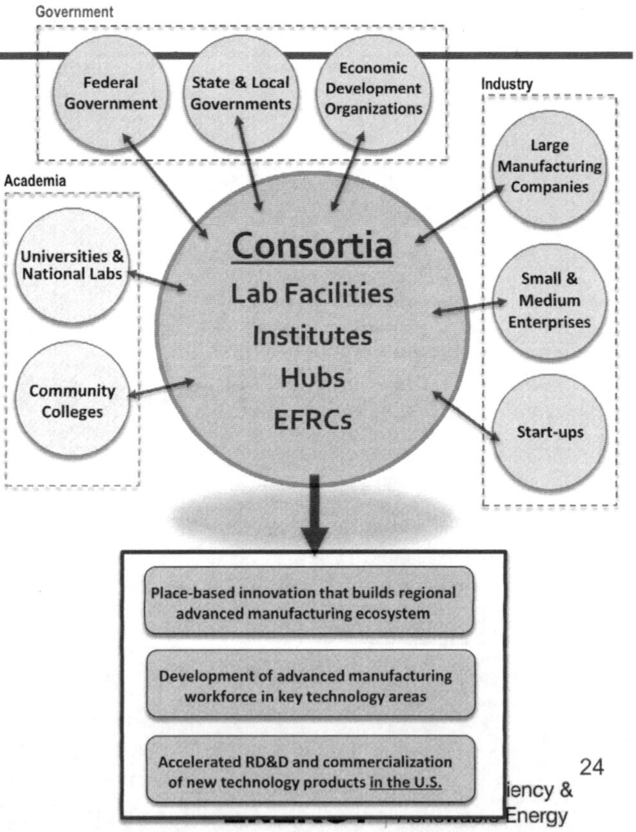

24

iency & Energy

24

DOE NNMI Institute: Modular Chemical Process Intensification

Objective: Develop a set of technologies that bring significant reduction in equipment size, process complexity, cost or risk reduction that will result in…

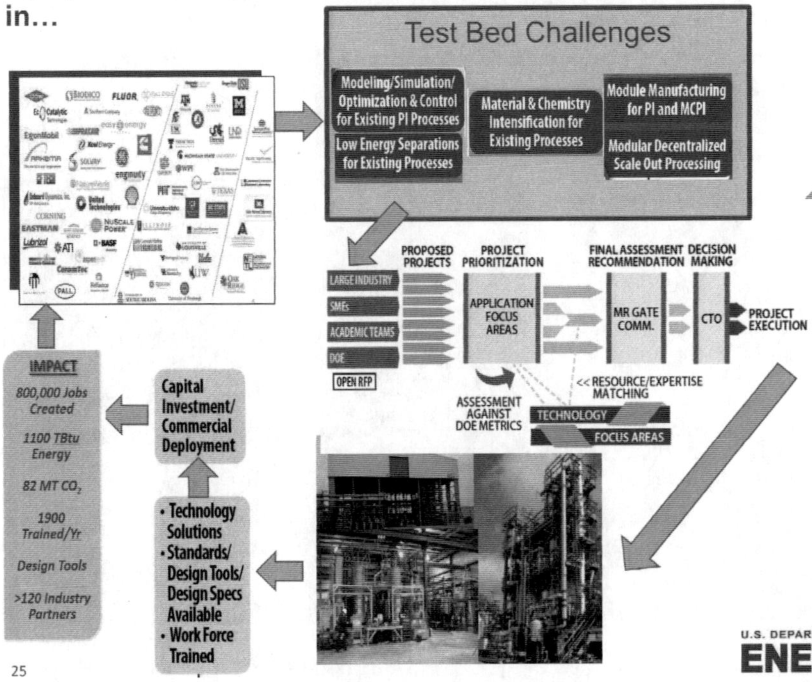

- 2x Improved energy efficiency - 20% Improved energy productivity

- 50% Decreased deployment cost

25

U.S. DEPARTMENT OF **ENERGY** | Energy Efficiency & Renewable Energy

DOE NNMI Institute #4 – RAPID (New York, NY)

$70M Federal + $70M Non-Federal Public Private Partnership

- 2x Improved energy efficiency - 20% Improved energy productivity

- 50% Decreased deployment cost

For more information about Membership, R&D Roadmaps, and project funding, go to: www.aiche.org/rapid

26

Manufacturing Demonstration Facility: National Lab Consortia

Supercomputing Capabilities

Spallation Neutron Source

MDF Manufacturing Demonstration Facility

America Makes

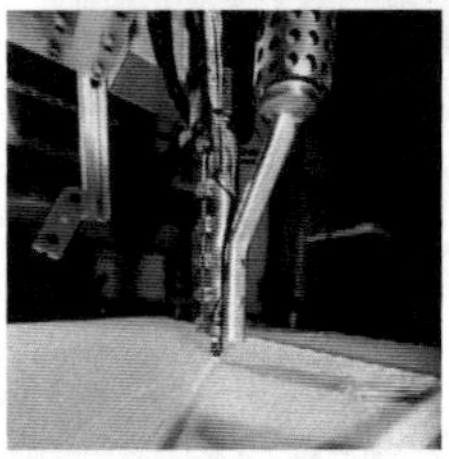

Additive Manufacturing

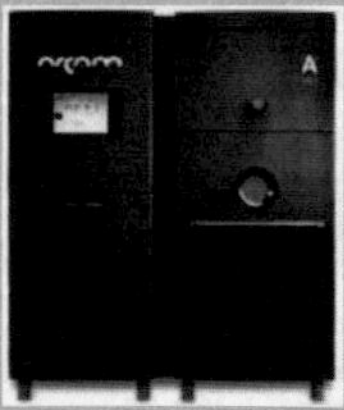

Arcam electron beam processing AM equipment

POM laser processing AM equipment

Research in partnerships at MDF can provide validation and feedback to further research in AM technologies utilizing various materials from metals to polymers to composites.

Other: Clean Water Production

Water Stress in the U.S.

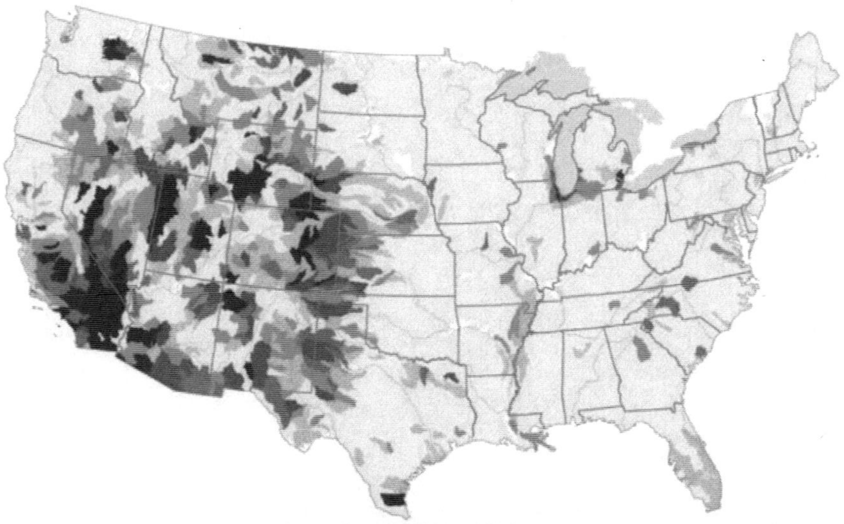

U.S. DEPARTMENT OF ENERGY | Energy Efficiency & Renewable Energy

30

What is 'Pipe Parity' for Clean Water

- Deliver Water with equivalent Economic & Energy cost
  - Price: Approximate \$0.50 / m3 (tonne)
    - Ranges from \$0.10 to \$1.00 nationally
  - Energy: Approximate: 1kWh / m3 (tonne)
    - 0.65 kWh (corresponding to 235m elevation change)
  - Environment: Approximate: 1lb / m3 (tonne)
    - Based on 0.69kg CO_2/kWh
  - Quality: 500 ppm TDS
  - Complimentary Cases: Desalination, Produced Water, Grey Water, etc.

U.S. DEPARTMENT OF ENERGY | Energy Efficiency & Renewable Energy

31

Framework Cost for Desalination in Clean Water

Goal = $0.50/m3

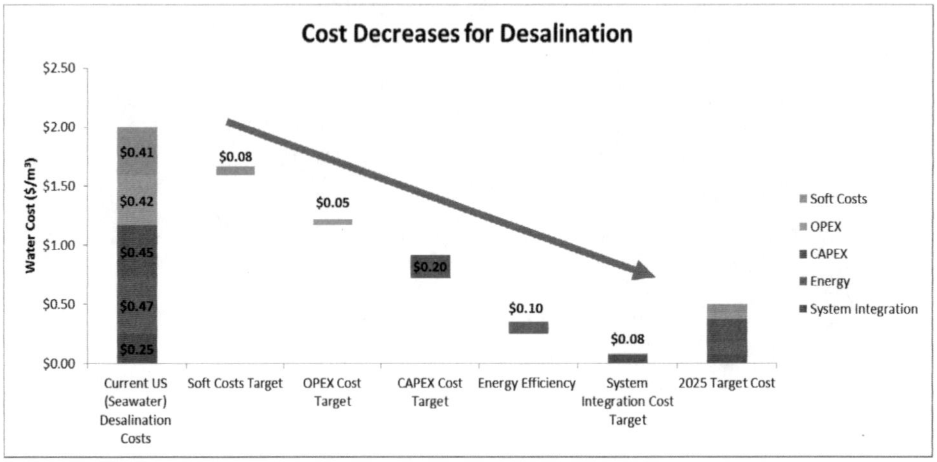

What are the technology R&D pathways that get us there?

U.S. DEPARTMENT OF ENERGY | Energy Efficiency & Renewable Energy

32

Some Possible Areas for Opportunity

- Operating Costs: Chemical additives (anti-bacterial, longer lasting membranes), Disposal / Post-processing of saline brines
- Capital Costs: Low-cost heat exchangers for thermal processes, Cost Effective membranes, Balance of Plant Equipment, Small Modular System Footprint
- Energy: Improve pressure energy recovery, utilize low-cost thermal energy
- System Integration: Intelligent design of water networks to minimize connection costs, Real-time Control and Sensor Systems
- Soft Costs: Workforce, Supply Chain, Expertise and Environmental Considerations

U.S. DEPARTMENT OF ENERGY | Energy Efficiency & Renewable Energy

33

Where are the possible gaps?

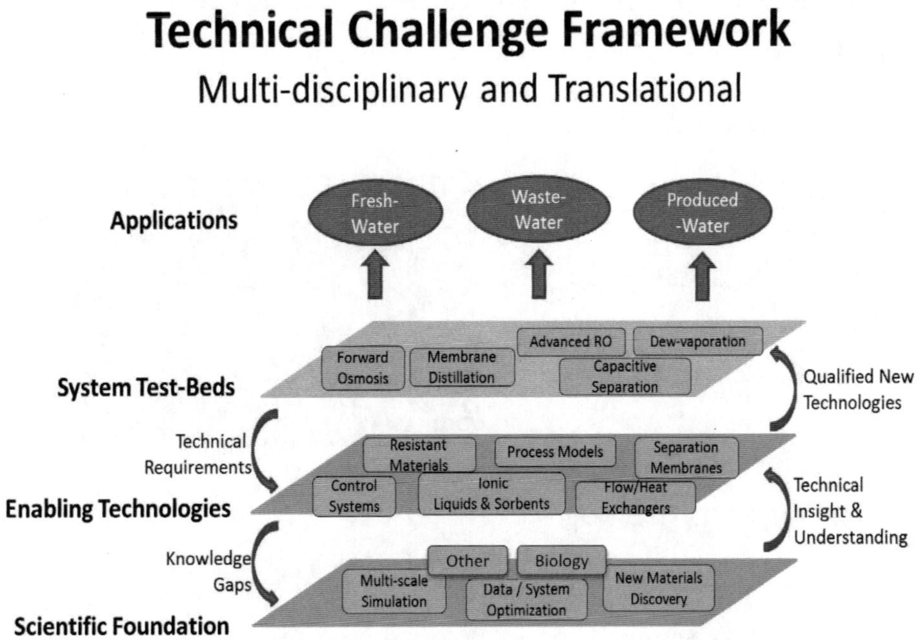

Other: High Performance Computing for Manufacturing (HPC4Mfg)

HPC4Mfg program

Brings the many benefits of high-performance computing to US Industry

- Accelerate innovation
- Lower energy costs
- Reduce testing cycles
- Reduce waste/reduce rejected parts
- Quality processes and Pre-qualify
- Optimize design
- Shorten the time to market

U.S. DEPARTMENT OF ENERGY | Energy Efficiency & Renewable Energy

36

Lowers barriers for industry use of world-class assets

- **Industry Status**:
  - Some larger companies use HPC, but struggle to stay current
  - Few small to medium companies use HPC
- **DOE Status**:
  - DOE labs possess 5 of the top 10 HPC systems worldwide; 2 of top 4 in Graph500
  - Broad expertise in the application HPC
  - Can be a challenge for industry to understand the best way to partner with DOE

HPC4Mfg creates partnerships that leverage DOE lab expertise and compute resources to address critical problems in the manufacturing sector

37

The HPC4Mfg program has a diverse portfolio

- $>13M technical portfolio
  - Executing on 41 projects with 33 industry partners and 6 labs
  - 40 demonstration; 1 capability
  - Seedling project funded by Office of Transportation
- Spring 2017 Solicitation
  - $3M available
  - New area of materials in severe environments
  - Other DOE offices informally involved
  - 57 Concept papers submitted; 26 moving forward to full proposals

8RIVERS
EATON Powering Business Worldwide
ACTASYS
PPG
ARCONIC
ACS Chemistry for Life
7AC
Harper International
SIERRA ENERGY
United Technologies Research Center
EWI We Manufacture Innovation
SHILOH
Ford
SORAA
TIMKEN
P&G
AGENDA 2020 TECHNOLOGY ALLIANCE
PURDUE UNIVERSITY CALUMET
LanzaTech
Carbon
SEPION TECHNOLOGIES
ROLLS ROYCE
ZoomEssence CREATING THE NEXT WAVE OF INNOVATION
GE
SAMSUNG SAMSUNG SEMICONDUCTOR USA
APPLIED MATERIALS make possible
ALZETA CORPORATION
lift
GM
U.S. DEPARTMENT OF ENERGY | Energy Efficiency & Renewable Energy

38

Extending scientific knowledge in different industrial sectors

HPC4 MANUFACTURING

Creating new lightweight alloys

Goal: Predict the strength of lightweight aluminum-lithium alloys produced under different process conditions; could save millions of fuel costs if used in aircraft design

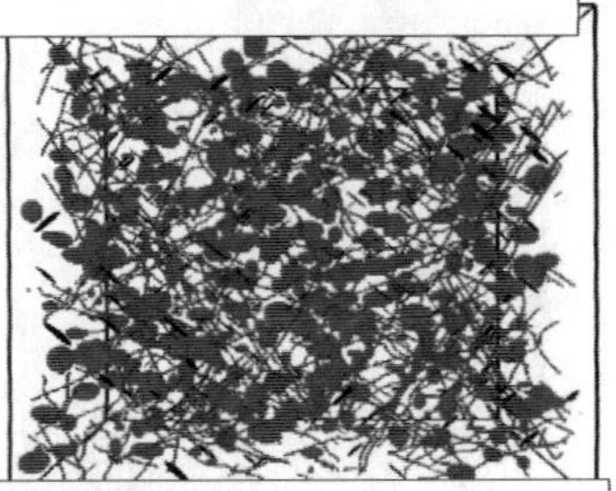

Results to date: Developed new dislocation mobility laws for Al-Li alloys; examining influence of different precipitate density; predicting yield strength for differing particle sizes

Team: LIFT with LLNL and Univ Mich.

Dendritic Growth in AM Parts

Goal: Use HPC to model multi-scale morphology of solidification microstructure of Nickel base 718

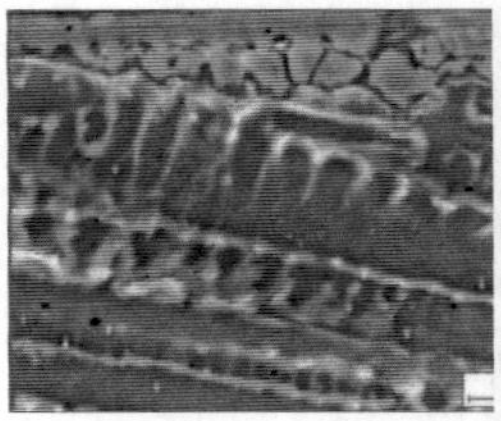

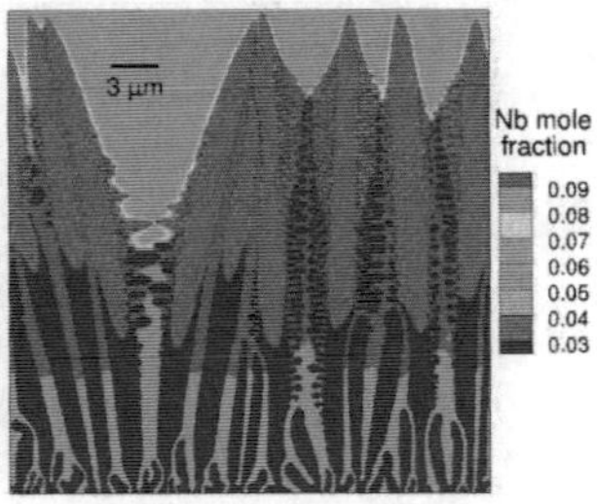

Results to date: Predicting crystal growth over large domains from multi-component alloys using phase field approaches; moving to new alloy systems and 3D

Team: UTRC with ORNL/LLNL

39

Improving industrial workflows and speeding up modeling time using HPC

HPC4 MANUFACTURING

Weld Predictor Tool

Goal: Develop an improved online welding software modeling application using advanced 3D models, more material hardening laws, and open source parallel codes

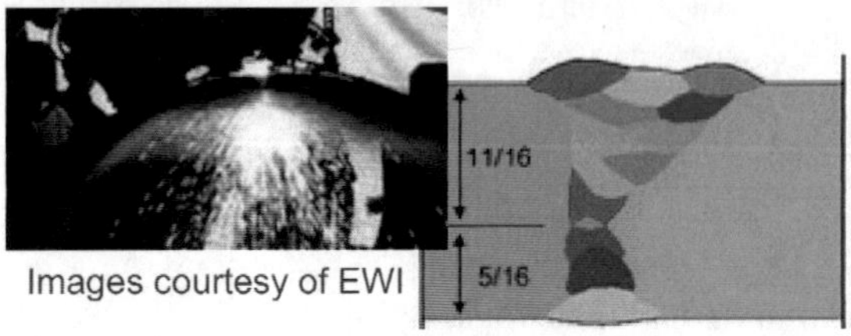

Images courtesy of EWI

Results to date: Developed new front end interface and automated meshing tools; working on new parallel simulation tools for thermal analysis, microstructure prediction, and mechanical analysis

Team: Edison Welding Institute with ORNL/OSU

Paper Towel Design

Goal: Use HPC to evaluate different microfiber configurations to optimize drying time while maintaining user experience

Results to date: New mesh tool reduces product design cycle by 2X cycle; additional cores by another 8X; largest non benchmark run of Paradyn code at LLNL

Team: Procter and Gamble with LLNL

40

Renewable Energy

Improving manufacturing operation in energy–intensive environments

HPC4 MANUFACTURING

Glass Making Operation

Goal: Optimize process operation and production of glass making using HPC; save 2 weeks of production per year per furnace; save 2.5 TBTUs of energy and avoid 130,000 metric tones on carbon dioxide emissions

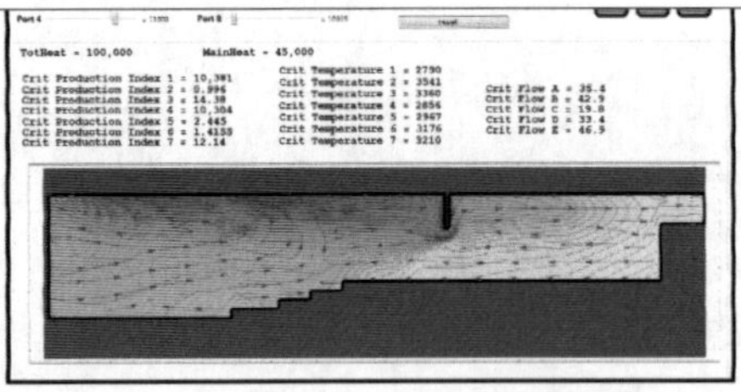

Results to date: Fully validated fluid model used for a parameterization study; statistical analysis and machine learning to create a reduced order glass furnace model for real time process adjustments

Team: Vitro Glass with LLNL

Blast Furnace Operation

Goal: Optimize blast furnace operation to reduce coal coke usage, energy consumption

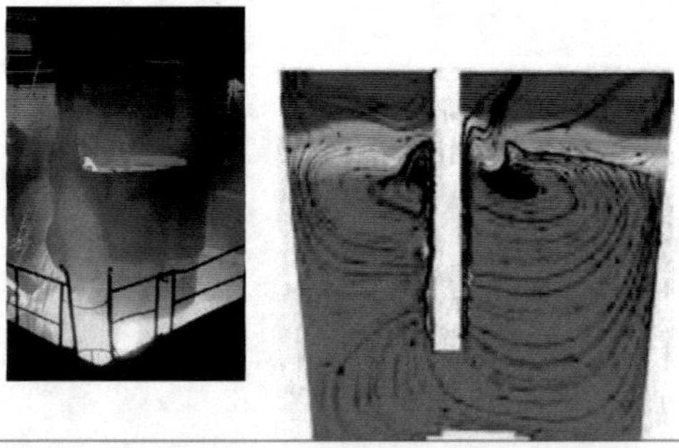

Results to date: Parallelized blast furnace code is 100X faster than previous code; studies conducted to evaluate coke usage as a function of operational parameters

Team: US Steel with Purdue and LLNL

41

Renewable Energy

Leading to significant energy savings in new products or processes

HPC4 MANUFACTURING

More efficient LED lightbulb

Goal: Model ammono-thermal crystal growth of GaN to scale up the process; reduce production costs of LED lighting by 20%

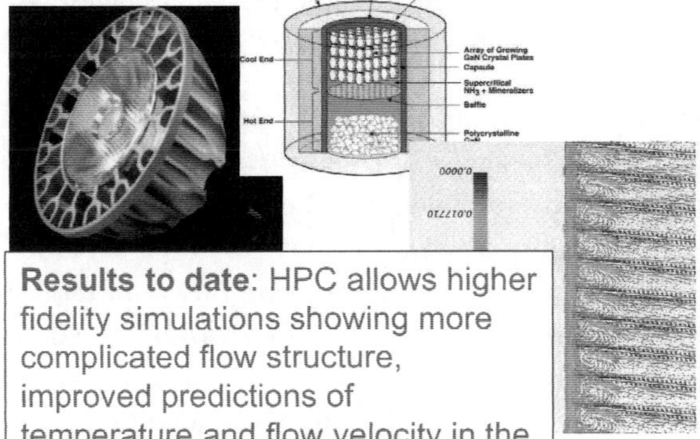

Results to date: HPC allows higher fidelity simulations showing more complicated flow structure, improved predictions of temperature and flow velocity in the reactor; now optimizing uniform growth of crystals

Team: SORAA with LLNL

Energy savings in paper making

Goal: Use multi-physics models to reduce paper rewetting in the pressing process; reduce 3rd most intense energy consumer in paper making; save 80 trillion BTU's per year

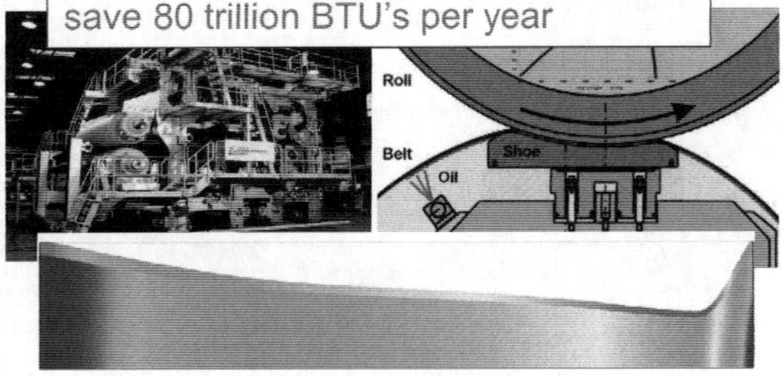

Results to date: Using both continuum and pore-scale approaches to determine how water flows through porous paper pulp; simulations can be used to optimize drying

Team: Agenda2020 with LBNL/LLNL

Thank You

A New Era for Alkylation

Southwest Process Technology Conference

Conference: Southwest Process Technology Conference

Year: 2017

Proceeding: 9th AIChE Southwest Process Technology Conference

Group: Southwest Process Technology Conference

Session: Catalysis & Reaction Engineering

Time: Thursday, October 12, 2017 - 11:00am-11:30am

Authors:

Carter, E., Honeywell UOP

In September 2016, Honeywell UOP introduced the refining industryâs first new liquid alkylation technology in 75 years. ISOALKY^{TM} technology offers refiners the ability to produce alkylate using a specialized ionic liquid catalyst with lower handling risks and comparable process and economic performance to traditional hydrofluoric or sulfuric acid technologies.

Researchers have assumed ionic liquids could be useful in industrial applications, but so far there has been limited success harnessing their potential as catalysts on a commercial scale. The ISOALKY process solution is successful on a commercial scale because the catalyst, a chloroaluminate ionic liquid with a proprietary composition, has finely tuned acidity to avoid undesirable cracking, isomerization, and heavy by-product formation, as well as properties that allow low inventory, low consumption, efficient separation, and easy on-site regeneration.

Chevron Corp., one of the worldâs leading integrated energy companies and Honeywell UOPâs alliance partner for ISOALKY technology, is converting the existing hydrofluoric acid alkylation unit at its Salt Lake City refinery into the first alkylation unit in the US based on ionic liquid technology. ISOALKY technology takes ionic liquids a step closer to becoming mainstream in industrial applications, and can have a significant effect on global production of clean fuels using a process with lower environmental impact.

Explosive Rate Increases Using Shaped Charges

AIChE Southwest Process Technology Conference 2017

Dr. Adrienne Van Kooperen

Technology Advances in Hydroprocessing

- Challenges of Increasing Hydrocracker Capacity
- Catalyst Technology Solutions
- Science Behind Shape & Size
  - Engineering Advanced Trilobe eXtra (ATX) Shaped Charges
- Case Study 1: Demonstrating Pressure Drop Reduction with ATX
- Case Study 2: Leveraging ATX Lower Pressure Drop to Maximize Throughput

Key Drivers for Debottlenecking Hydrocrackers

- Huge incentive for adding incremental processing capacity of heavy hydrocracker feeds especially in current market

- IMO restrictions on Bunker Fuel in 2020 create greater driver to upgrade additional heavy BBLs to low sulfur, lighter products

Challenge: maximum throughput within hydraulic limits often determined by reactor pressure drop

Grading Catalysts – the First Barrier to ΔP Build

- Tailored by size, shape, & activity to provide deep-bed filtration
- Design uses higher voidage, larger size materials followed by layers of decreasing size & shape to capture inert solids or highly reactive species
- Applies multiple materials with varying void fractions, structures, & shapes with varying degrees of activity

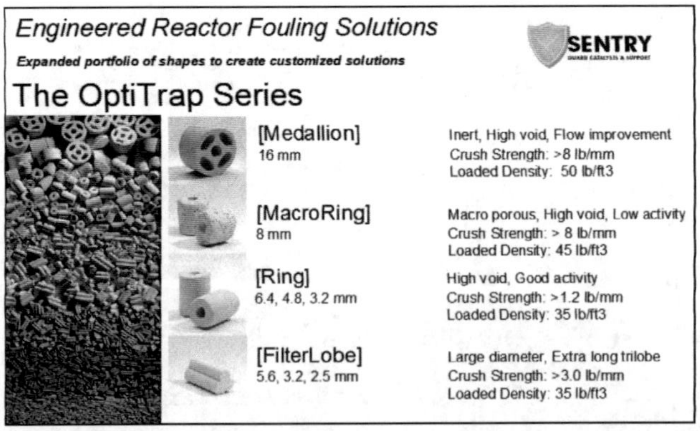

- Medallions
- Macroporous rings
- Specialty extrudates

Conventional Main Catalyst Shapes

- Historically main bed hydroprocessing catalysts consisted of cylindrical shapes which offer improved performance over spheres
- American Cyanamid Compony pioneered shaped-catalyst production in the early 1970's with Trilobe (TL) catalysts which remain the primary shape offered for the majority of hydroprocessing catalyst today

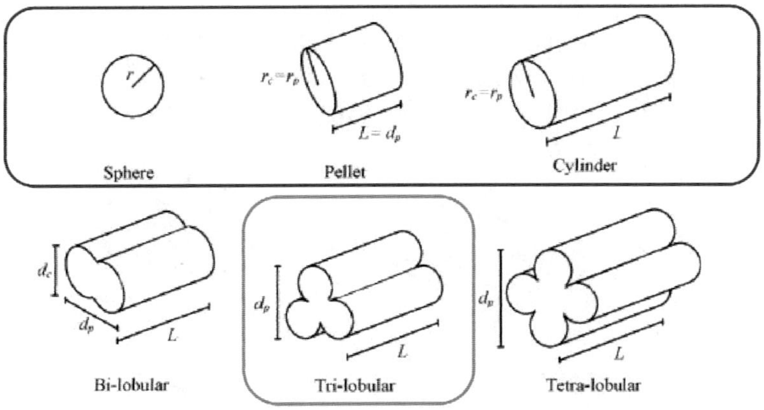

- Shape charges provide higher activity vs. cylinders due to enhanced utilization & improved ΔP reduction in packed beds → key to overall unit performance!

Advanced Trilobe eXtra (ATX) Catalyst Shape

- Criterion developed an advanced catalyst (ATX) in 2006
  - to reduce ΔP & improve catalyst activity
  - to meet the growing demand to operate existing hydrocrackers at increasingly higher capacities without additional capital investment or loss of performance

ATX Shape

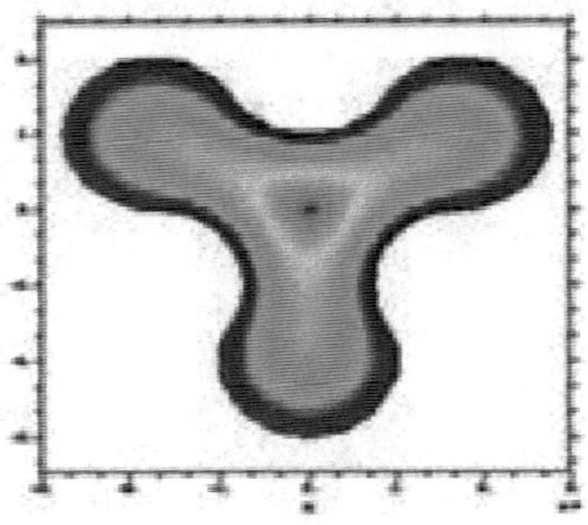

Multi-Lobed Catalyst Shapes

- Multi-lobed catalysts offer significant advantages due to the higher particle surface area (S_p)/ volume (V_p) ratios vs. spheres & cylinders
  - Increased S_p results in enhanced activity due to greater access to active sites on the catalyst surface & within the pore structure

- Shaped extrudates provide greater access for reactants to contact active sites in the pore structure close to the center by reducing diffusion limitations → improves selectivity by mitigating over-cracking to light ends

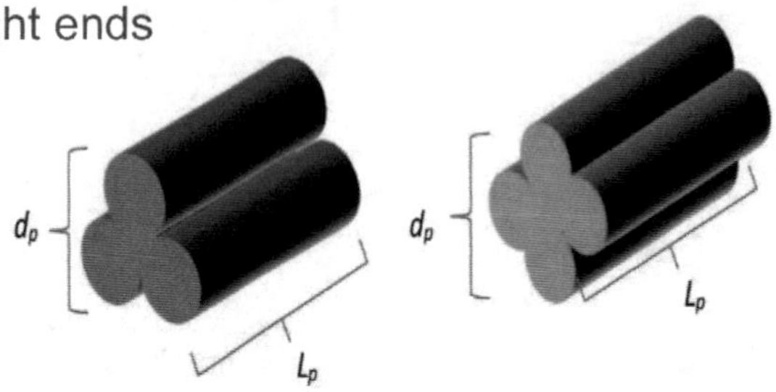

- Increased number of lobes ~ better utilization ~ higher activity / V_p

Catalyst Effectiveness Factor

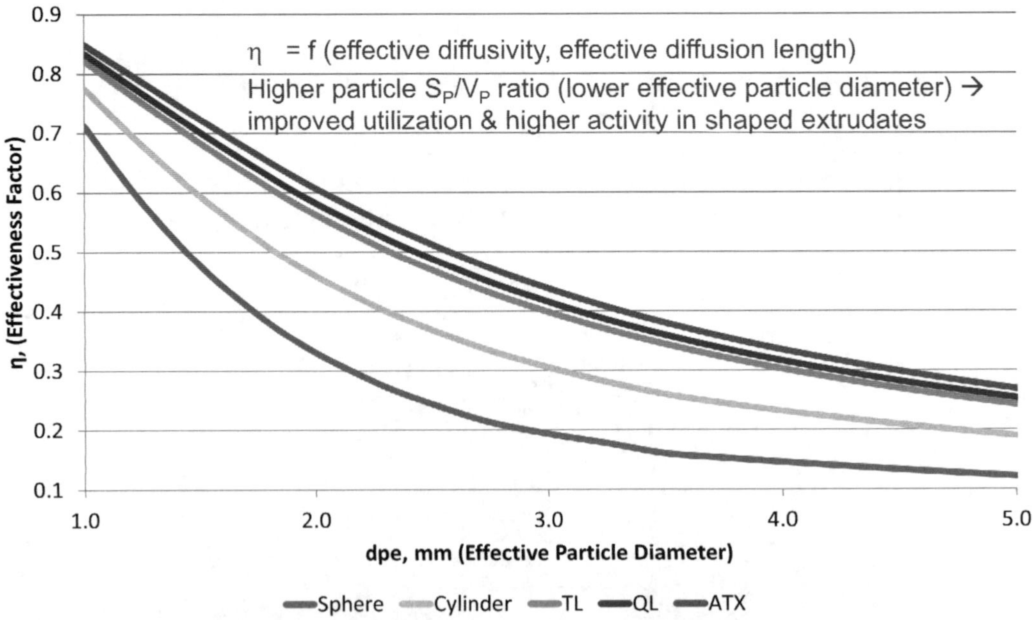

Shaped extrudates demonstrate smaller d_{pe} relative to actual physical dimensions which is a critical variable for ΔP across packed beds

Advanced Trilobe eXtra (ATX) Shape

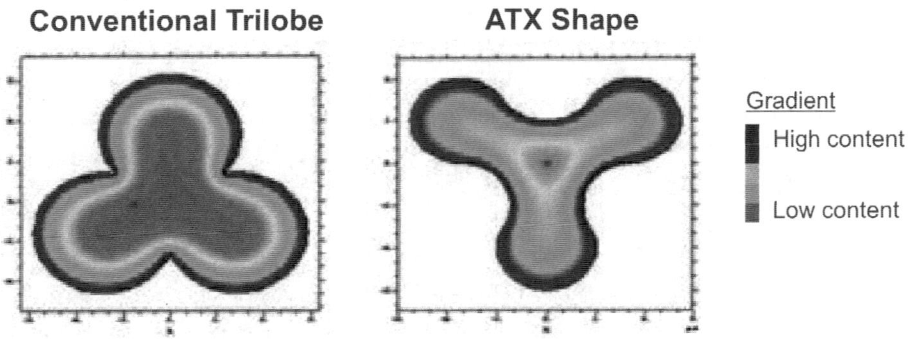

- Larger void fraction of ATX shape → higher particulate uptake & delayed onset of rapid ΔP build
- ATX commercially shows 15 to 20% lower SOR ΔP vs. conventional TL

Pressure Drop Relate to Particle Shape

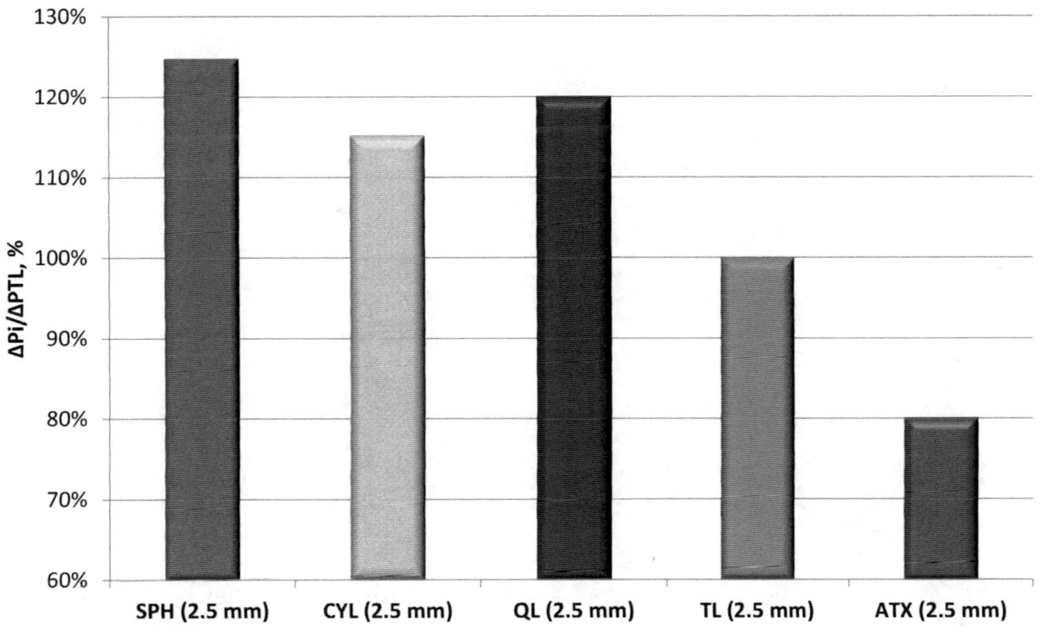

Advanced Trilobe eXtra (ATX) Shape

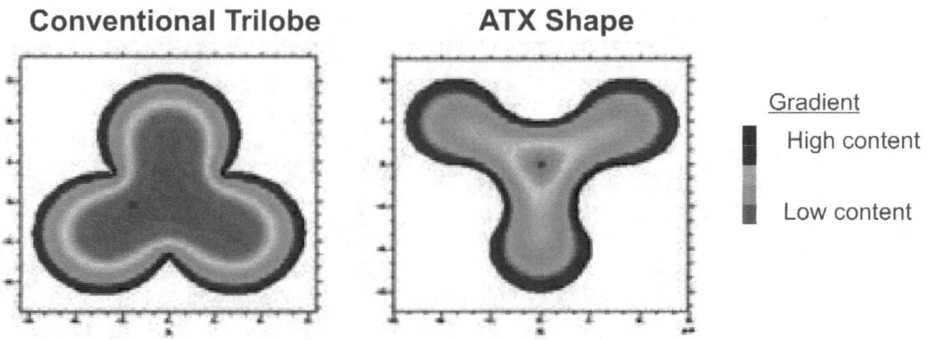

- Larger void fraction → higher particulate uptake & delayed onset of rapid ΔP build
- ATX commercially shows 15 to 20% lower SOR ΔP vs. conventional TL
- Improved diffusivity enhances catalyst utilization
- Better liquid yield due to shorter diffusion path → reduced over-cracking
- Lower CBD → lower fill cost (~10% lower $)

ATX Shape reduces ΔP without loss of Activity or Selectivity

Case 1: ΔP-Limited Upgrading of High Severity Feed

- Single-Stage HCU with recycle processing high severity blend of Synthetic (sweet) feed from oil sands upgrading & (sour) SR VGO from oil sands fractionation
- Historically, pressure drop limits feed rate
- Key driver → maximize throughput without loss of run length or product quality

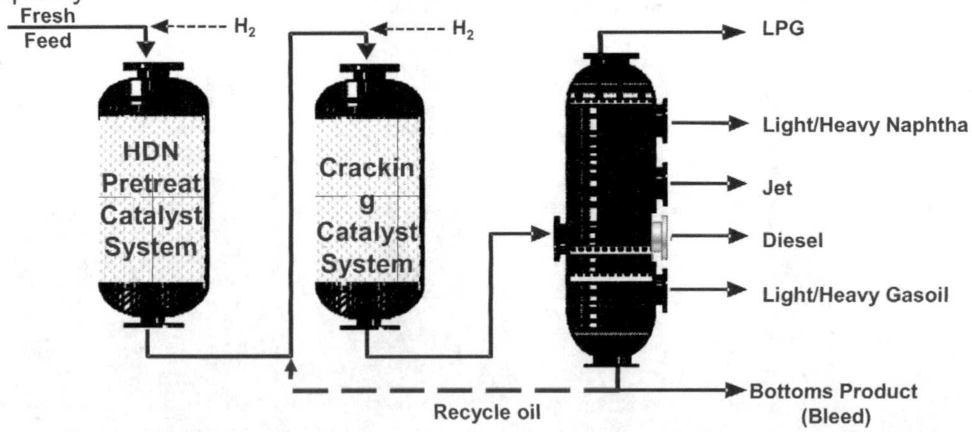

Case 1: Replacing TL with ATX = 15% Lower ΔP

- Demonstrated 15% ΔP reduction with ATX

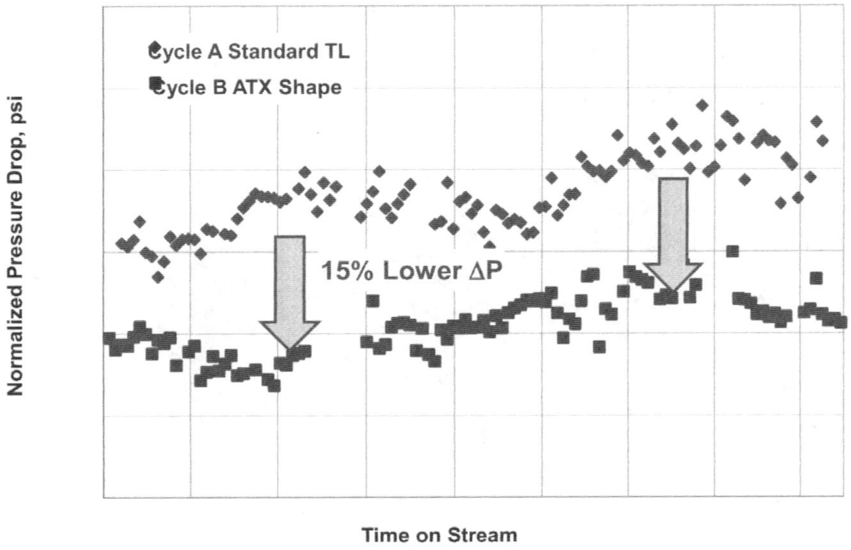

Replaced 60% of Total Catalyst Volume with ATX

Case 1…without loss of Cycle Length

- ΔP constraint alleviated with ATX & run length extended

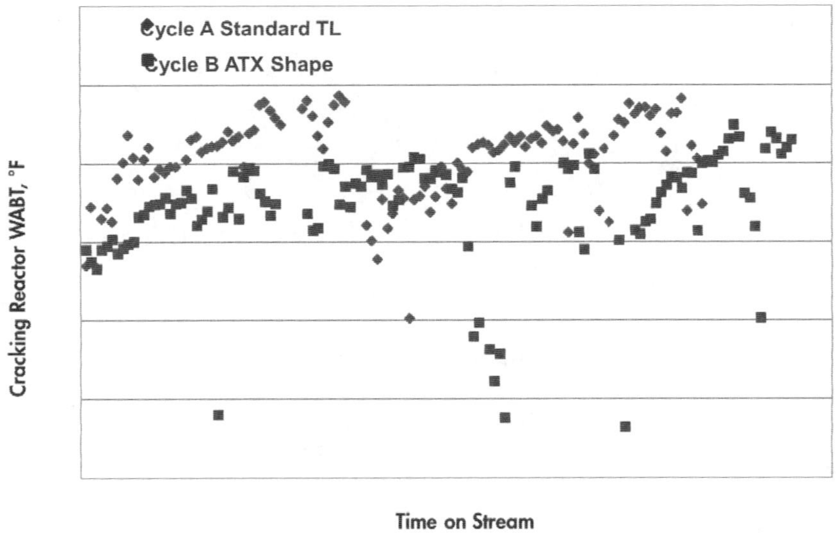

Replaced 60% of Total Catalyst Volume with ATX

Case 2: Increasing Rate in Largest Capacity HCU

- Single-Stage, OT HCU with two trains upgrading SR, Heavy & Coker Gas Oil to high quality distillate products & unconverted oil for FCC feed
- Cycle 1 optimized performance to increase throughput above design
- Identified greater capacity expansion for Cycle 2 by leverage catalyst advances to reduce ΔP with ATX-shaped charges

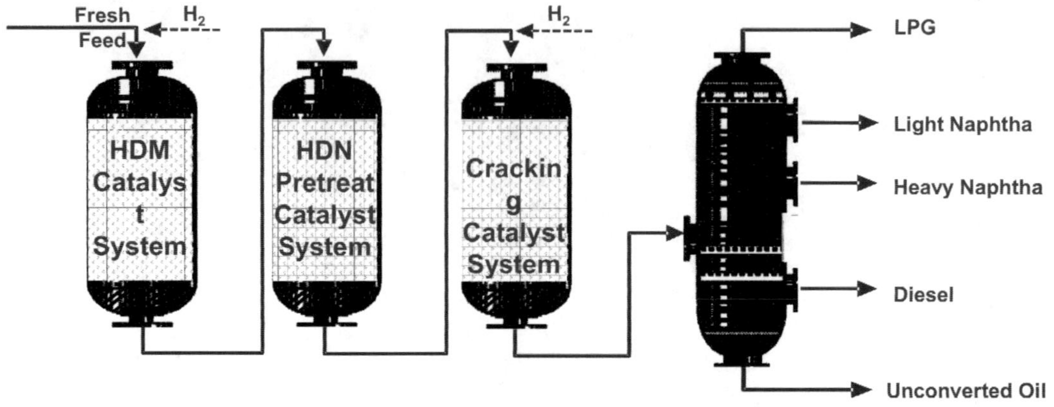

Case 2: Replacing TL with ATX = 20% Lower ΔP

- 20% reduction in Normalized ΔP with ATX Shape

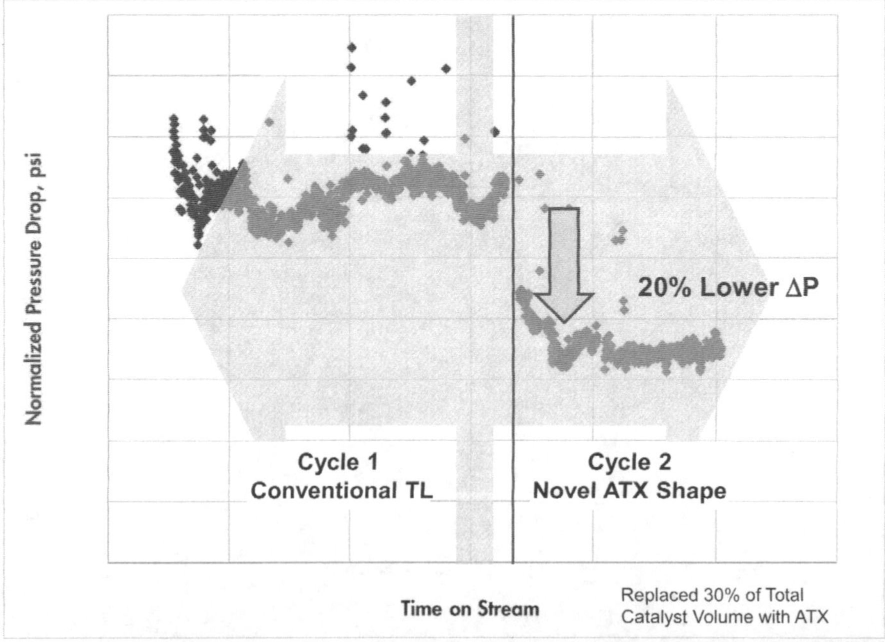

Case 2: Replacing TL with ATX = 35% Higher Rate

- Refiner leveraged ΔP reduction to increase throughput

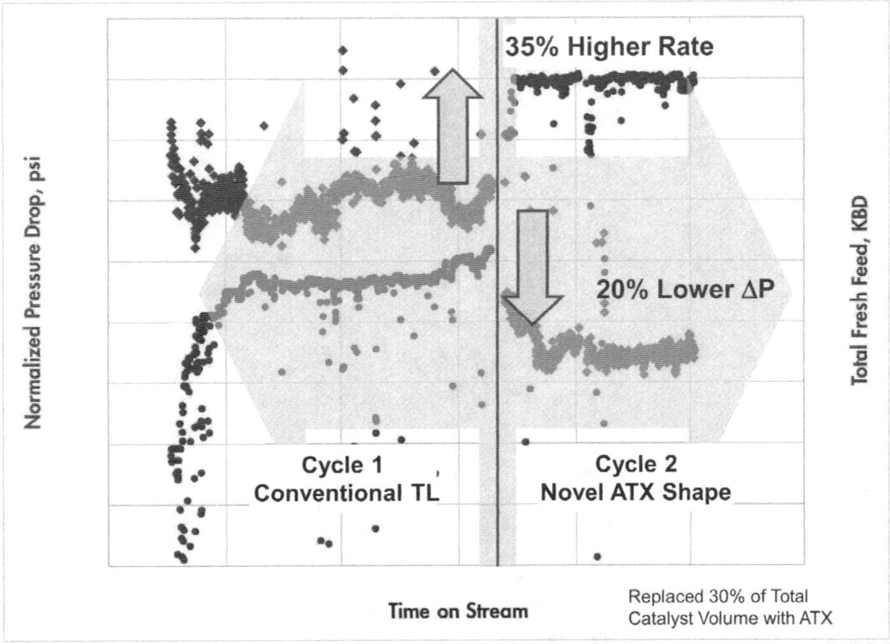

Case 2: ATX Shape Enables Increased Severity…

- Higher WABT with low deactivation despite rate increase

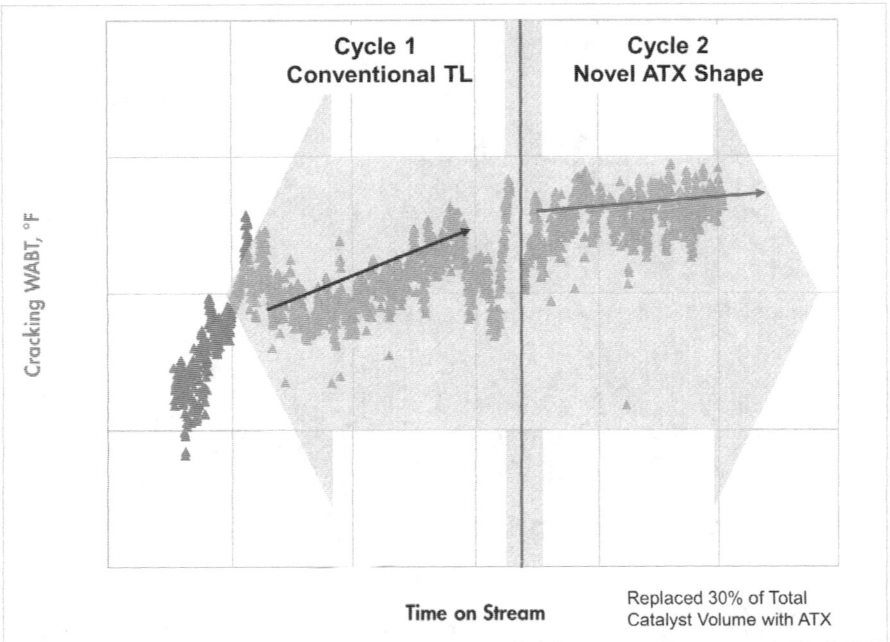

Case 2… with retention of High Gas Circulation Ratio

- Reduced ΔP enabled comparable increase in gas → same GOR

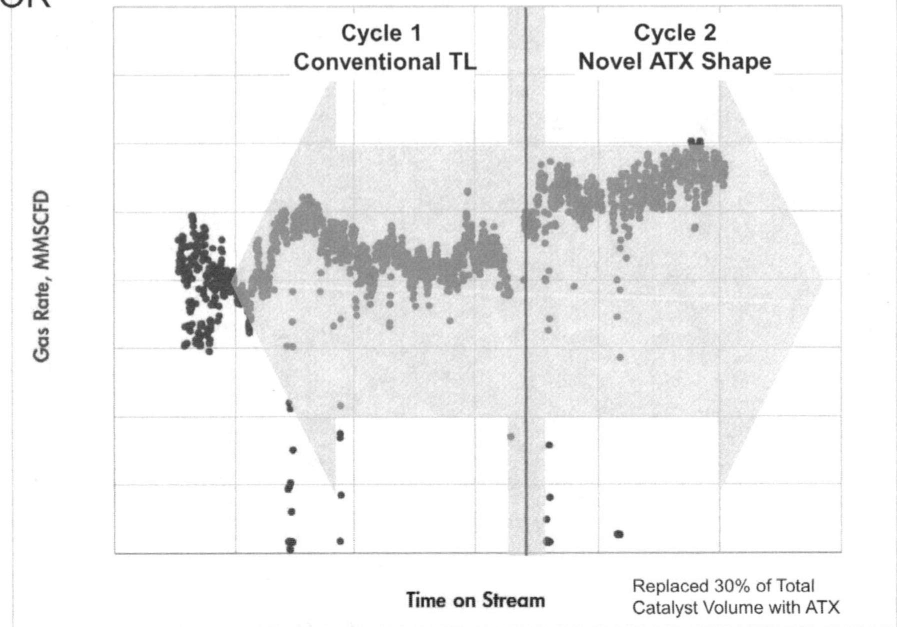

19

Case 2… without loss of Stability

- High GOR maintains uniform temperature profile = very stable control

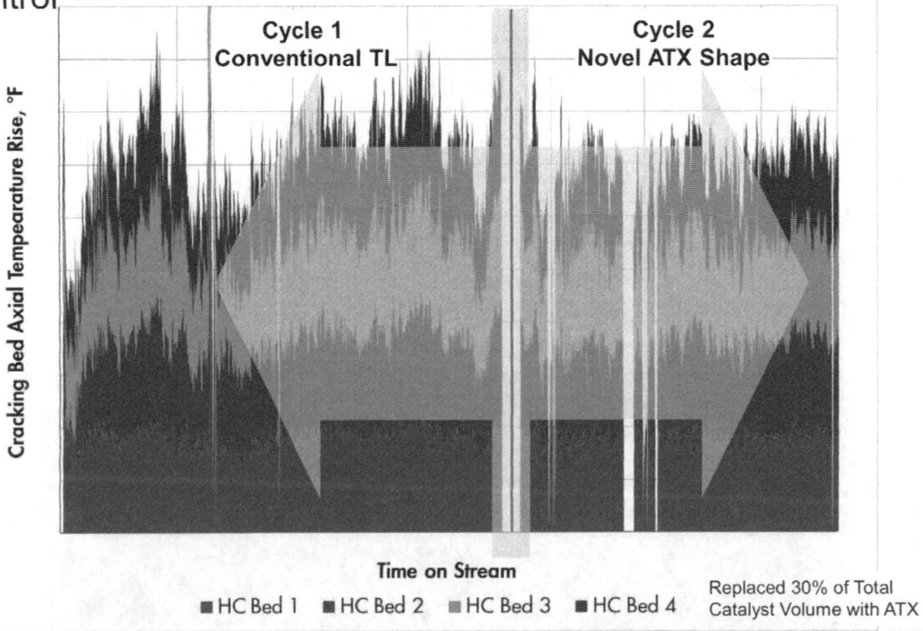

20

Case 2… without Loss of Total Liquid Product Yield

- Enhanced catalyst utilization with ATX maintains liquid yield

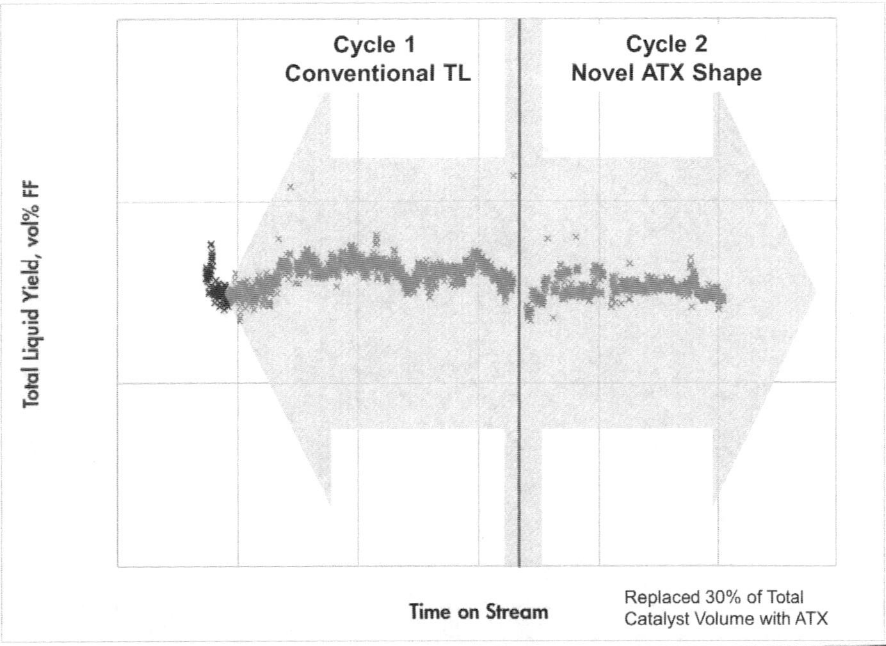

Case 2… or Added Light Ends Make

- While decreased diffusion pathway reduces over-cracking

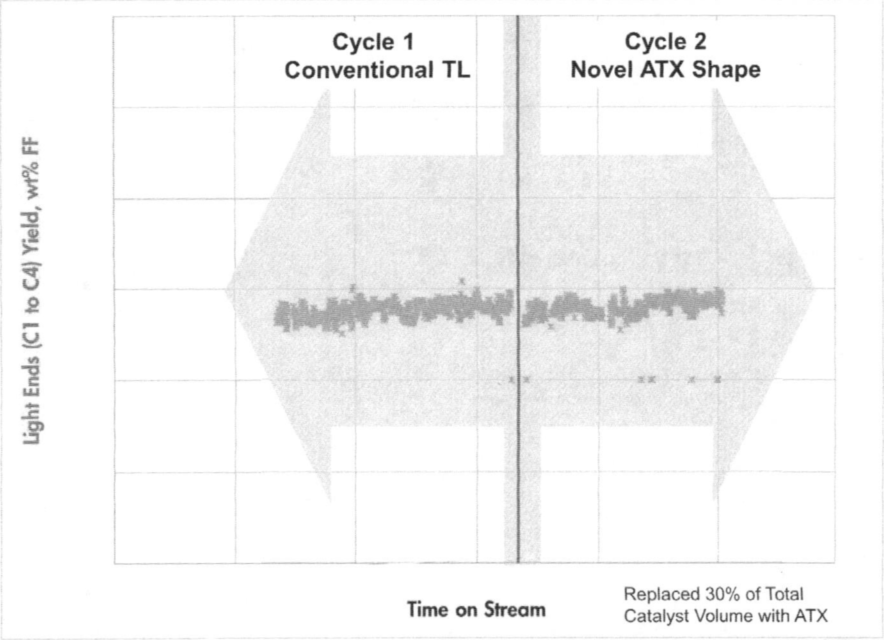

Technology Advancements Demonstrated

- Extrudate shape & size are critical factors for fixed bed reactors & have a significant impact on hydrocracking performance

- Criterion's Advanced Trilobe Extra shape demonstrates significant advantages over conventional TL
  - ATX enables large reductions in reactor system pressure drop
  - ATX demonstrates exceptional catalytic activity & stability
  - Continuous technology improvements supports massive rate increases in large-scale hydrocracking applications

Axial Dispersion in Fixed-Bed Reactors from Numerical Analysis

Nico Jurtz[1], Philipp Waldherr[1], Thomas Eppinger[2], Ravindra Aglave[3], Prof. Matthias Kraume[1], and Chandra Tourani[3]

[1]Technische Universität Berlin, Chemical and Process Engineering, Germany
[2]Siemens PLM Software, Nuremberg, Germany
[3]Siemens PLM Software, Houston, TX

nico.jurtz@tu-berlin.de

Outline

- Motivation
- Workflow
- How to efficiently generate packings of cylindrical particles?
- How to generate packings with a pre-defined bulk voidage?
- Numerical analysis of dispersion coefficient
  - Validation study
  - Impact of molecular Péclet number
  - Impact of tube-to-particle diameter ratio
  - Impact of particle shape
  - Limits of the Axial Dispersion Model
- Conclusion

Motivation

- Modeling of dynamic processes in fixed-bed reactors is of huge interest
- Particle-resolved CFD can predict accurately the steady-state conditions BUT is numerically demanding
- Simplified models like the axial dispersion model are efficient but rely on the knowledge on effective parameters like the axial dispersion coefficient D_{ax}

$$\frac{\partial c}{\partial t} = -u\frac{\partial c}{\partial z} + D_{\text{ax}}\frac{\partial^2 c}{\partial z^2} + \dot{R}$$

- No correlation available for
  - Complex particle shapes
  - Reactors with a low tube-to-particle diameter ratio
- Experimental determination is costly and time consuming

➢ **Use of CFD to determine D_{ax}**

Workflow

Methods and Workflow

1. **Packing generation using DEM**

Methods and Workflow

1. Packing generation using DEM
2. **Create CAD description based on DEM results**

Methods and Workflow

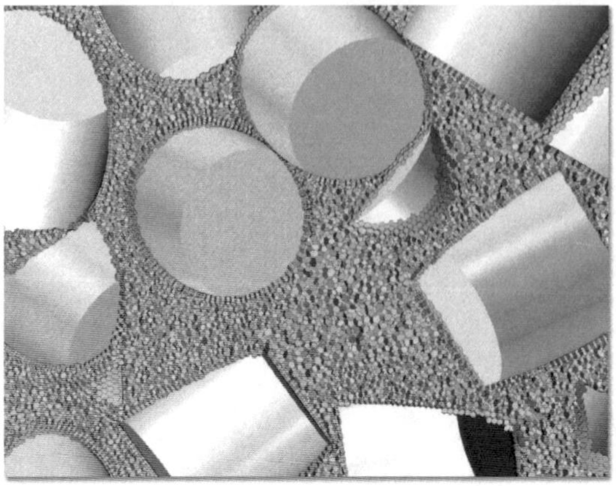

1. Packing generation using DEM
2. Create CAD description based on DEM results
3. **Meshing (local flattening near contact points)**

Methods and Workflow

1. Packing generation using DEM
2. Create CAD description based on DEM results
3. Meshing (local flattening near contact points)
4. **CFD simulation**

Methods and Workflow

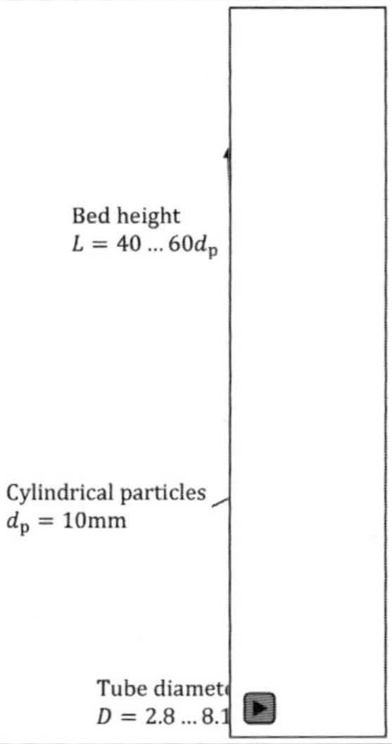

1. Packing generation using DEM
2. Create CAD description based on DEM results
3. Meshing (local flattening near contact points)
4. CFD simulation
5. **Particle tracking method to determine residence time distribution function**

Methods and Workflow

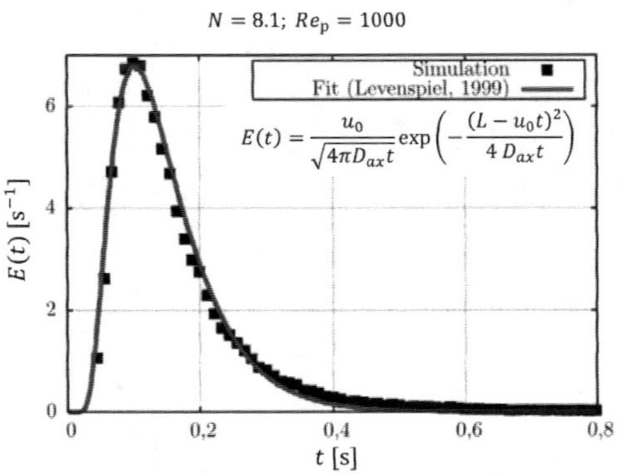

1. Packing generation using DEM
2. Create CAD description based on DEM results
3. Meshing (local flattening near contact points)
4. CFD simulation
5. Particle tracking method to determine residence time distribution function
6. **Determine D_{ax} by curve fitting**

Packing Generation

Chemical and Process Engineering
www.vt.tu-berlin.de

Slide 11

How to efficiently generate packings of cylindrical particles?

- Cylindrical particles were formerly modelled using the composite particle approach
  - Cylinders were approximated by a number of spherical particles
  - Large numerical overhead since all particles need to be tracked
  - Not all features could be resolved properly (e.g. edges)
- The new approach uses sophisticated contact detection algorithms to model genuine cylinders
  - All features can be resolved
  - Simulation time for packing generation is reduced significantly

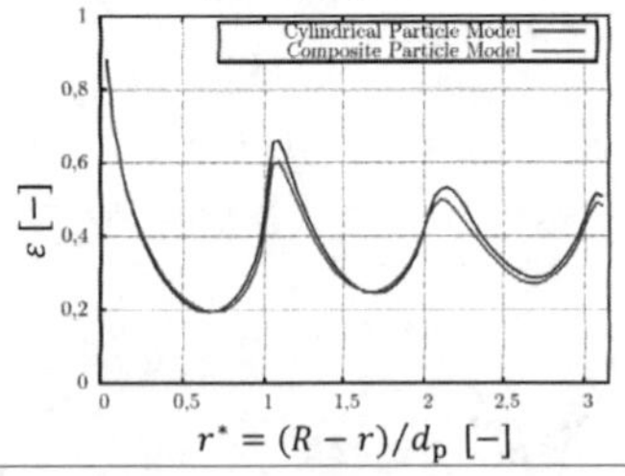

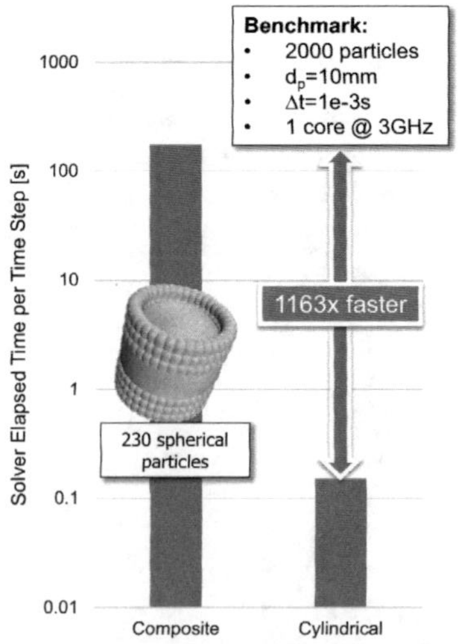

Chemical and Process Engineering
www.vt.tu-berlin.de

Slide 12

How to generate packings with a pre-defined bulk voidage?

- DEM-generated packings are looser than often found in experiments
- Due to wanted or un-wanted vibrational or shaking effects the particles re-order
- This leads to a decreasing bulk voidage
- Shaking or vibrations can be modeled explicitly but it's time consuming and hard to achieve a certain value
- The static friction coefficient can be used for fine-tuning the bulk voidage of the numerical packing

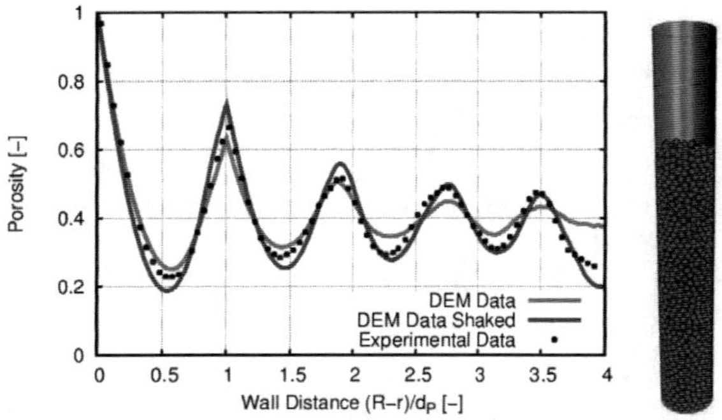

Effect of shaking on the radial void fraction distribution in a fixed-bed of spherical particles (N=7.99)

How to generate packings with a pre-defined bulk voidage?

- Virtual experiments using DEM
  - Cylindrical particles
  - Varying static friction coefficient and tube-to-particle diameter ratio
  - Bed height = $60d_p$
  - Evaluated bed length: $8d_p$-$56d_p$
- A new correlation is proposed as:

$$\varepsilon = f(D/d_{v,p}) \cdot g(\mu_{fs})$$

using:

$$g(\mu_{fs}) = \left(\frac{\mu_{fs}}{\mu_{fs,ref}}\right)^A$$

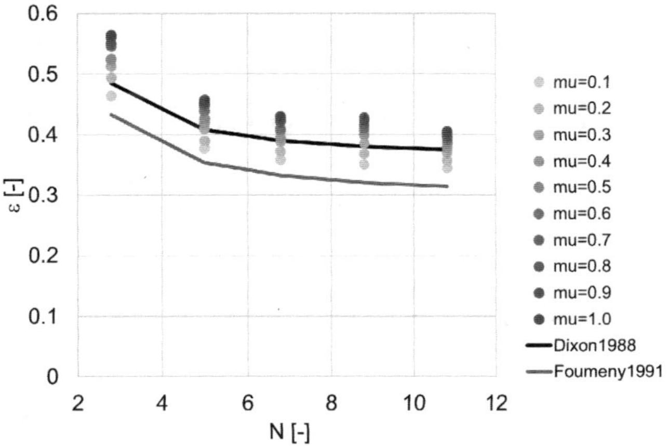

Bulk voidage as a function of tube-to-particle diameter ratio and static friction coefficient for cylindrical particles. Comparison of numerical data with correlations.

How to generate packings with a pre-defined bulk voidage? – Cylindrical particles

$$\varepsilon = \underbrace{\left(0.293 + 0.684 \cdot N^{-0.85} \cdot \frac{1}{\sqrt{1.837 \cdot N - 1}}\right)}_{\varepsilon = f(N) \rightarrow \text{[Foumeny1991]}} \cdot \underbrace{\left(\frac{\mu_{\text{fs}}}{0.0433}\right)^{0.0844}}_{\varepsilon = f(\mu_{\text{fs}})}$$

- Numerical Results

- Correlations of [Dixon1988] and [Foumeny1991] were tested for $\varepsilon = f(N = D/d_{\text{v,p}})$
- Correlation of Foumeny is in better agreement with DEM results
- Mean deviation of below 0.9%
- Maximum deviation of 3.4%

Method can be used to fine-tune μ_{fs} to meet a certain requirement for ε

[Dixon1988] Dixon, A. G. *Canadian Journal of Chemical Engineering.* 66 (5) 705-708. 198
[Foumeny1991] Foumeny, E. A. *Chemical Engineering Science.* 46 (9) 2363-2364. 1991

Numerical Analysis of Dispersion in Fixed-bed Reactors - Validation of Particle Tracking Method -

Validation

Experimental Set-Up [Abdulmohsin2016]:

- System: Helium/Air
- Spherical particles
- Tube-to-particle diameter ratio: *N*=6
- Concentration of He derived by measurement of thermal conductivity

Results:

- Very good agreement for low to moderate Re_p
- Only slight deviations for high Re_p → Neglecting turbulent dispersion

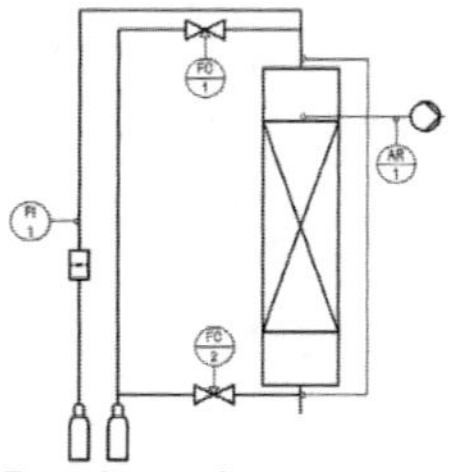

Experimental set-up

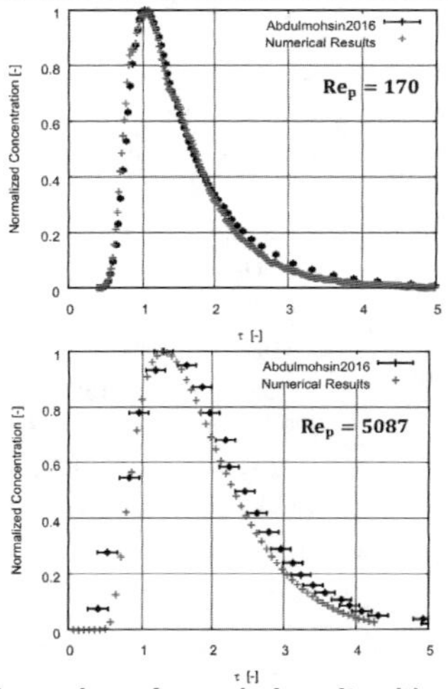

Comparison of numerical results with experimental measurements

[Abdulmohsin2016] Abdulmohsin, R. S. *Ann. Nucl. Energy.* 88 100 – 111. 2016

Numerical Analysis of Dispersion in Fixed-bed Reactors - Results for Cylindrical Particles -

Dispersion regimes

I. Diffusive regime
- Molecular diffusion controls the mixing almost completely

II. Transition zone
- Molecular diffusion still dominates the mixing but convection starts to contribute to it

III. Power law regime
- Mixing is almost driven by convection but molecular diffusion cannot be neglected entirely

IV. Purely convective regime
- Mixing is entirely dominated by convection

V. Turbulent regime
- Turbulent mixing additionally contributes to mixing

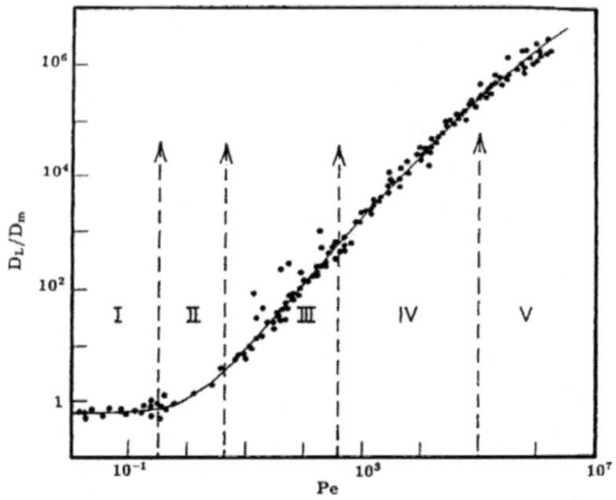

Dependence of longitudinal dispersion coefficient in the Péclet number, and the various dispersion regimes (taken from [Sahimi2012]).

[Sahimi2012] Sahimi, M. *Flow and transport in porous media and fractured rock.* VCH. 2012

Impact of Re_p on the radial velocity profile for cylindrical particles

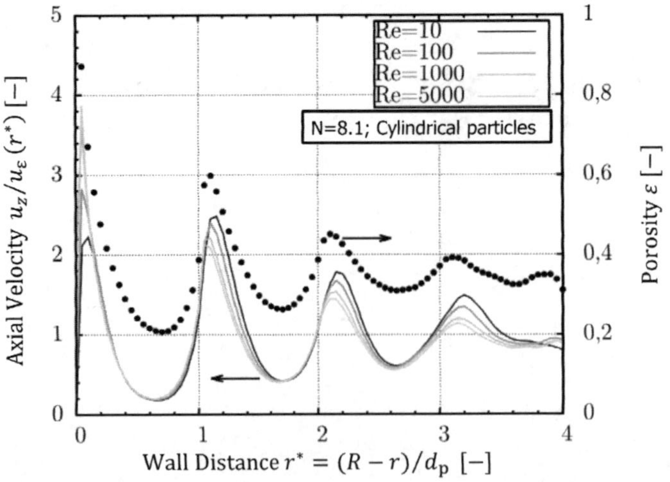

- The velocity field is directly linked to the void fraction profile
- Wall effect increases if Re_p ↑
- A more homogeneous velocity field in the core region can be observed
- A more plug-flow-like behavior can be assumed for higher Re_p

Dispersion coefficient for packings of cylindrical particles (N=2.0 - 8.1)

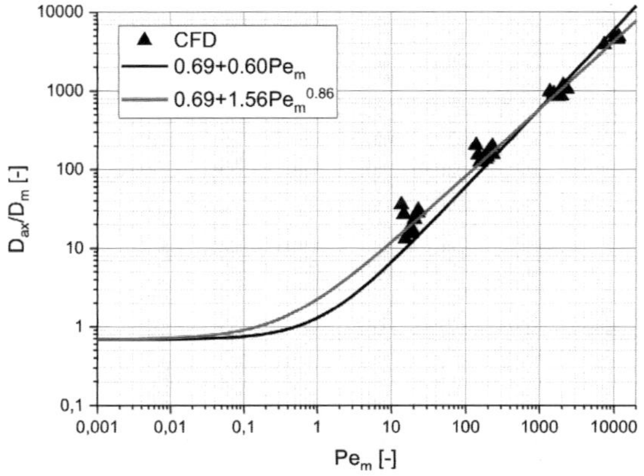

- Most correlations show the following form:
$$\frac{D_{\mathrm{ax}}}{D_{\mathrm{m}}} = \frac{D_{\mathrm{eff}}}{D_{\mathrm{m}}} + A \cdot Pe_{\mathrm{m}}^{B}$$
- Most authors report $B \approx 1$
- [Freund2008] reports $\frac{D_{\mathrm{eff}}}{D_{\mathrm{m}}} = 0.69$
- Simulation results are in good agreement when using $B = 1$
- However, the best fit is achieved with $B = 0.86$

[Freund2008] Freund, H. *Ortsaufgelöste Simulation von Transport-prozessen in durchströmten Festbetten.* SHAKER. 2008

Impact of tube-to-particle diameter ratio for cylindrical particles

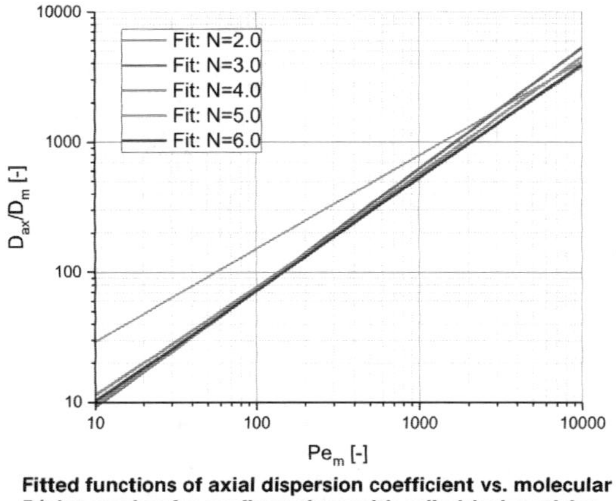

Fitted functions of axial dispersion coefficient vs. molecular Péclet number for configurations with cylindrical particles and different diameter ratios.

- For $N = 2$ significantly enhanced dispersion can be observed for low and moderate Pe_{m}
- For $N \geq 4$ almost no impact of N on $D_{\mathrm{ax}}/D_{\mathrm{m}}$ can be seen

Impact of particle shape

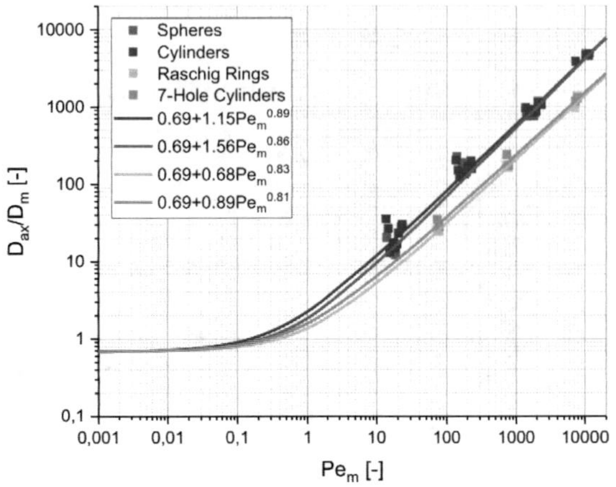

- Spheres and cylinders show very similar axial dispersion coefficients
- For spheres the exponent is closer to the theoretical value of 1
- Significantly reduced dispersion coefficient for particles with inner voids
- For all particle shapes an exponent $B < 1$ shows the best fit

Why do particles with inner voids show a reduced axial dispersion coefficient?

Radial profile of axial velocity for several particle shapes

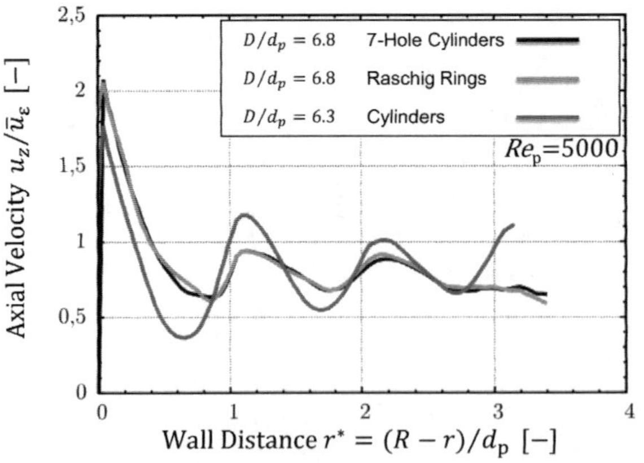

- More homogenous velocity distribution for particles with inner voids
- Closer to a plug-flow-like behavior → Decreasing dispersion coefficient

Be careful: Limitations of the axial dispersion model

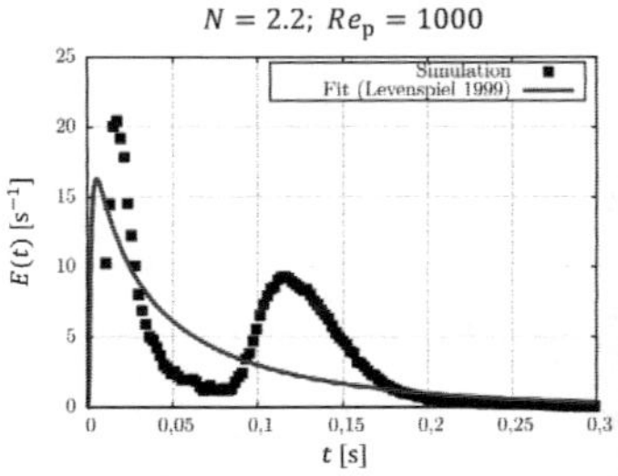

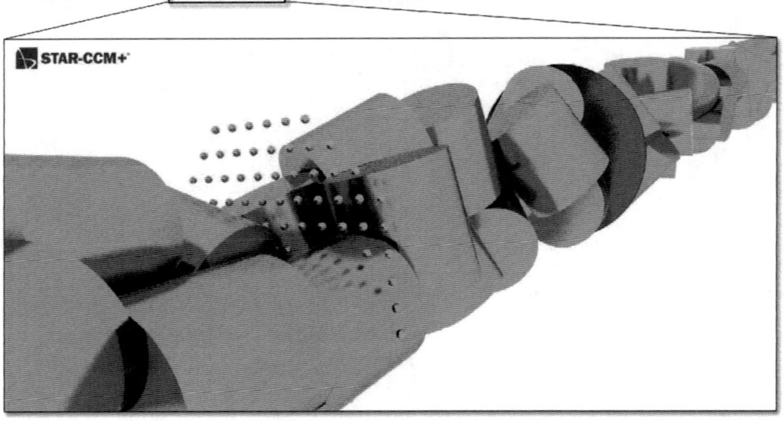

- Multimodal distributions are observed for some configurations
- Fit to Dispersion Model leads to non-meaningful results
- It's a result of back-mixing

Conclusion and Outlook

- Significantly faster bed generation by using sophisticated contact detection
- Static friction coefficient can be used to create packings of pre-defined bed voidage
- A fully automated workflow to determine the axial dispersion coefficient for spherical and non-spherical particles was presented
  - Cylindrical and spherical particles show similar dispersion coefficients
  - Particles with inner voids show significantly lower dispersion coefficients
- More validation work necessary
- Comparison of Particle-resolved CFD vs. Axial Dispersion for different particle shapes

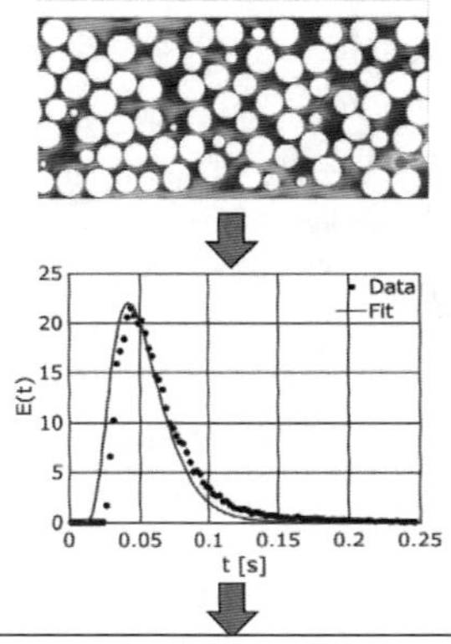

Thanks for your attention!

This work is part of the Cluster of Excellence "Unifying Concepts in Catalysis" coordinated by the Technische Universität Berlin. Financial support by the Deutsche Forschungsgemeinschaft (DFG) within the framework of the German Initiative for Excellence is gratefully acknowledged (EXC314).

Sustaining Economic Competitiveness for Industrial Catalyzed Processes

Southwest Process Technology Conference

Conference: Southwest Process Technology Conference

Year: 2017

Proceeding: 9th AIChE Southwest Process Technology Conference

Group: Southwest Process Technology Conference

Session: Catalysis & Reaction Engineering

Time: Thursday, October 12, 2017 - 9:30am-10:00am

Authors:

King, S., The Dow Chemical Company

The commercialization of catalyzed processes which produce hundreds of millions of pounds of product(s) annually is the result of a significant number of years of catalyst research, process research, and optimization of plant design. Despite this considerable investment in technology development, catalyzed processes still represent a unique opportunity for optimization through improved catalysts, enhanced product selectivity, extended catalyst life, or the use of cheaper feedstocks. In fact, many catalyzed processes, even those which have existed for many years in the chemical industry, must be continuously improved to sustain economic competitiveness. Numerous companies today maintain a major R&D investment in the development of improved processes as evidenced by the filing of numerous patent applications on processes which have been commercialized for 30 years or more.

This talk will give a high level overview of two well established catalyzed technologies: hydroformylation, and epoxidation. Included in the discussion are the major product(s) formed from the process, the catalysts employed, proposed mechanisms, and opportunities for technology improvements.

Improved Cogeneration Energy Efficiency through Gas Turbine Auto-Tuners and Optimizers

Heidi Holmes, Energy AC&O Technology Manager

2017 Southwest Process Technology Conference
October 12th, 2017

DOW®

DOW®

Why are we here presenting…

- To share a little about how our team came together to install auto-tuners/optimizers on our GE 7FA Gas Turbines in Plaquemine, LA
- To let our Energy peers know what kind of results are possible with this technology

What we had…

- All 4 of our GE 7FA GTs had GE CDMS systems installed with ability to see our combustion dynamics in our control room on our HMI.
- We had the ability to change the alert and alarm limits to make our CDMS display turn yellow and red when we exceeded limits.
- We had remote monitoring through GE so they could call and tell us when we had a problem.

What we did not have…

- Funds to bring in a tuner every time our CDMS system indicated that we had an issue.
- Procedural guidance to de-rate a unit if combustion dynamics were out of range.
- Plant knowledge of what kind of trouble each frequency band represented or what significance to place at what thresholds.

We installed Auto-Tuners / Optimizers

- In October of 2014, we installed EthosEnergy's Auto-Tuner & Optimizer, ECOMAX® Tru-Curve® on a single GE 7FA GT.
- After a full cycle of reliable run-time and a review of our combustion hardware, in 2016, we installed the same system on the remaining 3 GTs.

Without Ecomax®

With Ecomax®

Benefits of Auto-Tuning with ECOMAX®

- Allows for real-time auto tuning in both baseload and part load operations
- All knobs have max and min limits and pre-defined movement deltas set in the Turbine Controller Control Constants. We can change these limits at any time via the Turbine Controller utility.
- No tuner needed after CI, HGP or Majors
- Allows winter and summer max firing temperature adjustment while mitigating combustor dynamics
- Increases parts life by mitigation of Combustor Dynamics.
- De-rates GT if combustion dynamics can not be maintained below established limits.

Benefits of Tru-Curve®

- Tru-Curve® is a real-time optimizer for the power output of the gas turbine which safely monitors and manages operation once the CDMS and Auto-Tuning System are installed. Adjustments act as a safety measure against sudden variable changes that can cause dynamic or emission excursions.
  - User-defined emissions and dynamics tuning limits
  - All parameters have hard-coded maximum and minimum limits
  - TRU-Curve® reference temperature limit
  - This limit determines the maximum allowable reference temperature (TTRF1) for TRU-Curve®
- In short, because we are always tuned, we are able to run right up to the GT firing limit, year round.

ECOMAX® - Operator Interface

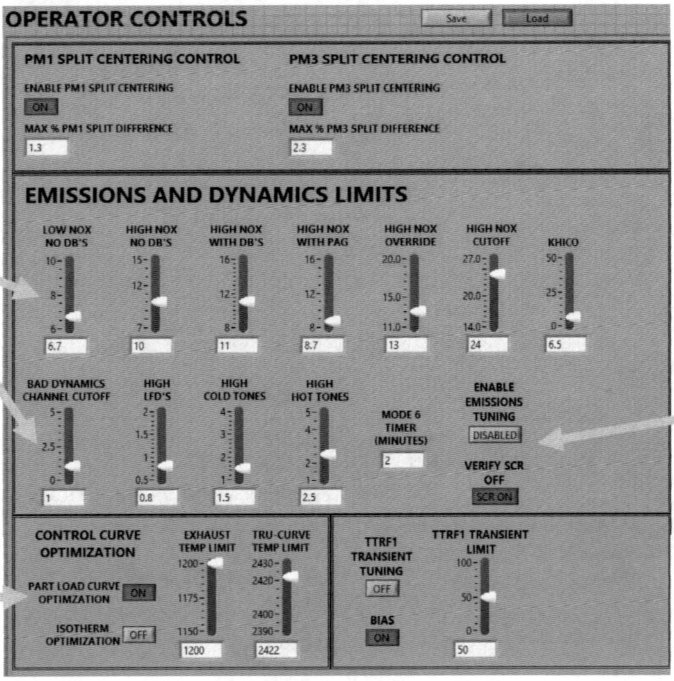

Slider bars allow easy modification of tuning limits

Manual settings to allow emissions tuning

Thermal optimization settings

DOW RESTRICTED - For internal use only

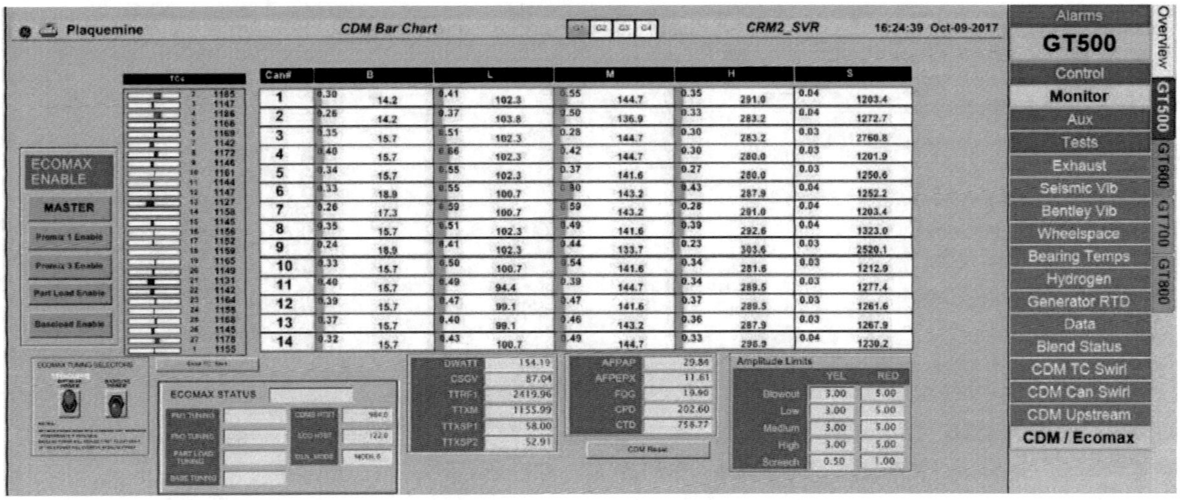

DOW RESTRICTED - For internal use only

ECOMAX® - NOx Emissions Tuning Requirements

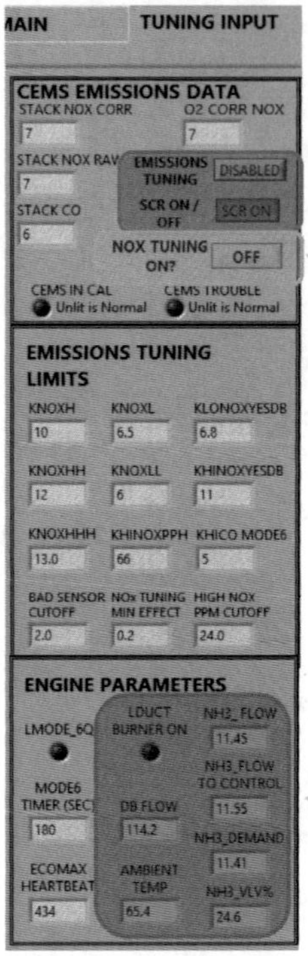

Manual settings to allow Emissions Tuning

Manual settings to allow Emissions Tuning

+

Analog parameters validating Ammonia Off

=

NOx Tuning Allowed

Analog parameters validating Ammonia Off

DOW RESTRICTED - For internal use only

Measurable, Reproducible Results

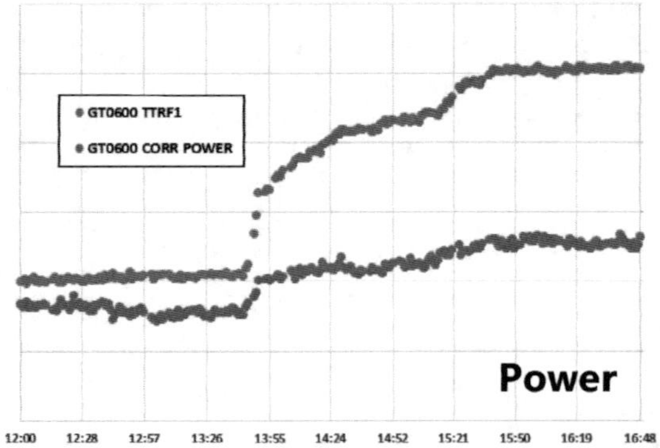

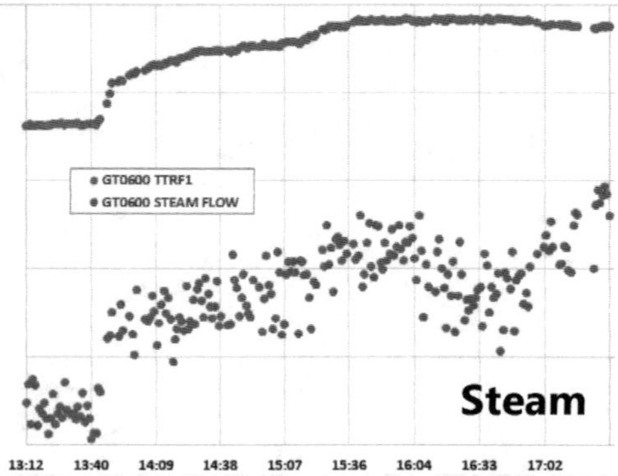

- For our 4 GTs, that was a move from 2400 to 2424 deg. F
  - We observed 1 MWe per each 10 deg. of firing temperature increase
    - 2.4 MWe per GT * 4 GTs or 9.6 MWe
  - 8 mlb/hr of HP Steam per each 10 deg. of firing temperature increase
    - 2.6 MWeqv. * 4 GTs or 10.4 MWeqv. of steam
  - 20 additional MWs for 98 mmbtus of additional fuel (4.9 INHR)

Fuel and CO2 savings

- Steam produced at GT incremental heat rate of 4.9 mmbtu/MWeqv. is steam that does not have to be produced with duct firing at a 10-11 mmbtu/MWeqv heat rate.
- Fuel savings:
  - ~60 mmbtu/hr for 10.4 MWeqv. of steam (~77 mlb/hr)
  - **~465K mmbtu/yr**
- CO_2 reduction for same steam:
  - **~27,200 tons/yr**
- Additional Power Sales for compelling ROI
  - 73,600 new MWs/yr produced at 4.9 mmbtu/MW

DOW PROUD!

Delivering strong on several of our Dow 2025 Sustainability Goals…

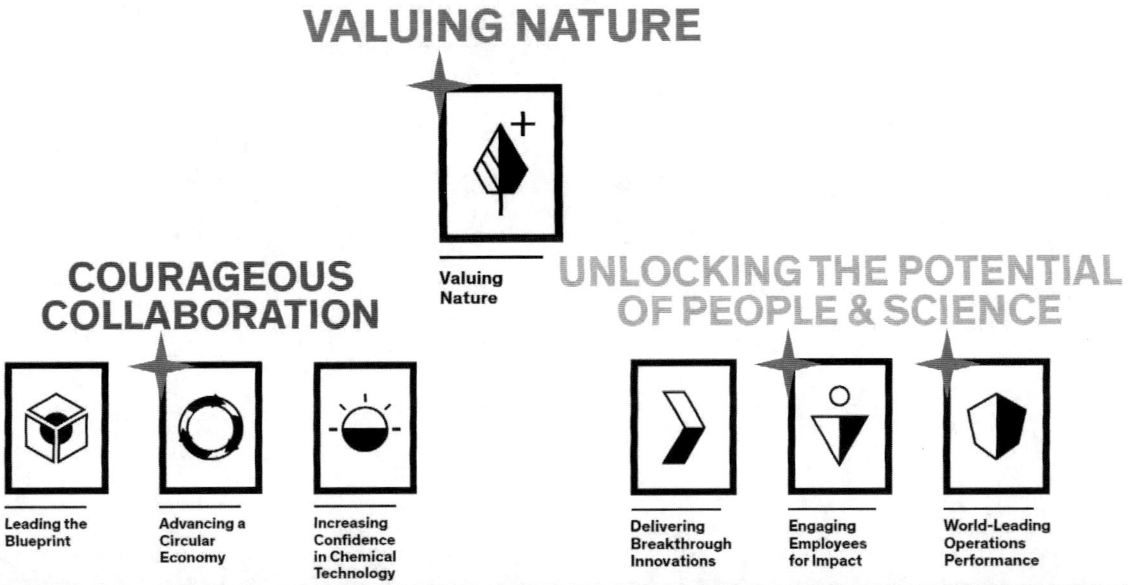

QUESTIONS?

ASCEND. PERFORMANCE MATERIALS

Thermal Oxidizer Optimization

Improving Fuel Economy While Maintaining Emissions Compliance

October 12, 2017

AIChE Southwest Process Technology Conference

Presenter

ASCEND. PERFORMANCE MATERIALS

Paul Pathasema
Process Control Sr. Engineer
Ascend Performance Materials, Chocolate Bayou Site

Introduction

Introduction

Ascend Performance Materials

Ascend is a global leader in proprietary technologies that are central to the production of nylon, plastics and synthetic fibers found in thousands of commercial and industrial products, including carpet, tires and apparel, as well as agricultural products, animal feed and personal care products.

Ascend Performance Materials operates the largest acrylonitrile production plant in the world at our Chocolate Bayou facility near Alvin, Texas.

Acrylonitrile (AN) is a high volume commodity chemical used as a monomer in the production of acrylic and modacrylic fibers, acrylic plastics and resins, specialty polymers, nitrile rubbers, and other organic compounds.

Introduction

Project Summary

Ascend operates a thermal oxidizer to destroy organics in process off-gas stream

- Business objective to optimize operation of this thermal oxidizer
  - Maintain environmental compliance
  - Reduce fuel consumption
  - Maximize heat recovery
- Model Predictive Control (MPC) successfully implemented
- Cost savings was realized through significant reduction in fuel gas
- Improved operational stability

Agenda

- MPC Control Basics
- Process Overview and Control Objectives
- Controller Design and Implementation
- Business Results

Model Predictive Control (MPC)

ASCEND PERFORMANCE MATERIALS

Model Predictive Control

ASCEND PERFORMANCE MATERIALS

MPC chosen for specific advantages over regulatory control

- Handles deadtime better than feedback control
- Inherently resolves multi-variable interactions
- Easier to incorporate feed-forward inputs
- Easier to incorporate override control
- Ideal platform for Constraint Optimization

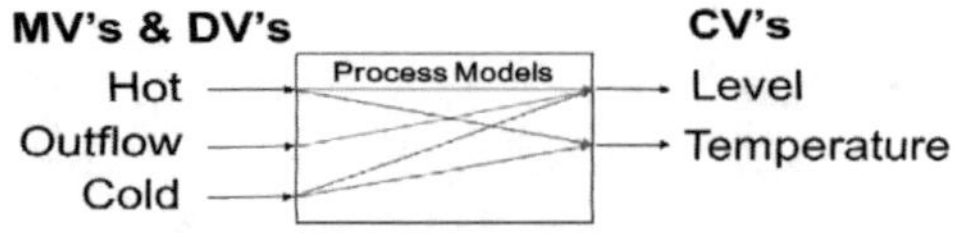

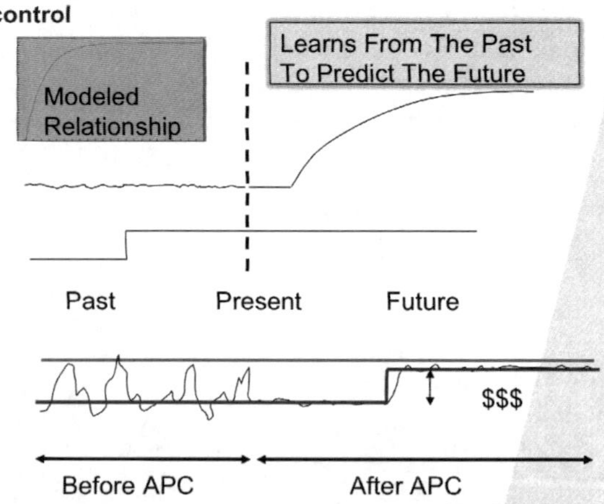

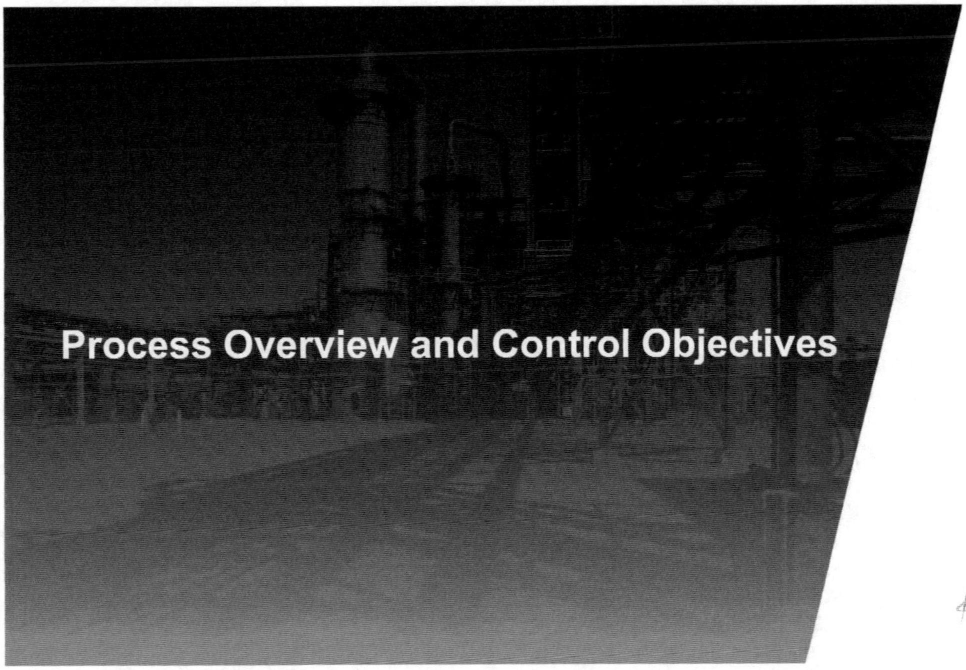

ASCEND.
PERFORMANCE MATERIALS

Process Overview

Incineration

Burner

- Burner combusts fuel gas to create heat
- Reducing air/fuel maintained with ratio controller
- Lead-lag controller enforces air and fuel cross-limits

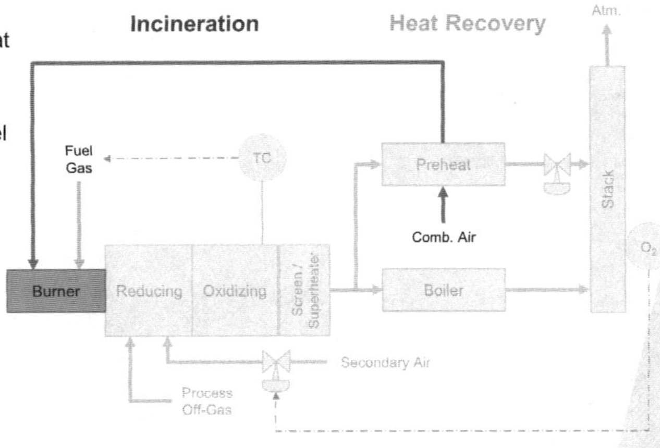

Process Overview

Incineration

Reducing Section

- Process off-gas fed to Reducing Zone; O_2-lean environment maintained
- Secondary air added downstream to compensate for changes in Process Off-Gas composition

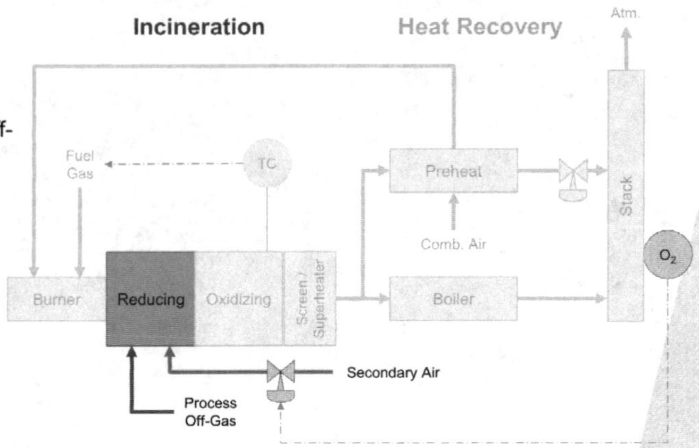

Process Overview

Incineration

Oxidizing Section

- Oxidizing Zone provides additional air (not shown) and residence time for completion of combustion
- Ox Zone is permit temperature
- Temperature maintained by modulating fuel gas

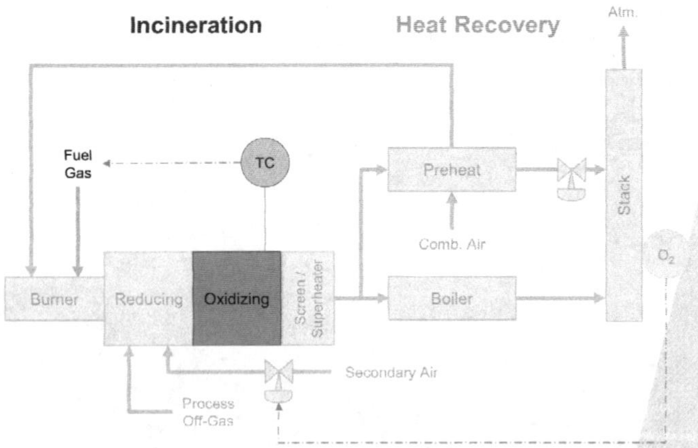

Process Overview

Heat Recovery

Boiler and Combustion Air Preheat

- Hot incinerator gas sent to Combustion Air Preheater and Boiler
- Split adjusted with outlet damper
- As damper is opened, Combustion Air temperature is increased, and steam generation decreased

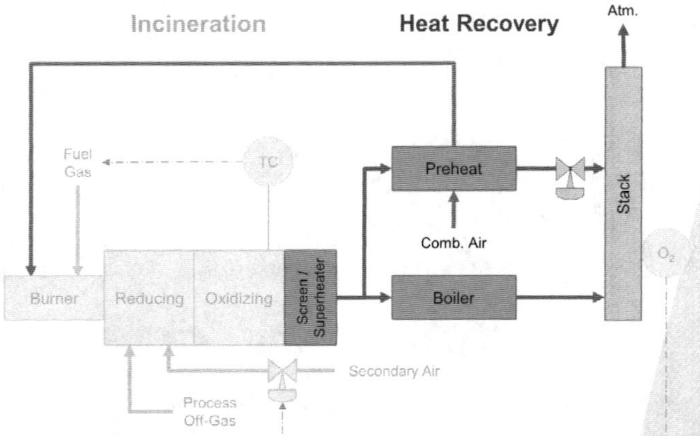

Control Objectives

Maximize Energy Efficiency

1. Control CO closer to permit limit – allows lower temperature and fuel
2. Maintain stack excess O_2; reduce if possible
3. Optimize heat recovery
4. Maintain differential temperature between the Reducing and Oxidizing zones

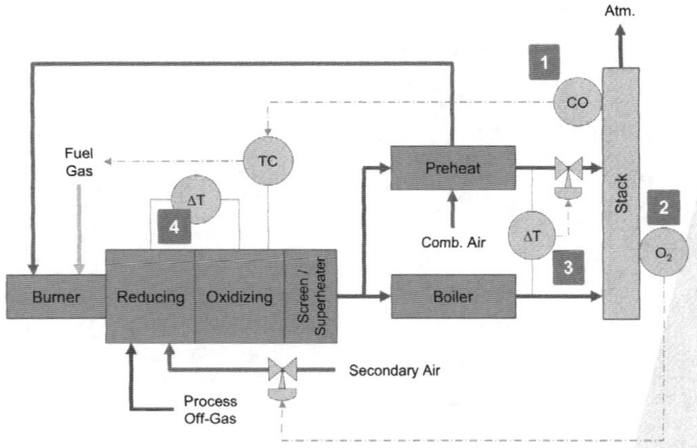

Controller Design and Implementation

ASCEND PERFORMANCE MATERIALS

Controller Design and Implementation

Initial Testing

Process Response

- Initial step testing of MVs provided some interesting insight
- Vastly different Steady-State times required two controllers
  - Controller 1: Flue Gas DT
  - Controller 2: CO and O_2

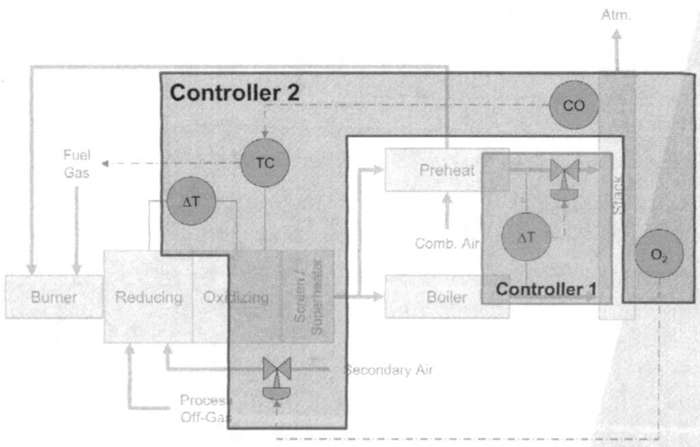

Controller Design and Implementation

Process Models and Configuration

Controller 1: Flue Gas ΔT

- Long steady-state time (5,400 sec)
- 1-MV, 1-CV (SISO) design
- 1 Constraint (Limit) Variable (LV) on preheated air Temp (protect preheater from damage)

CV1	Flue Gas ΔT (Boiler – Preheat)
MV1	Preheat Temp Controller (RCAS)
LV1	Preheated Air Temperature

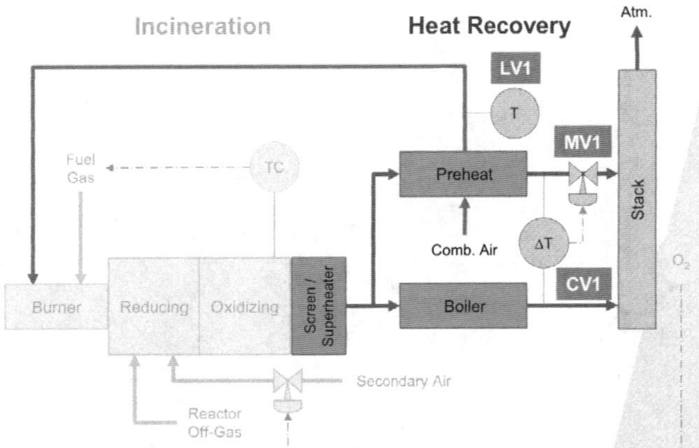

Controller Design and Implementation

Process Models and Configuration

Controller 1: Flue Gas ΔT

- Composite of two different dynamic effects
  - Initial change from outlet of Boiler (quick)
  - Secondary change from Combustion Air Preheater (very slow)

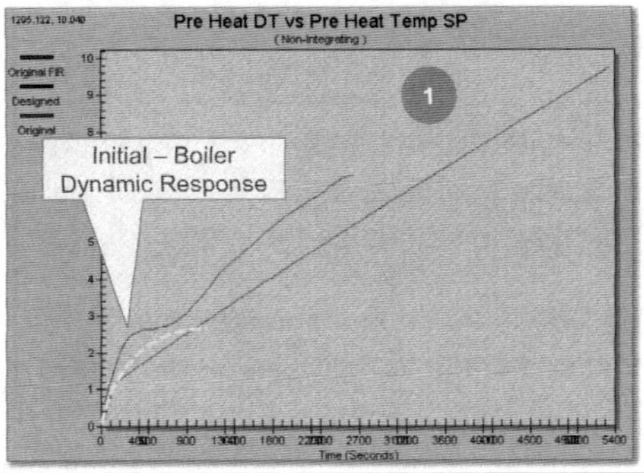

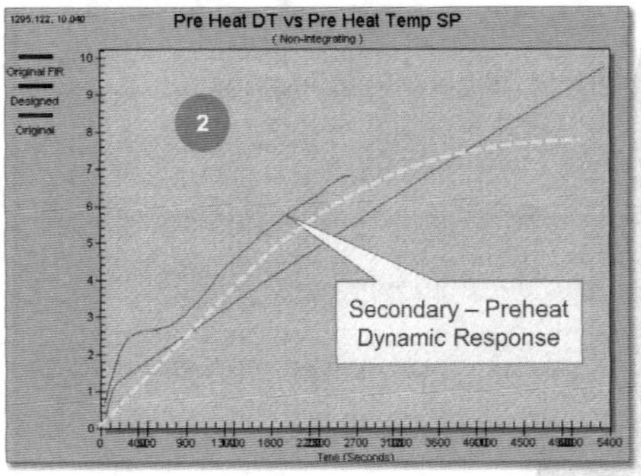

Controller Design and Implementation

Process Models and Configuration

Controller 1: Flue Gas ΔT

- Manually added points to ΔT response model due to unusual dynamics
- Achieved a reasonable fit to data

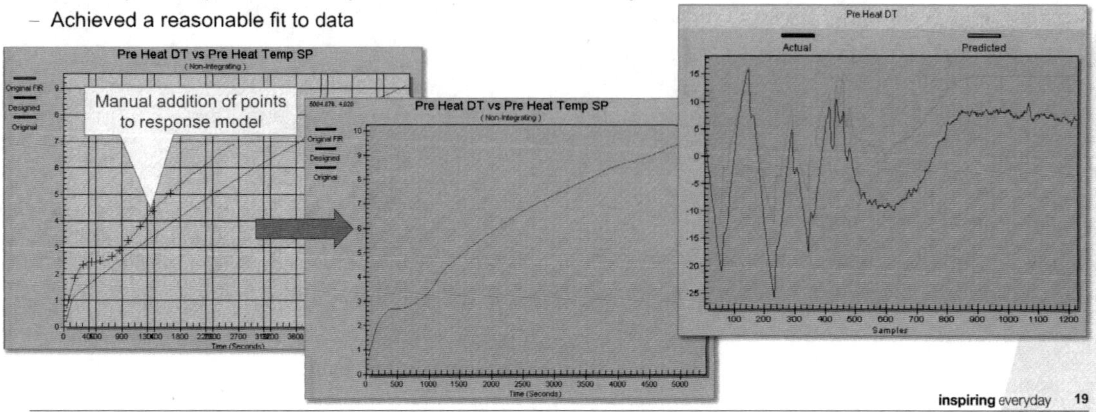

Controller Design and Implementation

Process Models and Configuration

Controller 2: CO and O_2

- 720 sec Steady-State time
- 2-MV, 2-CV design
- Constraint (Limit) Variable (LV) on ΔT between Reducing and Oxidizing Zone temperatures (stability) consideration

CV1	Stack CO (ppm)
CV2	Stack O_2 (%)
MV1	Ox Zone Temp (RCAS)
MV2	Sec. Air Valve Position (ROUT)
LV1	RedOx ΔT

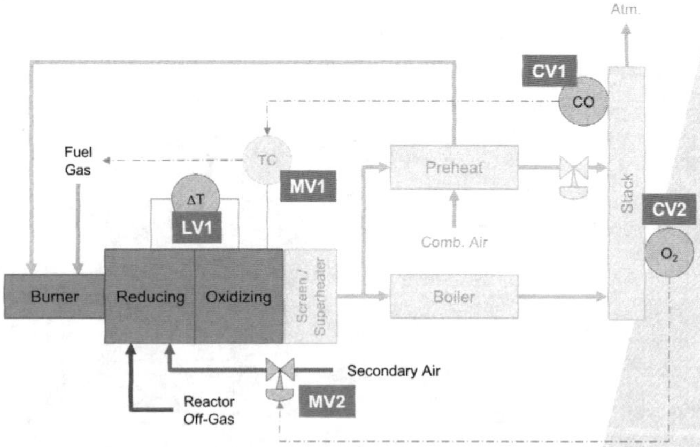

ASCEND. PERFORMANCE MATERIALS

Controller Design and Implementation

Process Models and Configuration

Controller 2: CO and O_2

- Made some reasonable adjustment to models
- Determined that expected disturbances had little or no effect
- Reasonable fit to data

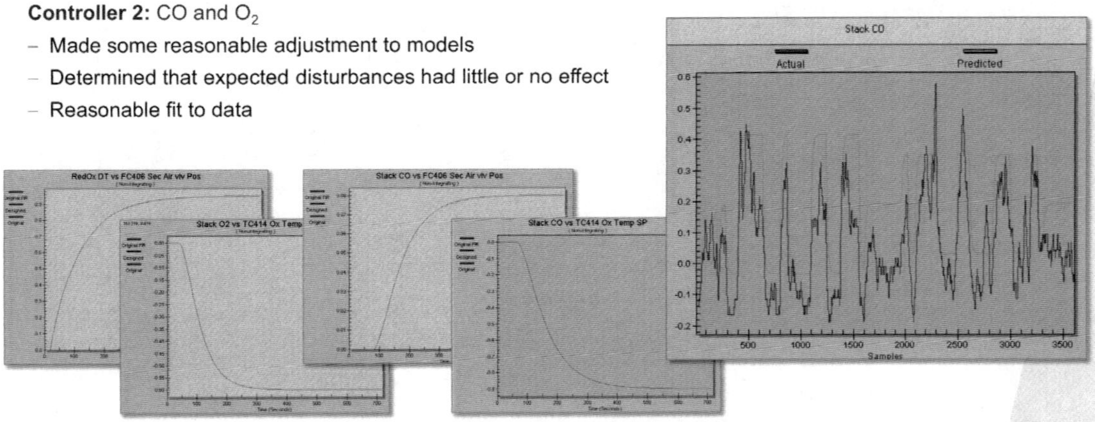

Business Results Achieved

ASCEND. PERFORMANCE MATERIALS

Business Results

Maintained CO limits

- MPC able to successfully maintain higher stack CO target
- Continued to operate well within permit limit
- No adverse affect to other emissions parameters

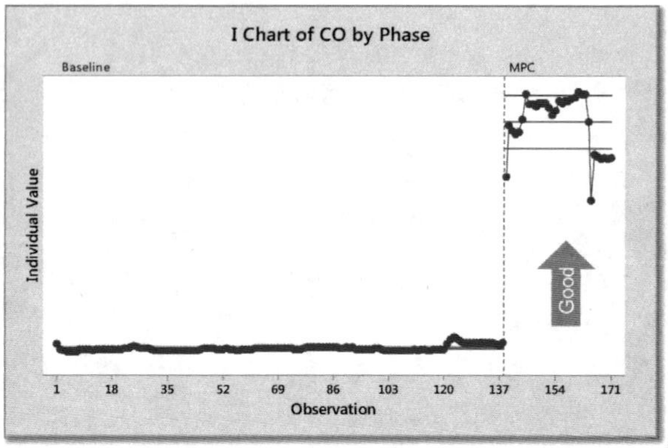

Business Results

Temperature

- Oxidizing Zone temperatures were reduced by 55°F
- MPC enforced RedOx ΔT constraint – stability maintained

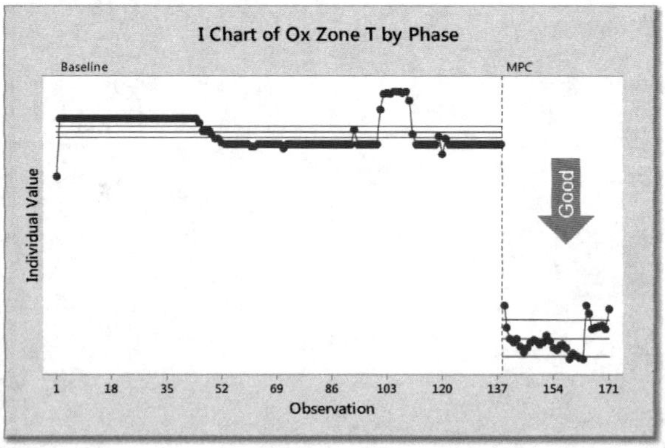

Business Results

Fuel Gas

- Realized significant reduction in fuel gas consumption for similar production rate
- If benefits sustained, performance reduces fuel consumption by ~12% annually

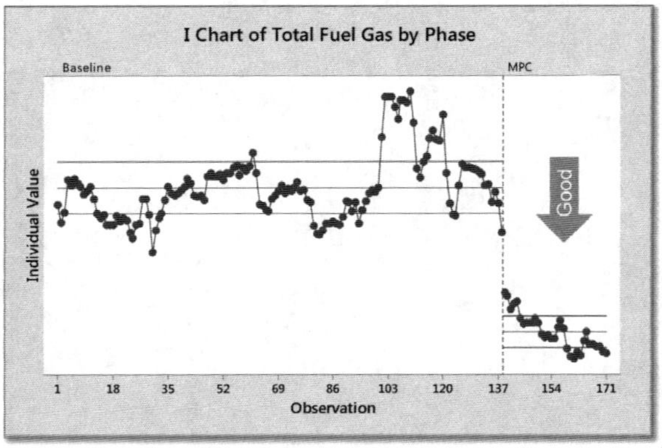

Business Results

Summary

- MPC successfully implemented on thermal oxidizer
- Objectives of reduced fuel usage and better heat recovery were met
- With sustained operation, expecting as much as 12% fuel savings annually

Questions?

ASCEND PERFORMANCE MATERIALS

ASCEND PERFORMANCE MATERIALS

North America
1010 Travis Street, Suite 900
Houston, TX 77002 USA
+1 713 315 5700

Europe
Watson & Crick Hill Park
11, rue Granbonpré - Bâtiment H
B-1435 Mont-St-Guibert
Belgium
+32 10 60 8600

Asia
Unit 3602, Raffles City
No. 268, Xizang Road (M),
Shanghai, China 200001
+86 21 6340 3300

inspiring everyday

ExxonMobil

Integration of Divided Wall Columns in a Tight Oil Processing Unit

Energy lives here™

Presented at:
9th AIChE Southwest Process Technology Conference
Session – Energy Efficiency
October 12-13th, 2017

Sundar Narayanan
sundar.narayanan@exxonmobil.com
Phil Melancon
phillip.s.melancon@exxonmobil.com
Kirtan Trivedi
kirtan.k.trivedi@exxonmobil.com

ExxonMobil
Research and Engineering

Global Outlook for Tight Oil
Liquids Supply

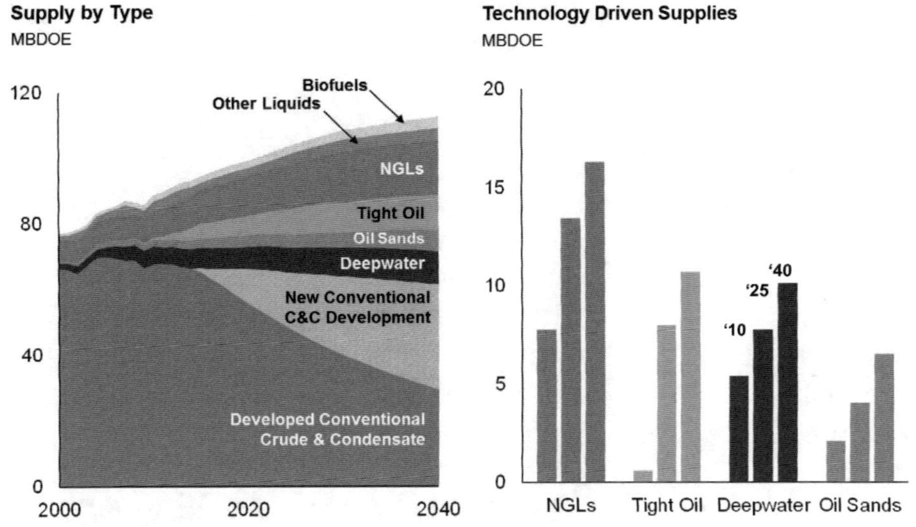

ExxonMobil Ref: XOM Energy Outlook 2040 2

Block Diagram of a typical Tight Oil Unit

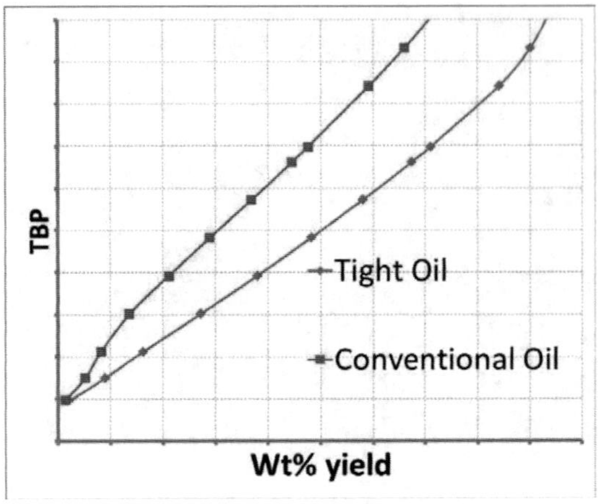

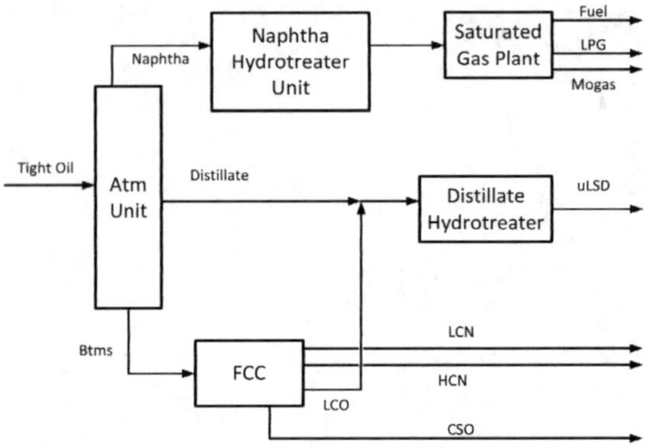

- Tight Oils contain higher distillate yields
- Low metal content Atmospheric tower bottoms can be directly sent to FCC unit
  - Helps in addressing FCC heat balance issues with low VGO yields
- This study assumes all new units except for FCC
  - Potential capacity debottleneck in existing units, DWC opportunity in separation sequence

ExxonMobil

3

Cooling Methods important in DWC processes

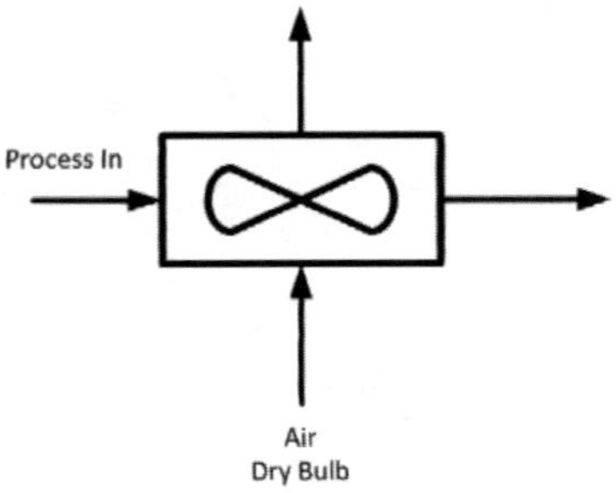

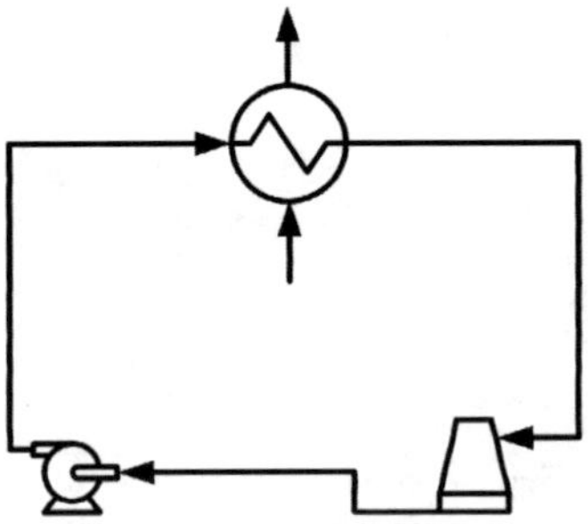

- Air Cooling
- Ambient Air
- Dry Bulb Temp
- Air side fouling
- **Tower Pressure - highest**

- Cooling Tower and water distribution facilities
- Approach to wet bulb temperature
- Potential CW fouling
- **Tower pressure - medium**

- Evaporative or Wet Surface Air Cooler
- Wet bulb temperature
- Low implementation in Refining
- Make up water, Quality, CoC
- **Tower Pressure – lowest; important parameter in DWC**

ExxonMobil

4

Saturated Gas Plant (SGP) stand-alone design

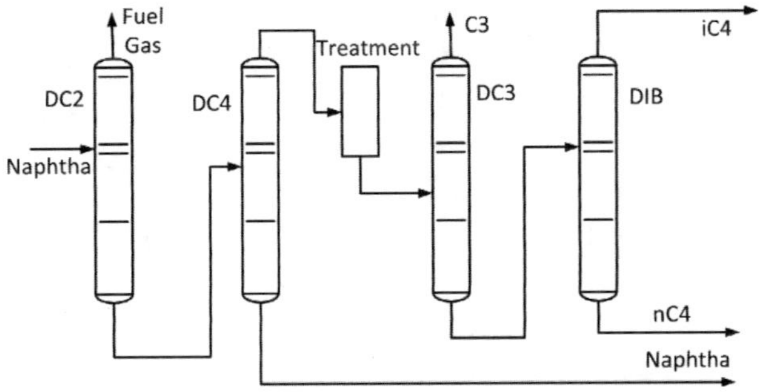

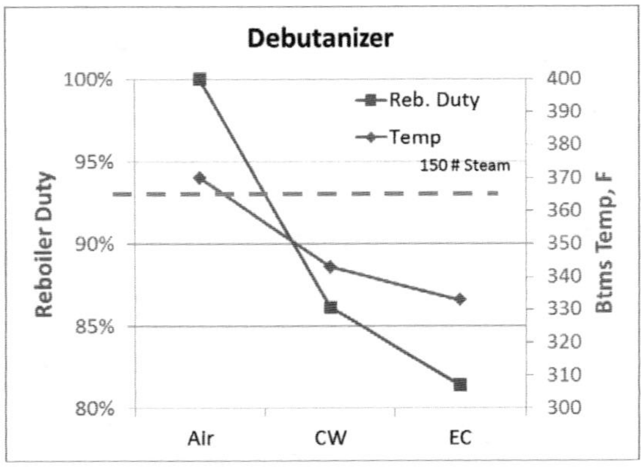

- Choice of cooling impacts separation, reboiler duty and hot utility type
- Decrease in pressure results in lower btms temp and lower reboiler duty
- With Air and CW, DC4 needs HP steam for reboiler
- With Evaporative cooling (EC), use of 150# steam is possible
- DC3/DIB lends itself to application of DWC

ExxonMobil

5

Energy Targeting for stand-alone SGP

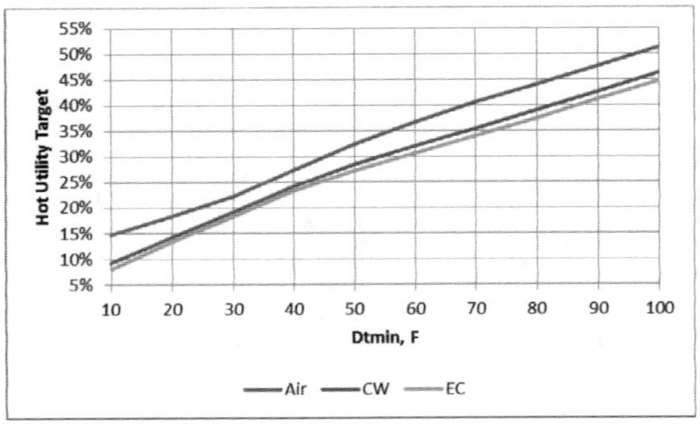

- Range targets enable capital / energy trade-off. 30-35% more energy is required at higher Dtmin
- Using CW or EC cooling can result in 5% savings in hot utility
- Design value of Dtmin changes the pinch determining stream
- Designer needs to be cognizant of this as the network topology will change

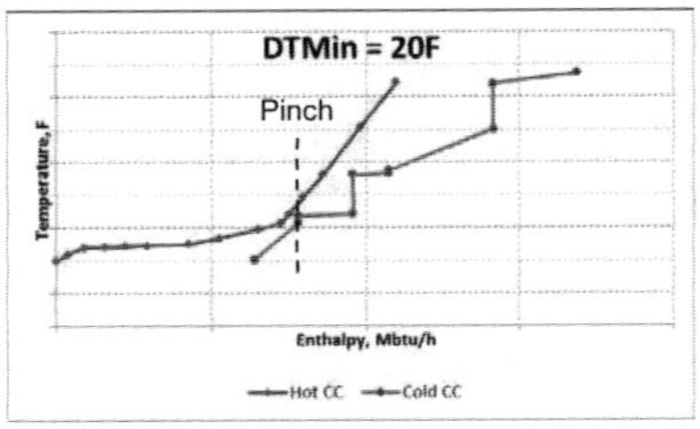

DIB Reboiler determines pinch point

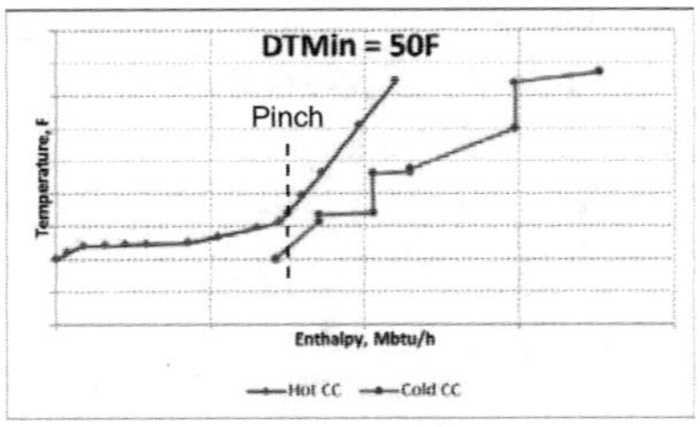

DC4 Condenser determines pinch point

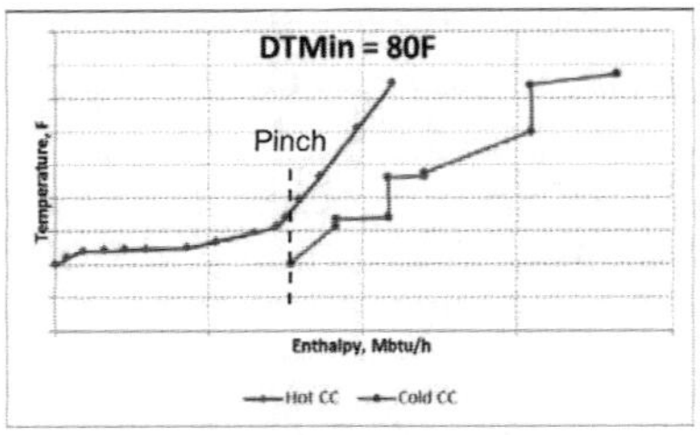

DC2 PA determines pinch point

ExxonMobil

6

Divided Wall Technology

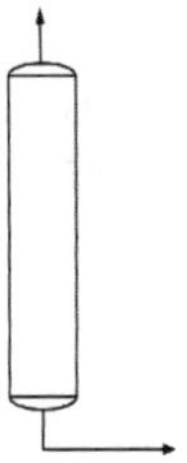

- Conventional distillation column
  - Leads to separation inefficiencies
  - Remixing of iC4 in DC3 column
- Petlyuk configuration and DWC address these inefficiencies
- DWC invented by R.O.Wright, Standard Oil, US Patent 2,471,134 (1949)
- Montz reports 120 applications
  - Unfixed dividing wall technology led to accelerated application of this technology
- BASF claims to have the largest number of commercial DWC installations

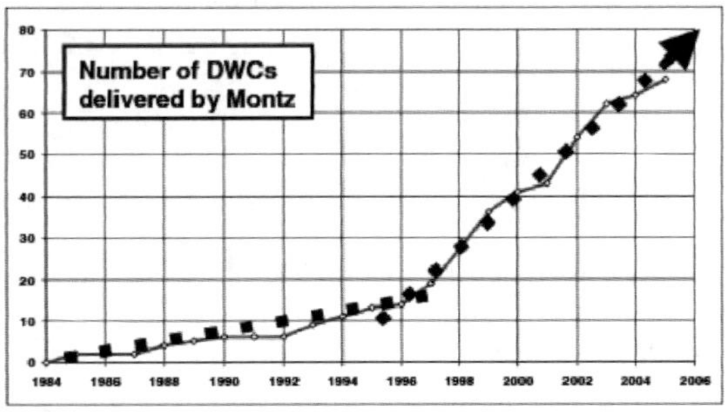

- Kaibel, Jansen, Olujic, " Unfixed Dividing Wall Technology for Packed and Tray Distillation Columns", Symposium Series no 152, IChemE, 2006
- http://montz.de/en/products/montz-dividing-wall-columns/, accessed 3/16/2016
- Yildirim, Kiss, Kenig, "Dividing Wall Columns in Chemical Process Industry: A review on current activities", Separation and Purification Technology 80, 2011, 403-417
- Schultz, Stewart, Harris, Rosenblum, Shakur, O'Brien, "Reducing Costs with Dividing-Wall Columns", CEP, May 2002

ExxonMobil

7

DWC column scheme

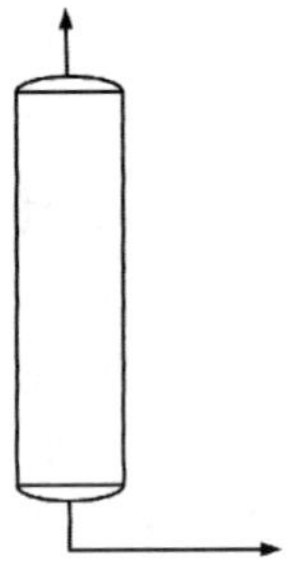

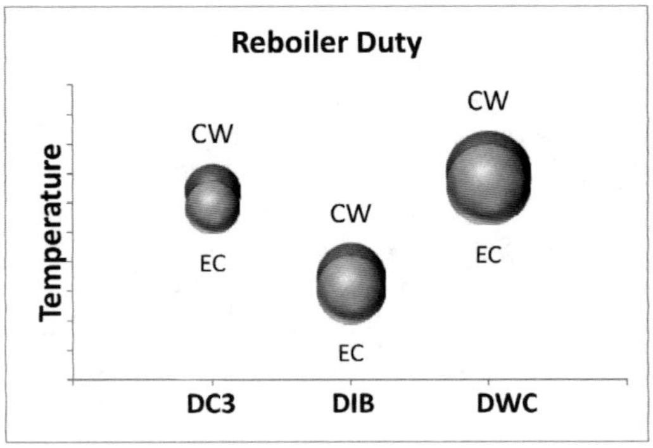

- Typical DWC application saves 20-30 %[1] in investment costs
  - Attractive with close-boiling mixtures (e.g., BTX separation)
- DWC for this case study motivated by plot-space, capital savings (equipment count)
  - DC3 / DIB separations challenge energy saving, especially with conventional cooling
- In this case study, energy savings is 5-10%
  - Lower than typical 25%[1] savings
- Less total energy required with DWC but partially at a higher temperature level
- Need to test DWC for feed flexibility, operability and control

1. http://montz.de/en/products/montz-dividing-wall-columns/, accessed 3/16/2016

ExxonMobil

8

Energy Impact at SGP Unit Level

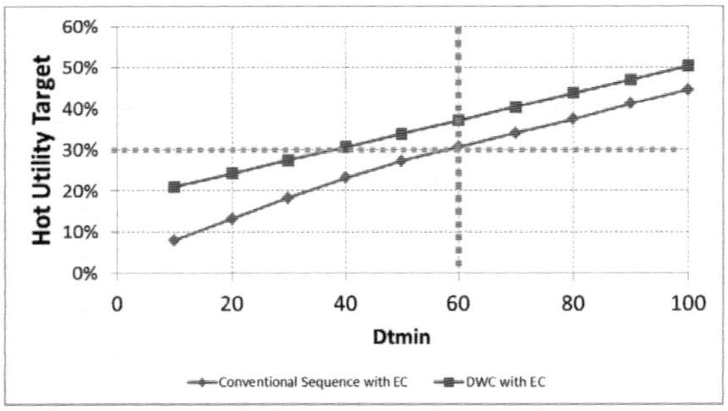

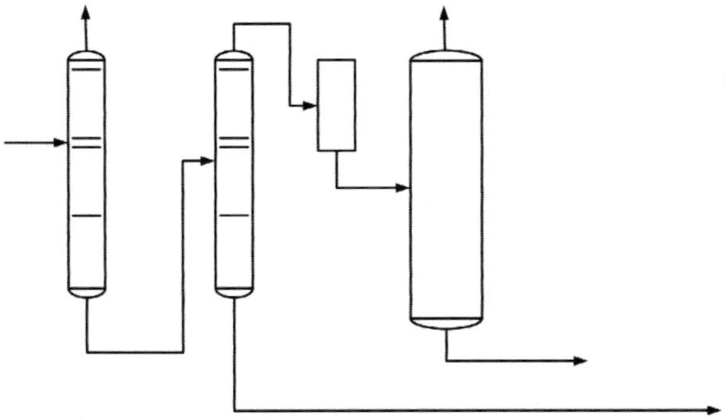

- At unit battery limits
  - For same Dtmin, total hot utility target is higher for DWC than conventional scheme !!!
  - Alternatively, for same energy target lower Dtmin required, higher capital
  - Need Process Integration tools to understand

ExxonMobil

9

Process Integration DWC vs Conventional

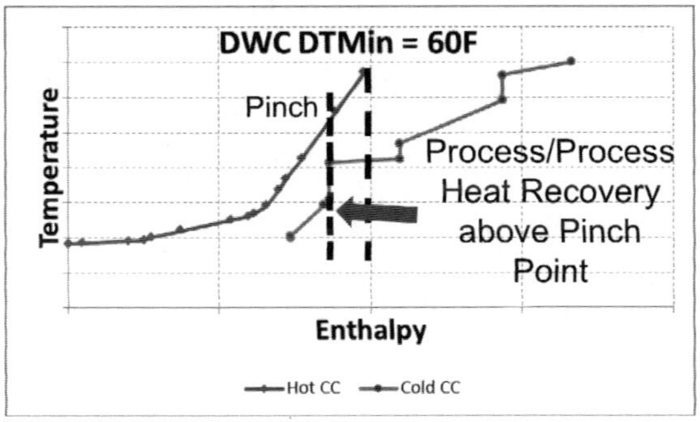

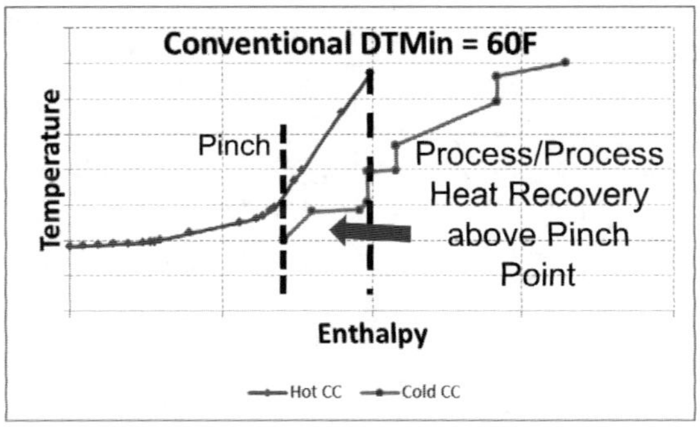

- DWC scheme – Pinch point is determined by DWC reboiler
  - Higher temperature
  - Process-to-process heat recovery is lesser
- Conventional scheme – Pinch point is determined by DC2 PA stream
  - Lower temperature
  - Process-to-process heat recovery is greater

ExxonMobil

10

Energy Efficient Refinery

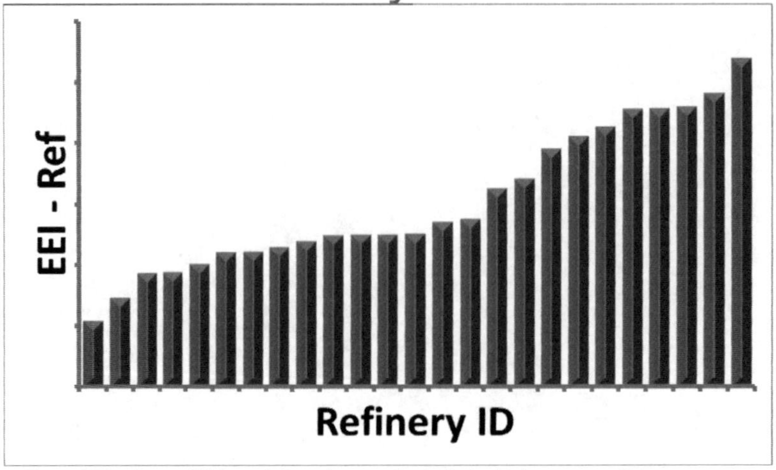

Note: - EEI – Energy Efficiency Index - Lower value indicates higher energy efficiency

- Energy Efficient Refinery has
  - ✓ High degree of inter-unit integration
  - ✓ Small number of large, efficient furnaces vs many small, inefficient furnaces
  - ✓ Less steam usage; Less steam generation in boilers
  - ✓ Less heat rejection to air and cooling water

ExxonMobil Silvestri and Trivedi, " Identifying Strategic Energy Opportunities Through Benchmarking", Presented at Texas Industrial Energy Manager's Forum, Houston, April 7th, 2016 11

Energy Impact at Site Level

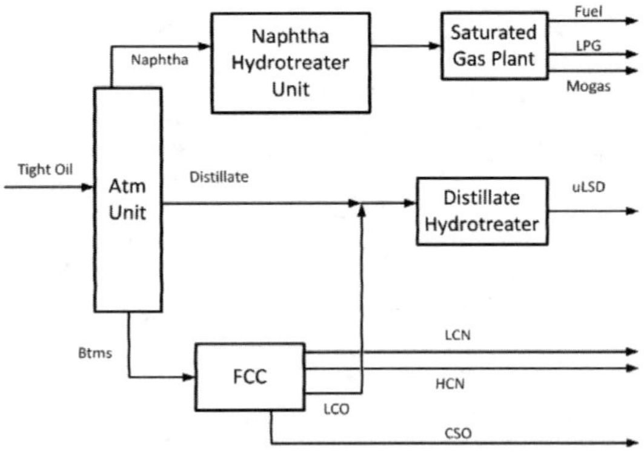

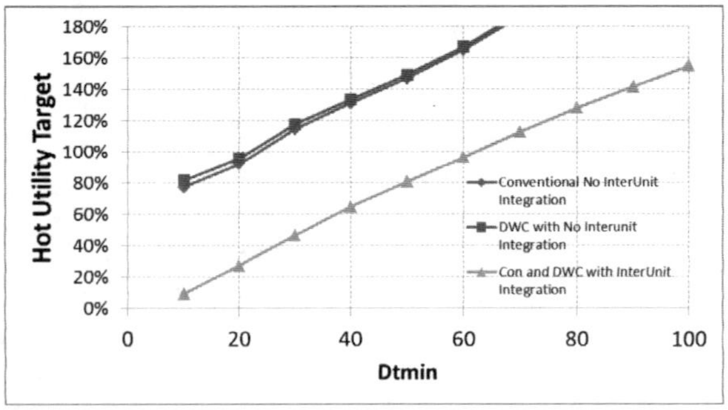

- From Energy target curves:
  - Without inter-unit integration, hot utility requirement
    - for DWC configuration > than for multi-tower configuration at lower Dtmin
    - gap between DWC & multi-tower configurations reduces at higher Dtmin
  - With inter-unit integration, hot utility requirement is similar for DWC and multi-tower configurations
- 60-75% hot utility saving possible by integrating all the new units
  - Operability, startup, shutdown issues should be evaluated

ExxonMobil 12

Site Wide Integration DWC vs Conventional

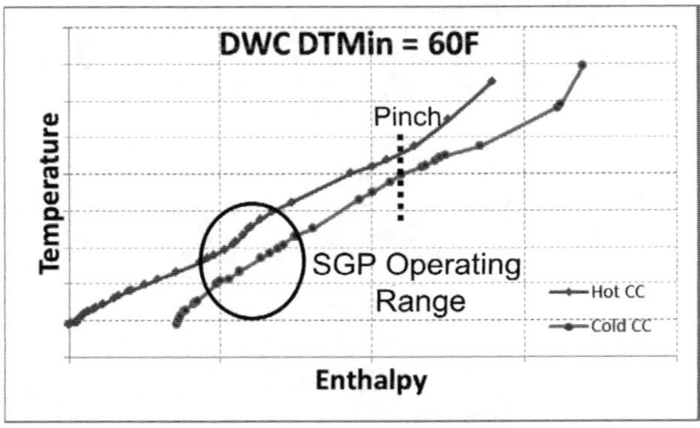

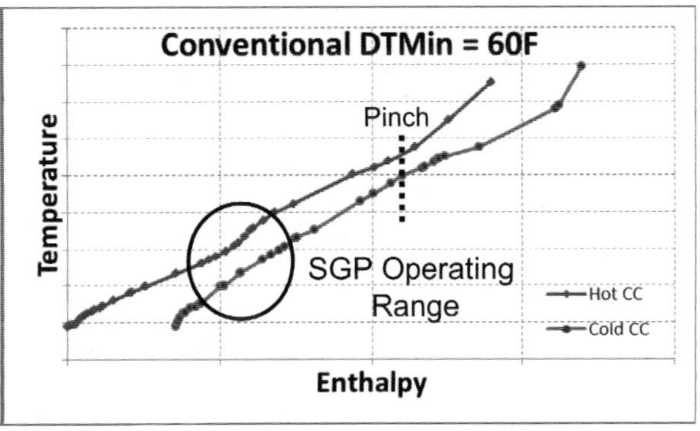

- In both cases the pinch point is determined by Distillate Hydrotreater (DHT) High Temperature Separator Liquid
  - Pinch point is towards the hot end of the process
  - In stand-lone SGP the pinch point was near the cold end of the process
- All SGP streams are below the pinch point with inter-unit integration
  - Energy differences between different SGP flow schemes will not impact overall hot utility requirement for the site

ExxonMobil 13

Summary

- DWC is a potential technology for capital savings
- DWC energy savings impact should be evaluated at site level
  - This study led to capital saving with no energy penalty on site level
  - Outcome would be different if considered on a stand-alone unit basis
  - Feed flexibility, operability and control issues of DWC should be addressed
- Inter-unit integration leads to energy efficient designs at site level
- Site-wide process integration studies need to be conducted to ensure economics are evaluated properly
- Pinch technology and process integration tools
  - Enhance the understanding of the thermodynamic effects governing the design
  - Evaluate range of options
  - Enable sensitivity studies

ExxonMobil 14

Smart Manufacturing Technologies and Data Analytics for Improving Energy Efficiency in Industrial Energy Systems

Robert W. Ivester, Ph.D.
Deputy Director
Advanced Manufacturing Office, DOE

Joe Cresko
Advanced Manufacturing Office, DOE

Sachin Nimbalkar, PhD
Oak Ridge National Laboratory

AIChE Southwest Process Technology Conference
Thursday, October 12,

ORNL is managed by UT-Battelle
for the US Department of Energy

OAK RIDGE National Laboratory | MANUFACTURING DEMONSTRATION FACILITY

AMO* Strategic Analysis Team

Joe Cresko
Advanced Manufacturing Office, DOE

William Morrow, Arman Shehabi, and Prakash Rao
Lawrence Berkeley National Laboratory

Alberta Carpenter
National Renewable Energy Laboratory

Diane J. Graziano
Argonne National Laboratory

Sujit Das and Sachin Nimbalkar
Oak Ridge National Laboratory

** Advanced Manufacturing Office (AMO)*

OAK RIDGE National Laboratory | MANUFACTURING DEMONSTRATION FACILITY

Why Energy Efficiency in Manufacturing Facilities is Important

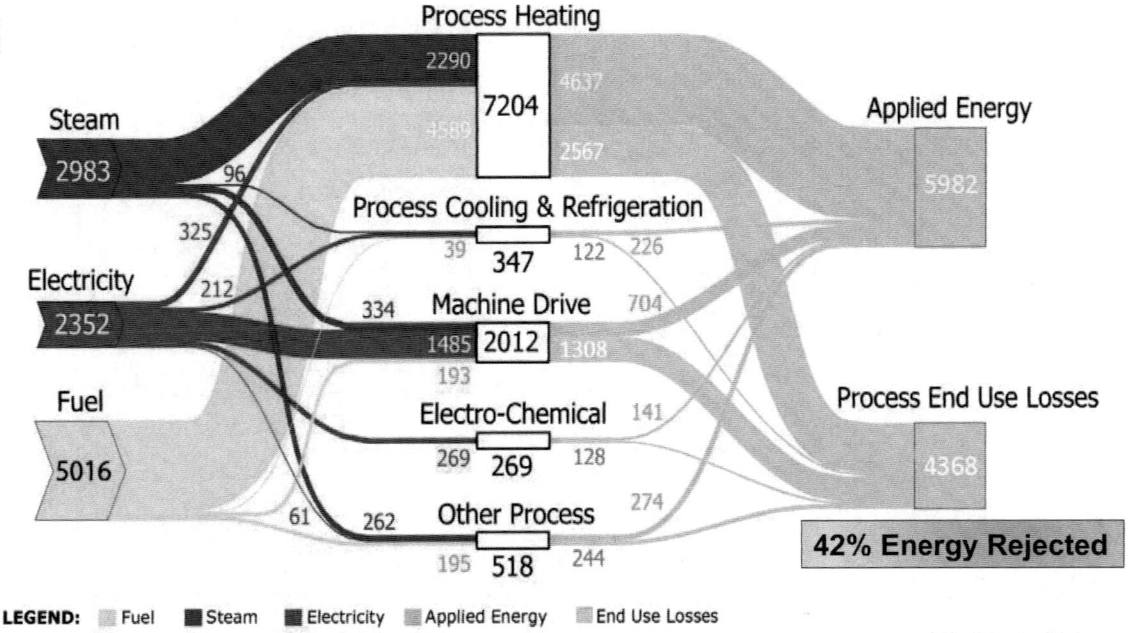

Source: Manufacturing Energy & Carbon Footprints, DOE

3 OAK RIDGE National Laboratory

Smart Manufacturing Technologies to Improve Energy Efficiency – Intelligent Efficiency

- NSF - Smart Manufacturing is about **Data → Information → Knowledge →Wisdom**" (DIKW).
- ACEEE 1 - The **integration** of **all facets of manufacturing** through the use of information and communication technologies
- ACEEE 2 – **Intelligent Efficiency** is the system level energy efficiency made possible by the use of **information and communications technologies and data analytics**

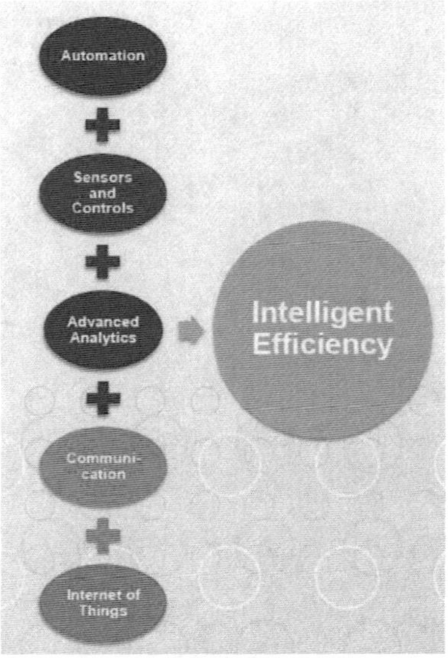

- NSF (National Science Foundation). 2015. NSF Workshop on Research Needs in Advanced Sensors, Controls, Platforms, and Modeling (ASCPM) for Smart Manufacturing. Atlanta, GA.
- ACEEE 1 - Elliott, R. N. 2016. Energy Efficiency and Industry: The National Trend. Washington DC: American Council for an Energy-Efficient Economy. http://aceee.org/blog/2015/12/energy-efficiency-and-industry
- ACEEE 2 – Ethan Rogers (ACEEE), Better Buildings Summit, Smart Manufacturing Session, May 16, 2017

4 OAK RIDGE National Laboratory

Typical Vs Intelligent Efficiency

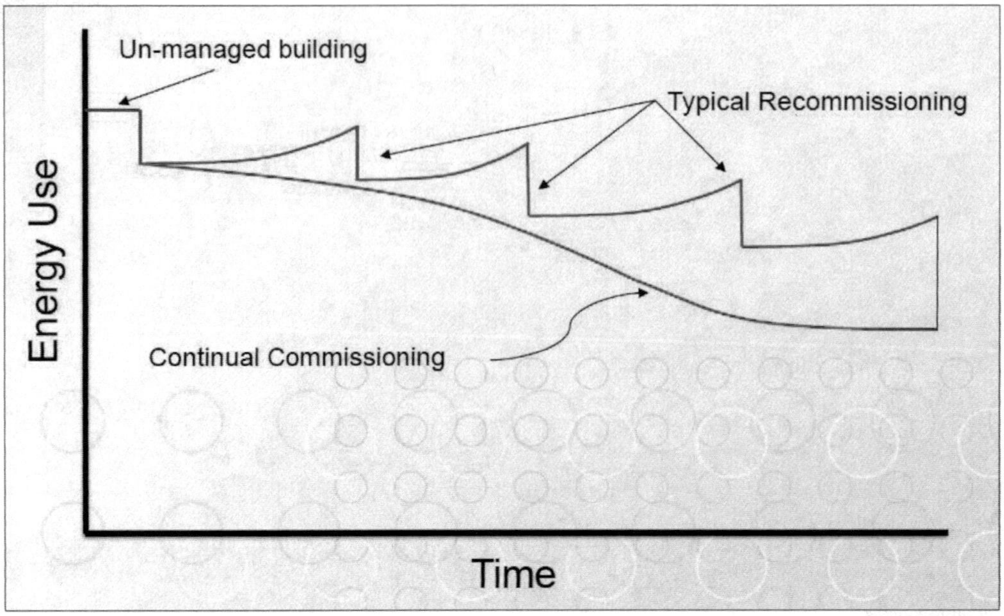

- ACEEE 2 – Ethan Rogers (ACEEE), Better Buildings Summit, Smart Manufacturing Session, May 16, 2017

5

OAK RIDGE National Laboratory

Implementation of Smart Manufacturing Technologies - Retrofitting vs Built-in Smart

A vibration sensor helps monitor a milling machine in a GE plant and transmits information wirelessly (Source – GE, WSJ 2016).

Smart manufacturing at the center of the fourth industrial revolution (Source - Ken Moir 2016)

GE, WSJ 2016 - Mary Catherine O'Connor, Bringing Smart Technology to Old Factories Can Be Industrial-Size Challenge
https://www.wsj.com/articles/bringing-smart-technology-to-old-factories-can-be-industrial-size-challenge-1465351322

Ken Moir 2016 - Smart manufacturing at the center of the fourth industrial revolution
https://www.nicelabel.com/blog/2016-09-06/smart-manufacturing-center-fourth-industrial-revolution/

6

OAK RIDGE National Laboratory

Smart Manufacturing Projects Hierarchy

- **Level Zero**
  - Project adding control systems without communication
    - Valuable, but not new
- **Level One**
  - Control systems exist
  - Project adds communication of control systems' status (includes sensor)
- **Level Two**
  - Control systems are added or modified
  - Project adds communication of control systems' status
- **Level Three**
  - Control systems include non-local information in decision making

Source - Muller, M. 2016. Smart Manufacturing and Industrial Analytics. US Department of Energy IAC Centers Training Materials.

OAK RIDGE National Laboratory

7

Explaining Smart Project Hierarchy Through an Example – Air/fuel Ratio Control

Air-fuel ratio control – **Level 0 Smartness** (Source: DOE 2015)

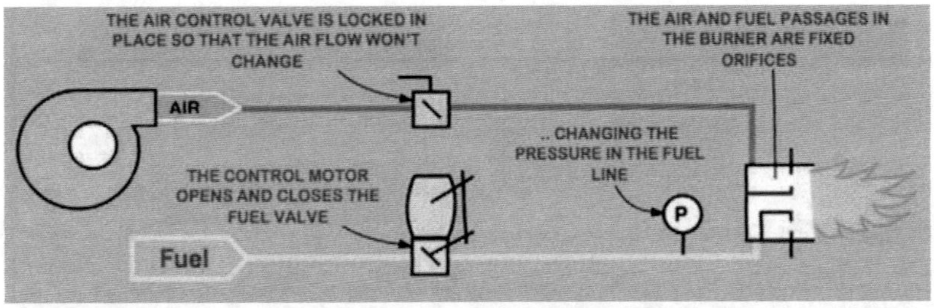

Fully metered mass flow control system with integrated oxygen sensors - **Level 3 Smartness** (Source: DOE 2015)

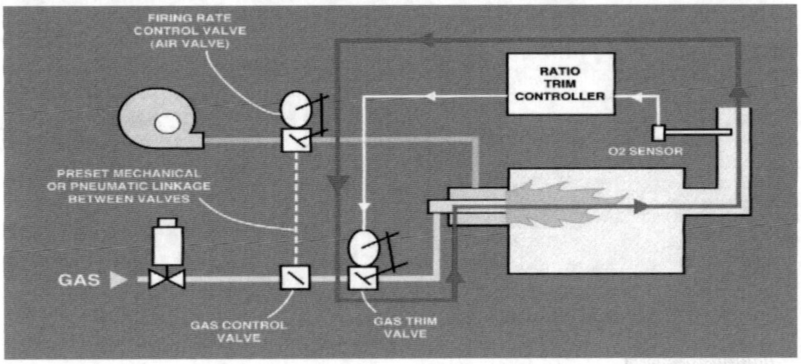

DOE (Department of Energy). 2015. Energy Efficiency Improvement for Process Heating Equipment, Workshop on Industrial Process Heating Systems – Qualified Specialists Training.

OAK RIDGE National Laboratory

8

One more Definition ...

Smart Manufacturing is the ability for **everyone in the organization** to have the **actionable information** they need, **at the time they need it**, so that they can contribute to the **optimal operation** of the enterprise through informed, data-based decision making.

Ethan A. Rogers The Energy Savings Potential of Smart Manufacturing - 2014

9 OAK RIDGE National Laboratory

Example 1 - General Mills Managing Energy as an Ingredient (Metering & Data Analytics)

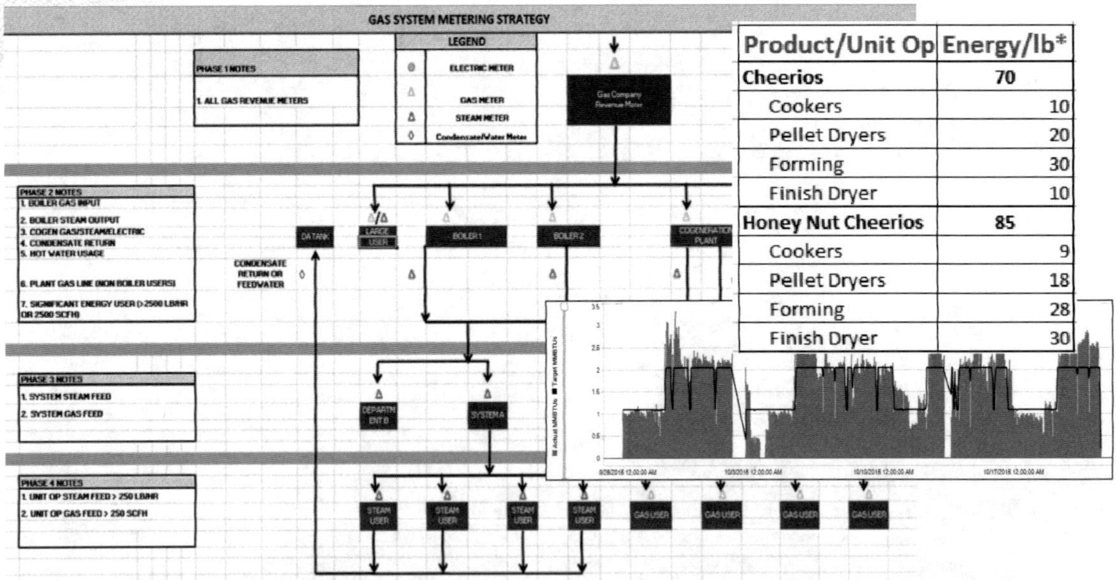

Product/Unit Op	Energy/lb*
Cheerios	**70**
Cookers	10
Pellet Dryers	20
Forming	30
Finish Dryer	10
Honey Nut Cheerios	**85**
Cookers	9
Pellet Dryers	18
Forming	28
Finish Dryer	30

Real-time energy monitoring at General Mills (General Mills 2016).

General Mills. 2016. Transforming Manufacturing: Smart Manufacturing and Advanced Data Analysis Drive Energy Savings. Washington DC: 2016 Better Building Summit.

10 OAK RIDGE National Laboratory

Example 2 - Celanese Energy Dashboards

Energy Dashboards to Engage Operators in Energy Optimization (Celanese 2016)

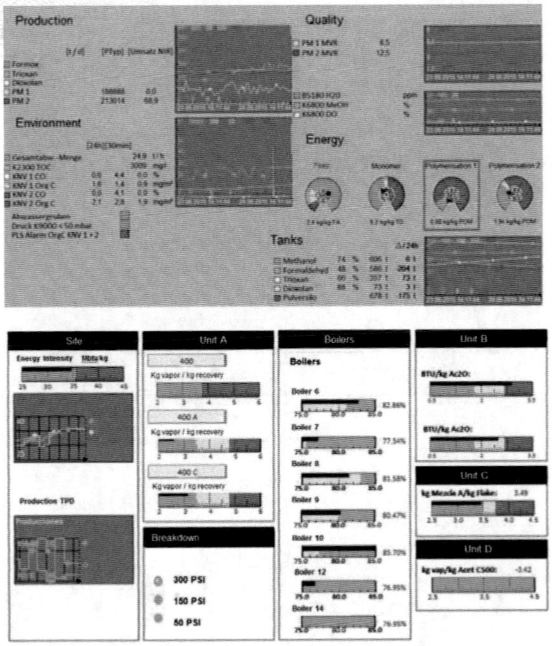

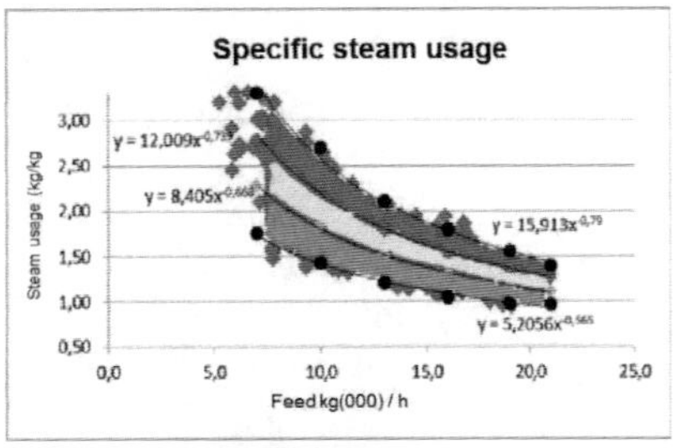

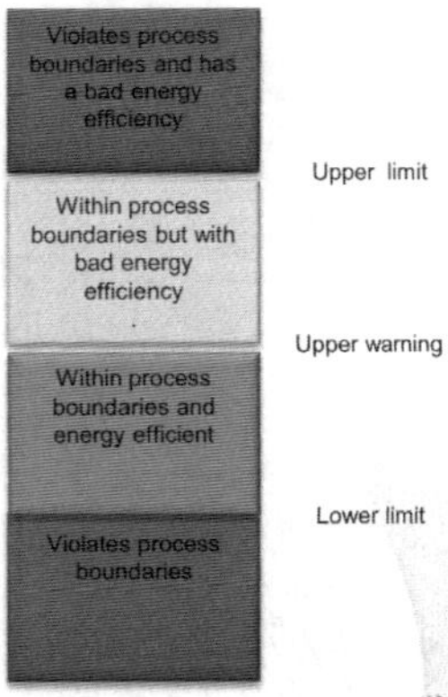

11

Celanese. 2016. Energy Dashboards—Driving Operator Engagement in Energy. Washington DC: 2016 Better Building Summit.

OAK RIDGE National Laboratory

What is the Potential Energy Impact? Economic Impact?

As per Clean Energy Smart Manufacturing Innovative Institute:

"Energy savings were estimated conservatively at 12.5% for each of first 5 years with steadily increasing market penetration of CESMII's SM technologies to 5% by year 5. In years 5 through 10, energy savings of 20% were assumed with market penetration steadily increasing from 5% to 15%."

There is a Need for the Sophisticated Energy Impacts Analysis

Source: Ray Collett, CEO, CESMII – AMO Peer Review June 2017.

12

OAK RIDGE National Laboratory

Smart Manufacturing Energy Efficiency Technologies Implemented During IAC Assessments That Could Be Upgraded to Higher Levels of Smartness

Energy Efficiency Opportunity	Energy System	No. of times Implemented (2000-16)	Avg. Savings % of Total Plant Source Energy	Avg. Savings % of Total Plant Energy Cost	Avg. Payback (Year)
Improve combustion control capability	Process Heating	36	2.6%	2.4%	2.2
Install automatic stack damper	Process Heating	10	0.9%	0.8%	1.2
Enhance sensitivity of temperature control and cutoff	Process Heating	12	2.6%	3.6%	0.7
Use power factor controllers	Motor	90	0.3%	3.5%	1.6
Install motor voltage controller on lightly loaded motors	Motor	10	1.5%	1.5%	2.2
Use multiple speed motors or adjustable-frequency drives (AFDs) for variable pump, blower and compressor loads	Pump	391	2.1%	2.0%	1.9
Install timers on light switches in little used areas	Lighting	73	1.0%	1.0%	0.8
Use photocell controls	Lighting	267	1.0%	0.9%	0.7
Install occupancy sensors	Lighting	1183	0.7%	0.6%	1.2
Use computer programs to optimize heating, ventilation, and air conditioner (HVAC) performance	HVAC	29	4.5%	4.2%	0.5
Install outside air damper/economizer on HVAC unit	HVAC	72	4.1%	3.4%	1.5
Install timers and/or thermostats	HVAC	275	2.3%	2.0%	0.6

Most of the energy efficiency projects implemented based on DOE ESAs and IAC assessments were **level 0 or 1** (simple controls or control devices with sensors)

13

OAK RIDGE National Laboratory

Industry Level Impacts of Level 0 to 1 Smart Manufacturing Energy Efficiency Projects Implemented during the IAC Assessments

Industry	No. of times Implemented (2000-16)	Avg. Savings % of Total Plant Source Energy	Avg. Savings % of Total Plant Energy Cost	Avg. Payback (Year)
Food	443	0.9%	1.8%	1.2
Wood product	173	2.1%	2.4%	1.4
Paper	197	1.3%	2.1%	1.2
Chemicals	202	1.4%	2.1%	1.1
Plastics and rubber product	427	1.4%	2.2%	1.3
Primary metal	231	1.1%	1.4%	1.3
Fabricated metal product	379	1.9%	2.5%	1.2
Computer and electronic product	323	1.7%	1.9%	0.9
Elec. equip. appliances, components	153	1.8%	2.1%	1.4
Transportation equipment	194	1.6%	1.6%	1.3

14

OAK RIDGE National Laboratory

Conclusions

- SM technologies and data analytics approaches can **improve energy efficiency and reduce energy costs** in process-supporting energy system
- Many manufacturers are **already taking advantage of the increasing capabilities of technologies** such as energy management systems, advanced metering and sub-metering solutions, and internet-connected sensors
- Most energy efficiency technologies implemented based on DOE ESAs and IAC assessments were **level 0 or 1** (simple controls or control devices with sensors)
- Implementing higher-level technology levels (level 2 or 3) is **likely to increase energy savings** and could make savings sustainable
- There is a need for the **sophisticated energy impacts analysis**

15

OAK RIDGE National Laboratory

Questions

Robert Ivester (DOE)
Robert.Ivester@ee.Doe.Gov

OAK RIDGE National Laboratory | MANUFACTURING DEMONSTRATION FACILITY

Intensified Applications in Centrifugation

Southwest Process Technology Conference

Conference: Southwest Process Technology Conference

Year: 2017

Proceeding: 9th AIChE Southwest Process Technology Conference

Group: Southwest Process Technology Conference

Session: Process Intensification

Time: Thursday, October 12, 2017 - 11:00am-11:30am

Authors:

Pomaville, T., B&P Process Equipment and Systems
Paz, G.

Recent interest from the chemical industry for modular process intensification equipment has motivated another look at the potential for picking up where previous work was done on the topic to develop a centrifugal vapor liquid contactor. Features of such a unit which offer unique advantages in the field of fractionation are high capacity with a large number of theoretical plates, very low pressure drop, and short contacting times. Some preliminary testing of an acetone/water system has been done, the results of which are reported herein along with the analytical method that was used to arrive at experimental conclusions. The approach itself is intended to provide a qualitative view of an entire spectrum of fractionation applications, characterized in terms of one number, the flow factor. There are conditions where the centrifugal contactor appears to be efficient. Because this appears to occur at low flow factors that correspond to vacuum and steam stripping operations, the results are encouraging. At the time this work was done, a capacity for the unit had not yet been determined. It is, however, presumable that the optimum efficiency could be achieved by varying rotor speeds.

Intensified Conversion of Natural Gas Using Proton-Conducting Ceramics

Southwest Process Technology Conference

Conference: Southwest Process Technology Conference

Year: 2017

Proceeding: 9th AIChE Southwest Process Technology Conference

Group: Southwest Process Technology Conference

Session: Process Intensification

Time: Thursday, October 12, 2017 - 9:30am-10:00am

Authors:

Kjølseth, C., University of Oslo,

Today, more than 50 million ton of hydrogen is produced annually, predominantly through steam methane reforming (SMR) with subsequent water gas-shift combined with pressure swing adsorption (PSA) for product separation. This technology is, however, not suitable for small scale on-site reforming systems due to high capital costs, moderate energy efficiency and large system size. Further, methane is considered to be an alternative source for synthesis of valuable products currently obtained by crude oil processing. Methane dehydroaromatization (MDA) under non-oxidative conditions over cation-modified zeolites has received considerable attention as a promising route for direct conversion of methane into highly value-added chemicals such as aromatics and hydrogen. The MDA process suffers however from several shortcomings that impose serious limitations for commercialization. The reaction is limited by thermodynamics and fast deactivation of the catalyst on-stream due to accumulation of carbon deposits on the catalyst surface.

For both processes we demonstrate that a galvanically driven catalytic membrane reactor based on a ceramic proton conductor enables process intensification through integration of catalytic reactions, heat management and separation in one process step. A tubular membrane with electrolyte composition BaZr0.7Ce0.2Y0.1O3-ï¤ (BZCY72) (25-30 Î¼m) on a Ni-BZCY72 support is utilized with Ni and Cu as the anode for SMR and MDA experiments, respectively.

ExxonMobil

October 12th, 2017

Reverse Flow Reactors: Process Intensification for High Temperature Reactions

Energy lives here

Federico Barrai[2], Dhaval Bhandari[1], Frank Hershkowitz[1], Charlie Chun[1], James R. Lattner[2], Elizabeth Mahoney[2], Brian Weiss[1]

[1]ExxonMobil Research and Engineering Company, [2]ExxonMobil Chemical Company

9th AIChE Southwest Process Technology Conference
Process Intensification Session
Galveston, TX

Chemicals - An Essential and Growing Part of Life

Chemicals Demand by Region*

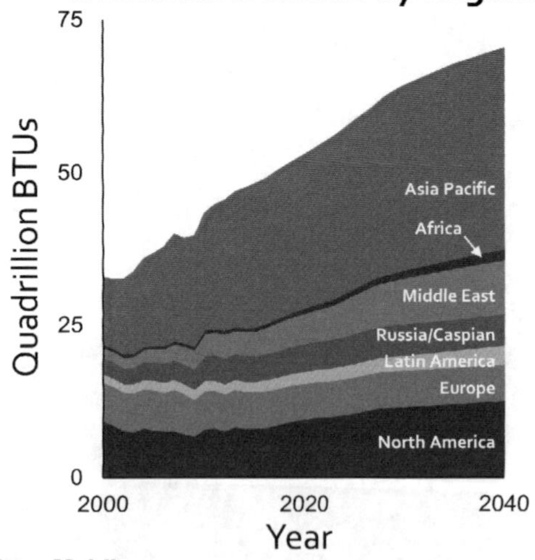

Growing Middle Class

Rapid Urbanization

New Advantaged Feedstock

ExxonMobil

*2017 ExxonMobil Energy Outlook

2

Endothermic Reactions Vital for Chemicals Production

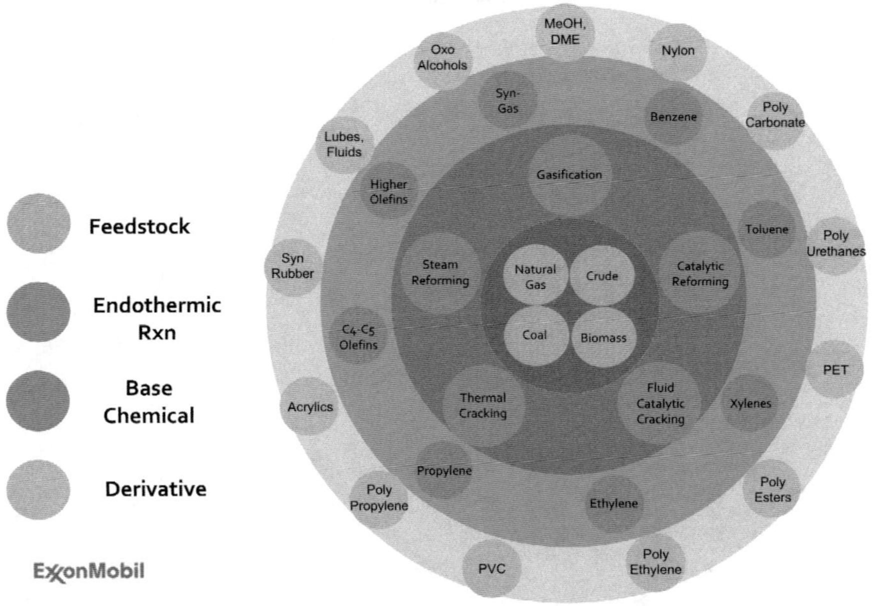

ExxonMobil

3

Key Endothermic Reactors – A Short Review

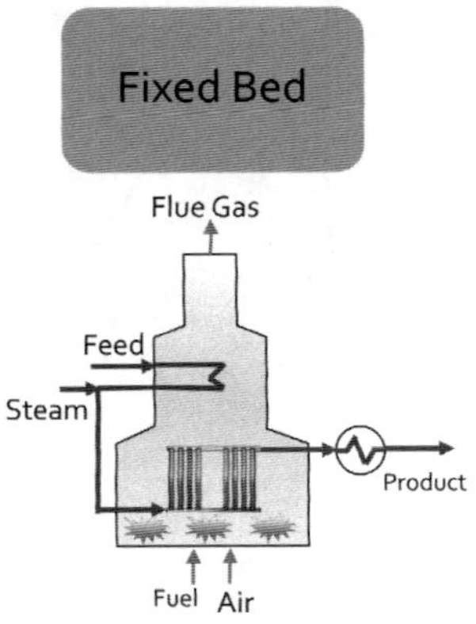

Typical Rxn: Steam cracking, steam reforming

Temperature: 700-1000 °C

Feed res. time: 0.1-0.5 s

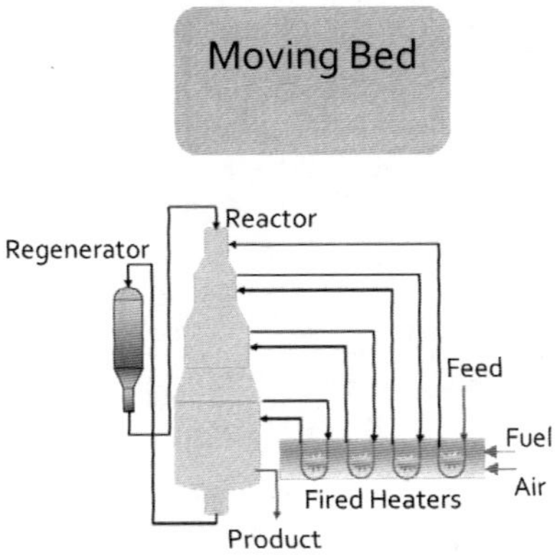

Typical Rxn: Naphtha reforming, catalytic dehydrogenation

Temperature: 400-600 °C

Feed res. time: 10-30 s

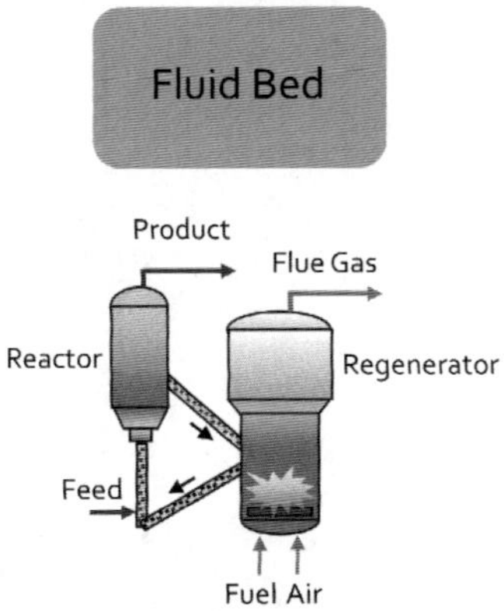

Typical Rxn: Fluid catalytic cracking, fluid coking

Temperature: 400-600 °C

Feed res. time: 2-5 s

ExxonMobil

4

Can PI Concepts be Applied to Endothermic Reactors?

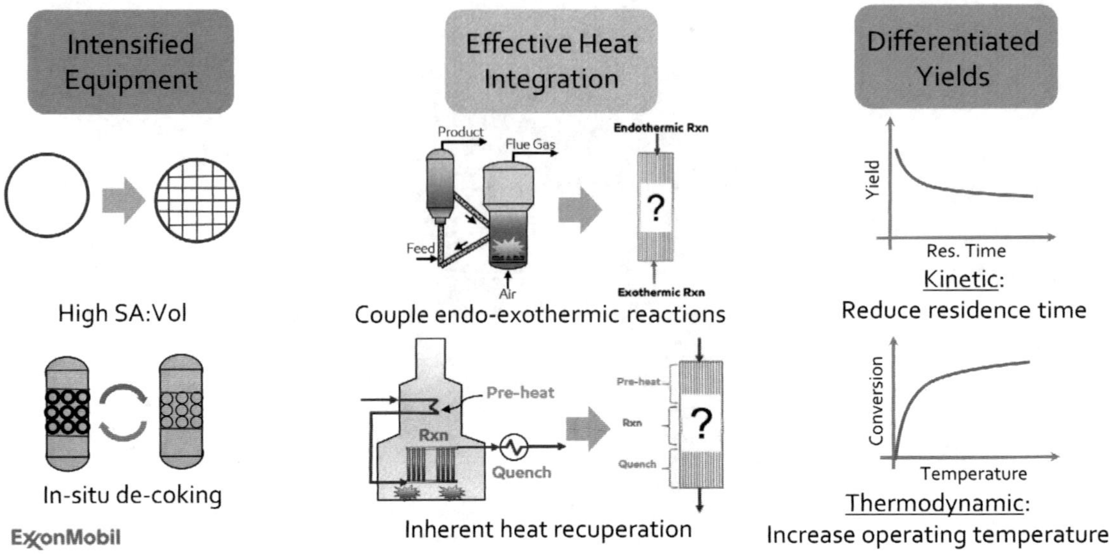

ExxonMobil

5

Reverse Flow Reactors (RFR) – An Efficient Way to Execute Endothermic Reactions

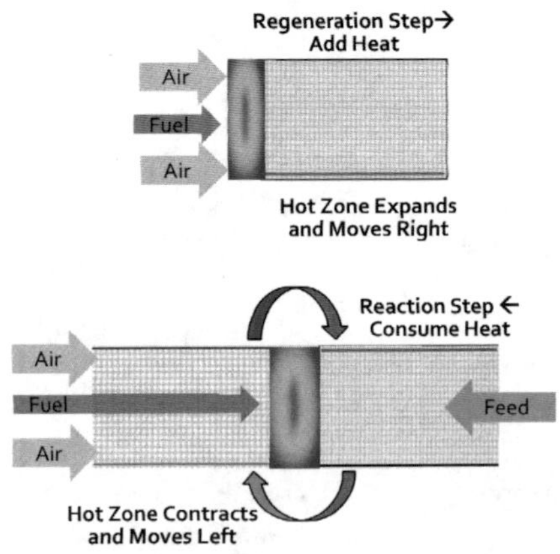

Consider a packed bed
- Combust in 1st step to add heat
- Gas/solid heat exchange creates temperature wave

Add a second packed bed
- Reverse flow and
- Rxn in 2nd step to consume heat

Cycle every 3–60 seconds
- Establish steady state temperature profile
- Very hot in middle, cool at ends

ExxonMobil

6

Reverse Flow Reactors Technology - Background

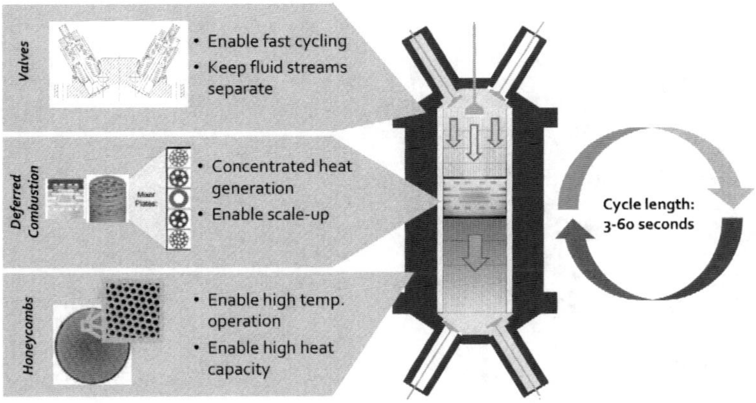

Heat input
Fuel Combustion
(Coke burn-off)

ExxonMobil

Reverse Flow Reactor for Hydrogen Production

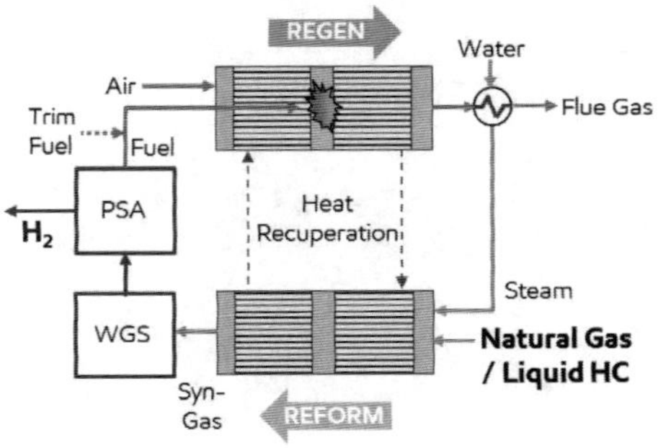

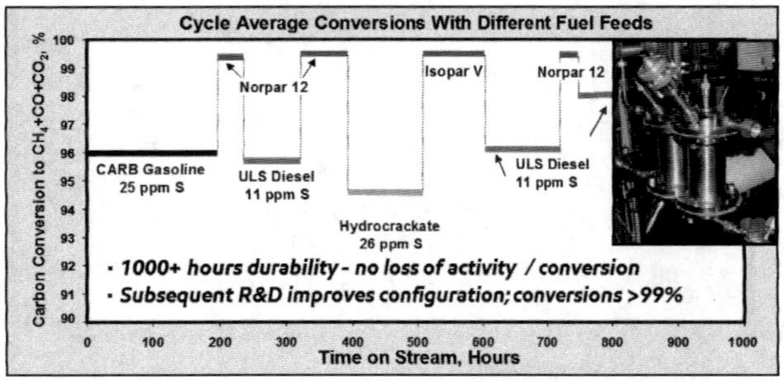

Technical Challenge
Metallurgy limit
Air separation cost
Reforming at High P
Energy efficiency

RFR Features
High temperature ceramics
Self-cleaning/decoking
High pressure operation
Inherent heat recuperation

Impacts
Improved efficiency
Eliminates air separation
Lower CAPEX
Reduced energy, CO_2 emissions

ExxonMobil

8

An Integrated Approach to Advance PI Technologies

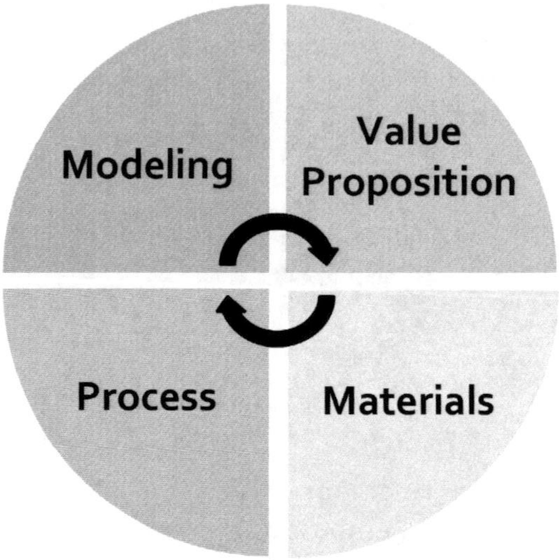

ExxonMobil

9

An Integrated Approach to Advance PI Technologies

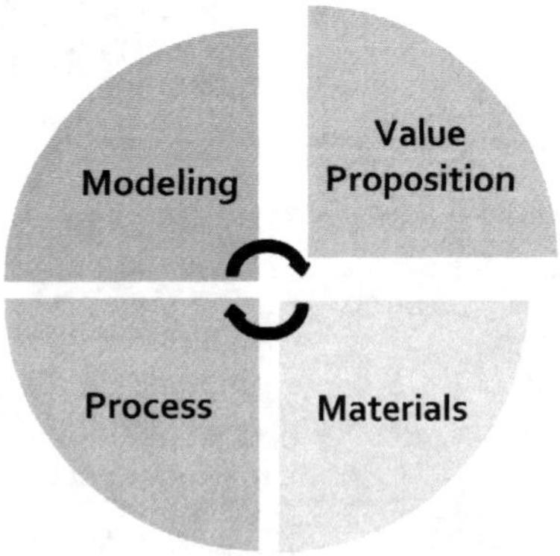

Value Proposition

CAPEX	OPEX	Yield
Footprint	Operability	Sustainability

- What is the key incentive for new development?
- Opportunity identification and assessment (Ideation → Define → Screen → Prioritize)
- Develop broader and iterative metrics → NPV, energy, differentiated yields, CO_2

ExxonMobil

10

An Integrated Approach to Advance PI Technologies

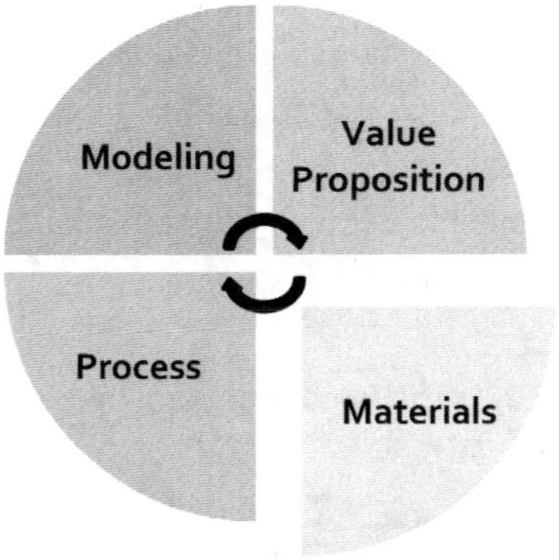

Materials and Internals

- Materials know-how, advanced manufacturing and characterization tools are all critical
- Applications will be challenging. Material durability and internals scale-up not trivial

ExxonMobil 11

An Integrated Approach to Advance PI Technologies

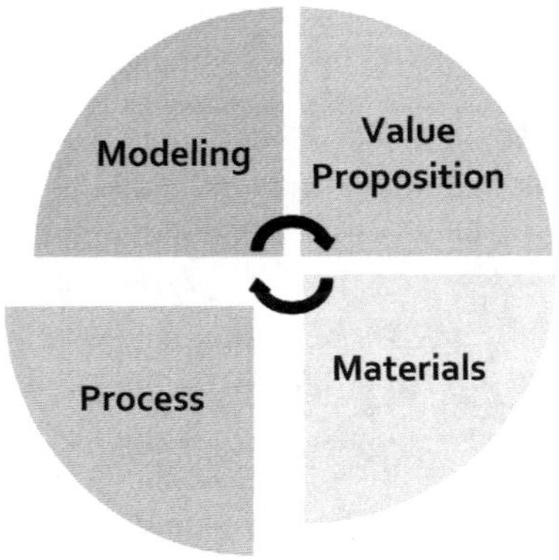

Process Development

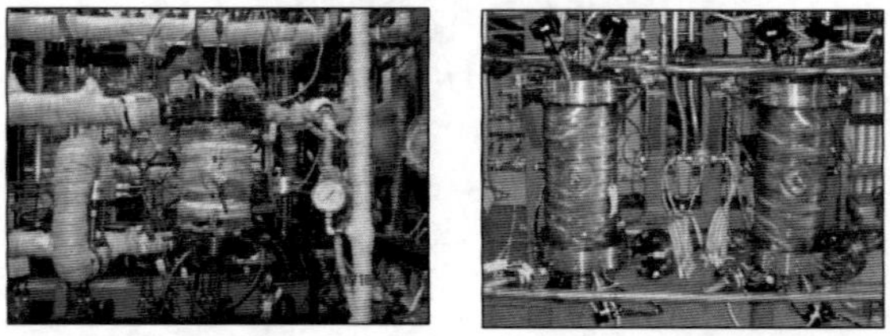

- Identify key technical, commercial risks. Develop pilot units tailored to meet application needs, scale and complexity
- Close coordination between stakeholders as technology progresses (Concept → Development → Deployment)

ExxonMobil 12

An Integrated Approach to Advance PI Technologies

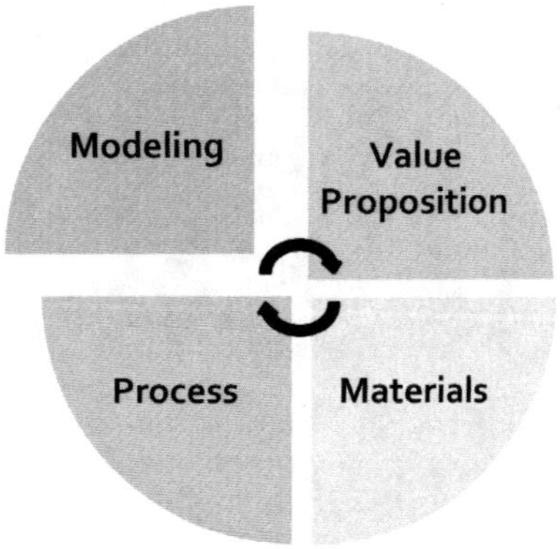

Modeling and Fundamentals

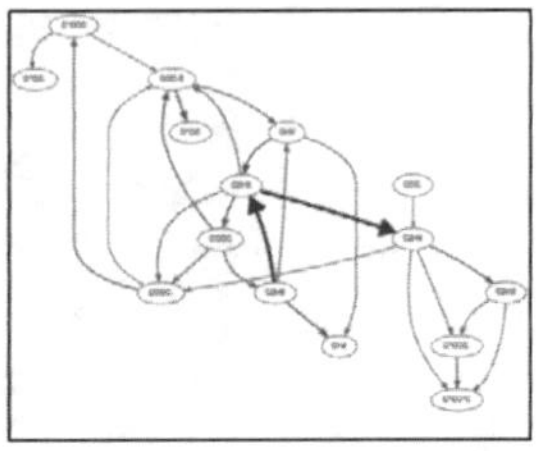

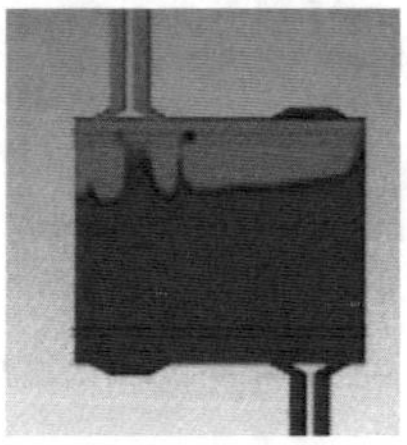

- Kinetic, dynamic, CFD, flow sheet models are all important to fully understand technology
- Models must provide guidance for material selection, process optimization

ExxonMobil

Some Thoughts

- High temperature endothermic reactions are a key part of modern chemical industry
- Highly developed, integrated nature of conventional reactors paired with high CAPEX and slow replacement rates make it a tough challenge!
- Reverse flow reactors can be an advantageous manufacturing tool on several fronts → Capital, Energy, Yields, CO_2...
- PI is not just a buzzword. Number of commercial applications are in practice, however several barriers exist
- A multi-disciplinary effort is required to advance any new PI technology

ExxonMobil

ExxonMobil

The Linde Ammonia Concept (LAC™)

Simplifying NH_3 Production for Improved Economic Feasibility

Rameshwar Hiwale and Carrie Lalou

October 12, 2017

THE LINDE GROUP

Linde

Agenda.

THE LINDE GROUP

Linde

1. Major Characteristics of the Linde Ammonia Concept
2. The Linde Ammonia Concepts
3. Key Elements
4. Experience References

Linde Ammonia Concept (LAC™)
Characteristics

- Generation of an inert-free ammonia synthesis gas in the right H_2 to N_2 ratio
- Adjustable CO_2/NH_3 Product Ratios
- Design feedstock flexibility from light hydrocarbon feedstocks (LAC.L) to heavy hydrocarbon feedstocks (LAC.H)
- Flexibility for valuable by-products, extension, and integration

3

Linde Ammonia Concepts (LAC)
Overview

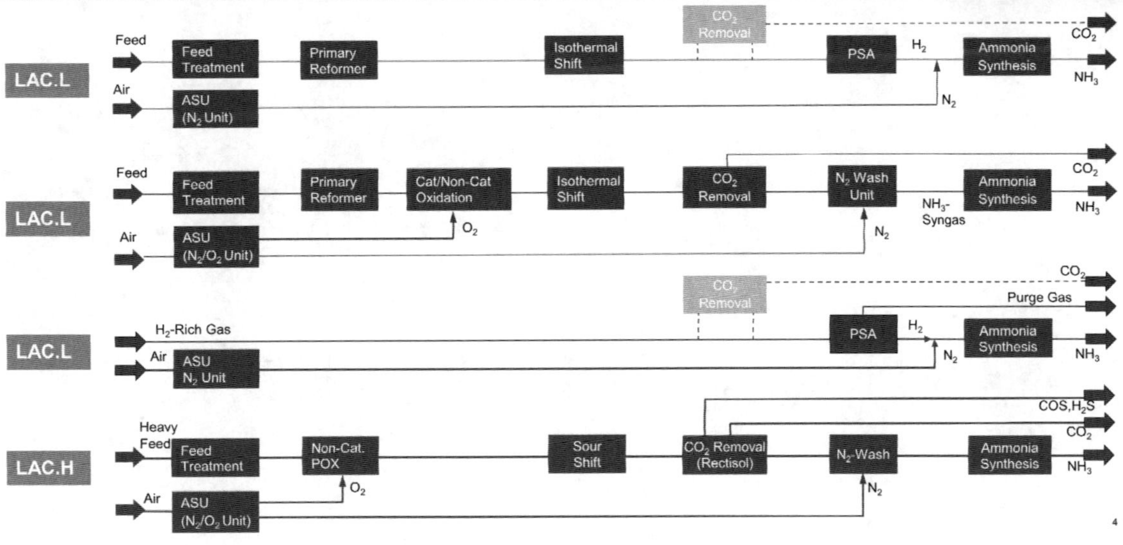

4

Linde Ammonia Concept (LAC_L)
Light Hydrocarbon Feedstocks

A combination of proven technologies

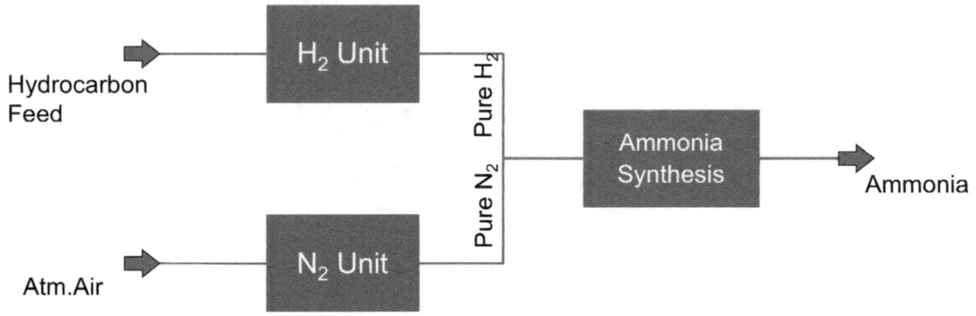

5

Linde Ammonia Concept
Light Feedstocks LAC.L1

The first plant went into operation in 1996 in India for GSFC with a capacity of 1350 mtpd NH_3; it is currently operating at 1475 mtpd NH_3.

- Natural gas feedstock
- Still one of the most energy efficient ammonia plants
- Steam reforming-based H_2 plant with Isothermal shift reactor
- Nitrogen generation unit with argon by-product
- Inert-free syngas

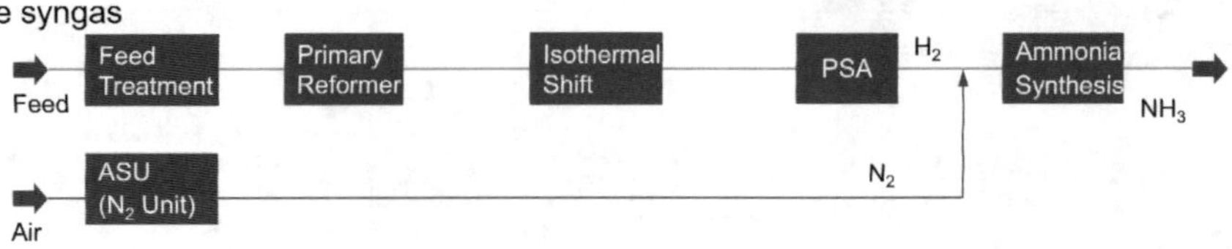

6

Linde Ammonia Concept
Comparison of LAC.L1 process with conventional scheme

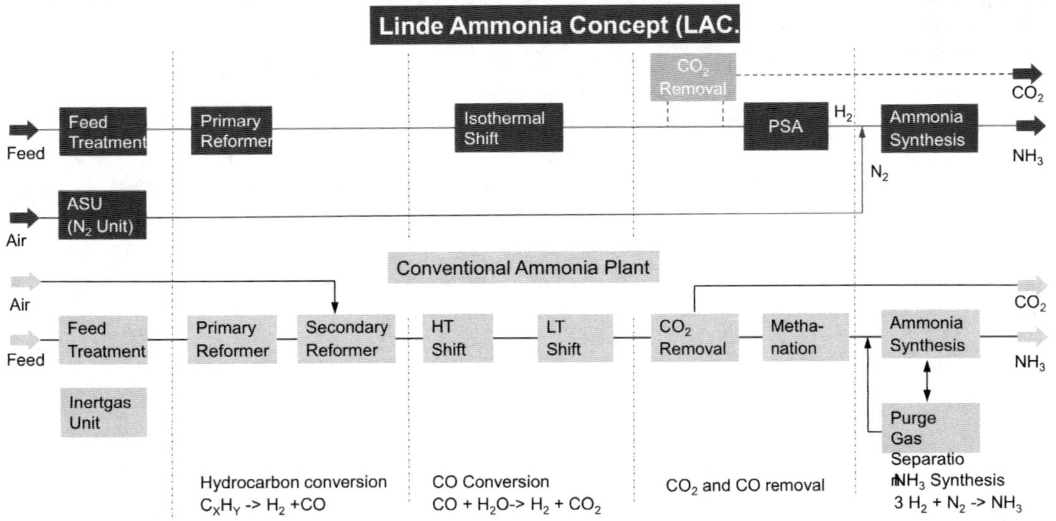

7

Linde Ammonia Concept (LAC.L1)
Flexibility in plant design and operational reliability

- CO_2 removal is eliminated for pure NH_3 production
  If CO_2 is needed, size of CO_2-removal is adjusted according to CO_2 demand
  CO_2 removal can be integrated at a later stage with no pre-investment
- Pure hydrogen and pure nitrogen are directly available from process streams
- Other potential by-products such as O_2, Ar, CO, CO_2 and MeOH can be easily integrated
- High flexibility with regard to steam and driver concepts allows perfect integration into existing sites or downstream units
  Process steam system is separated from export and driver steam system
- Provision of a sophisticated gas purification system by pressure swing adsorption (PSA), which has proven and unmatched reliability and even allows NH_3 production without shift conversion in operation

8

Linde Ammonia Concept (LAC.L1)
Cost and energy savings

THE LINDE GROUP Linde

- Elimination of three catalytic process steps; reduces the total catalyst volume to approximately 50% of that in a conventional plant
- N_2 is added at the back end of the process resulting in smaller and more efficient syngas cooling
- The generation of an oxygen and inert-free synthesis gas, giving important savings in the synthesis loop and eliminating a purge gas purification step
- Process condensate treatment is eliminated by routing the process condensate directly to the isothermal shift reactor for the production of process steam
- Simplified flowsheet also results in a reduced pressure drop through the system, reducing compressions requirements
- LAC is a much more direct route to ammonia resulting in reduced start-up times and important lifecycle cost savings

9

Linde Ammonia Concept (LAC.L1)
Summary

THE LINDE GROUP Linde

Overall simplification of the classical process route, resulting in

- Savings in investment, catalyst replacement and spare part costs, construction time, site area, and maintenance
- Flexibility for extension and integration
- Increased reliability

10

Linde Ammonia Concept (LAC.L2)
Development from LAC.L1 to LAC.L2

THE LINDE GROUP Linde

Development drivers:

- Increase flexibility with respect to NH_3 to CO_2 ratio
- Increase energy efficiency
- Achieve higher plant capacities

Modifications of Linde Ammonia Concept (LAC.L2) compared with (

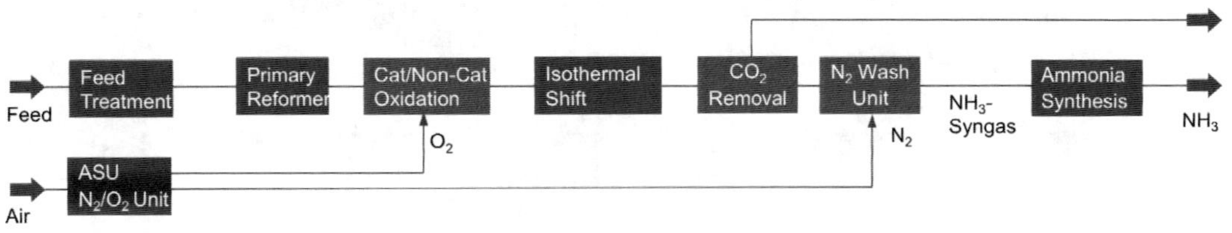

11

Linde Ammonia Concept (LAC.L3)
Spin off: Ammonia from H_2 – rich gases

THE LINDE GROUP Linde

Option 1

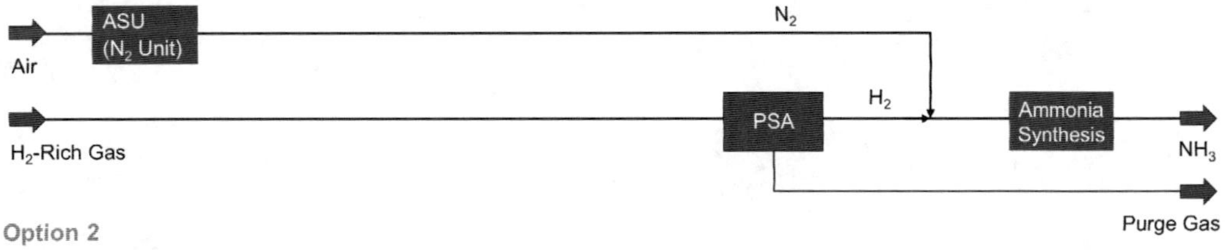

Option 2

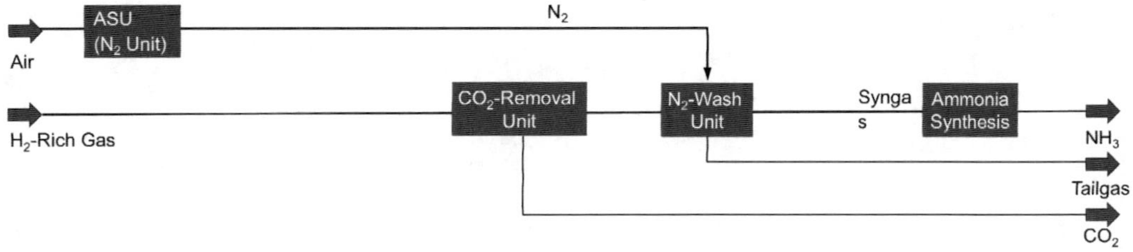

12

Linde Ammonia Concepts
Heavy Feedstock LAC.H

Ammonia/UREA plant, started up in 1981, designed for 1350 mtpd NH_3; revamped to 1750 mtpd NH_3; running at 1850 mtpd NH_3

- High pressure gasification of heavy fuel oil with oxygen
- Rectisol® wash system for syngas clean up and acid gas removal
- Liquid nitrogen wash system for an inert-free synthesis gas with optimum H_2/N_2 ratio
- Syngas from the plant is also used for other products such as methanol (MeOH), CO, etc

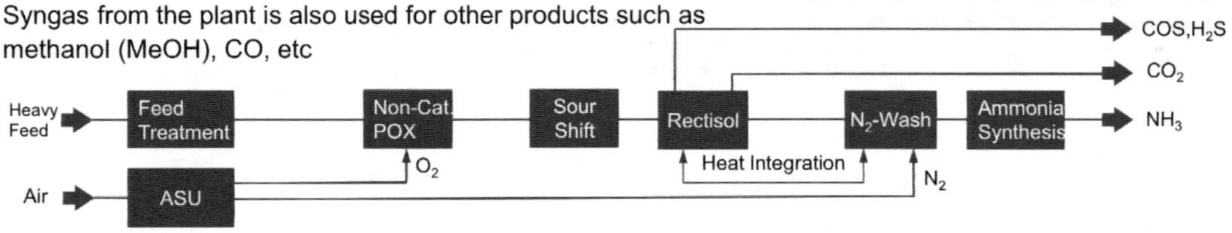

13

Agenda.

1. Major Characteristics of the Linde Ammonia Concept
2. The Linde Ammonia Concepts
3. Key Elements
4. Experience References

Linde Ammonia Concept (LAC_L)
Key elements

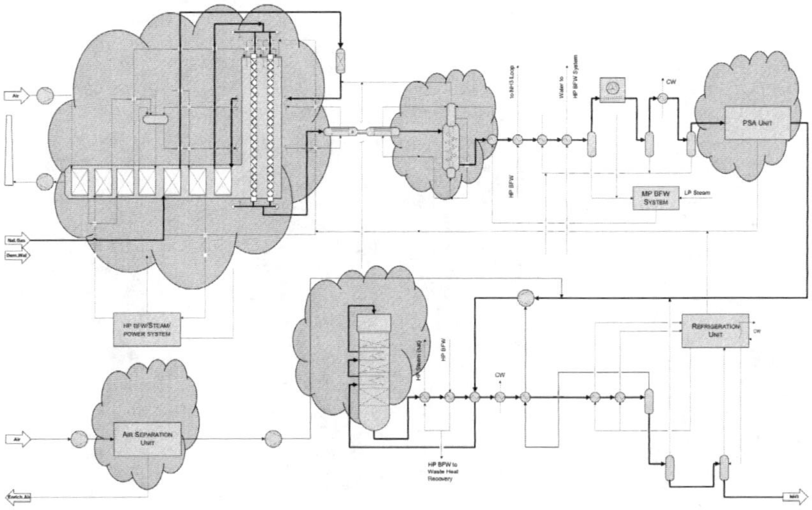

15

Linde Ammonia Concepts (LAC)
A combination of proven technologies

Linde-owned technologies

Air Separation | Rectisol Wash unit | Isothermal Shift Conversion | Nitrogen Wash Units

Reforming Technologies | Pressure Swing Adsorption | Catalytic Oxidation

16

Linde Ammonia Concept (LAC)

A combination of proven technologies

Licensed technologies/equipment

Partial oxidation/
non-catalytic oxidation

Activated amine-based
CO_2 removal system

Ammonia synthesis
converter

17

Agenda.

1. Major Characteristics of the Linde Ammonia Concept
2. The Linde Ammonia Concepts
3. Key Elements
4. Experience References

Linde Ammonia Concepts (LAC)
Reference summary LAC.L Plants

THE LINDE GROUP Linde

Country	NH_3 Capacity, mtpd	Year of Start-up	Scope
India	1350	1996	EP
Australia	230	1999	EP
Australia	600	1999	EPC
China	250	2005	EP
Saudi Arabia	615	2016	EPC
Russia	1340	2016	EPC
USA	540	2016	EPC
Confidential	1000	2019	Basic Engineering

19

Linde Ammonia Concept (LAC)
Reference summary LAC.H Plants

THE LINDE GROUP Linde

Country	Capacity mtpd	Year of Start-up	Scope
India	1350	1981	EP
China	1000	1996	EP
China	1000	2003	EP
USA	1100	2012	FEED

20

Linde's recent Experience in LAC.L Projects:
HyCO + NH_3 Plant for the Sadara Complex

THE LINDE GROUP

Linde

Linde's largest HyCO facility and first NH_3 supply scheme in the Middle East
Products supplied to Sadara Chemicals Company LLC (Saudi Aramco/Dow JV)

Technology
- 2 x 65% SMR HyCO trains
- 1 x Ammonia loop
- 20kt storage tank

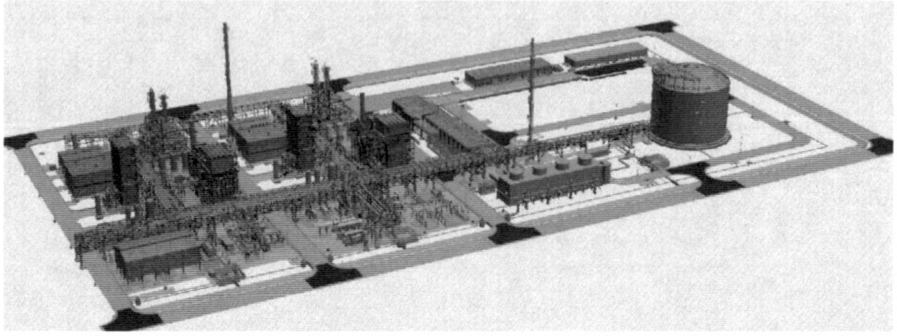

Capacity
- Syngas: ~ 130,000 Nm3/h
- CO: ~ 21,320 Nm3/h
- H_2: ~ 91,000 Nm3/h
- NH_3: ~ 615 tpd

Feedstock
- Natural gas

Design and EPC of the plant executed by Linde Engineering
Facility owned and operated by Linde with successful start up in first half of 2016

21

Linde's recent Experience in LAC.L Projects:
HyCO + NH_3 plant for the Sadara Complex

THE LINDE GROUP

Linde

22

Acknowledgments

Klemens Wawrzinek

Process Engineering HS
Pullach, Germany

Bernd Kandziora

VP Business Development and Technology
Hydrogen and Synthesis Gas Plants
Pullach, Germany

23

Collaborate. Innovate. Deliver.

Thanks for your attention.

Shell ConSep™ Trays Improve Diesel Recovery and Naphtha Quality in a DHT Stripper

Gregory A. Cantley – Marathon Petroleum Company LP, Russell, KY
Waldo De Villiers – Shell Global Solutions (US) Inc., Houston, TX
Eric Vos – Shell Global Solutions International BV, Amsterdam, Netherlands
Grzegorz Kamyszek – Sulzer Chemtech USA, Tulsa, OK

2017 AIChE Southwest Process Technology Conference, Galveston, TX, (Unpublished)

DHT History

- Originally built in mid-70s
- Revamped to Low Sulfur Diesel operation in early 90s
- Revamped to accept Coker Naphtha in early 00s
- Rates on LP Stripper were fairly constant until the Coker Naphtha revamp when overhead rate increased from 1% to about 7.5% of charge rate
- Charge Rate has crept up about 30% from the original design to the Pre-Revamp rates

2

DHT Stripper Pre-Revamp

- Prior to revamp the dP in the top 6 trays showed that the rectifying section was flooded at 0.20 psi per tray.
- Lab analysis showed the D3710 98% endpoint was around 440°F.
- The overhead naphtha was routed back to the crude unit to be reprocessed.

3

DHT Stripper Pre-Revamp

- After a pre-revamp catalyst change, the unit charge was increased 5%.
- The dP in the top trays increased to 0.22 psi per tray.
- The D3710 98% endpoint was approaching 460°F.
- A plan was needed for the next catalyst change!

4

DHT Stripper

- A 7'-0" diameter rectification section with 6 floating round valves trays.
- The open area on the tray was 9.6%
- The downcomers are sloped.
- 24" Tray Spacing

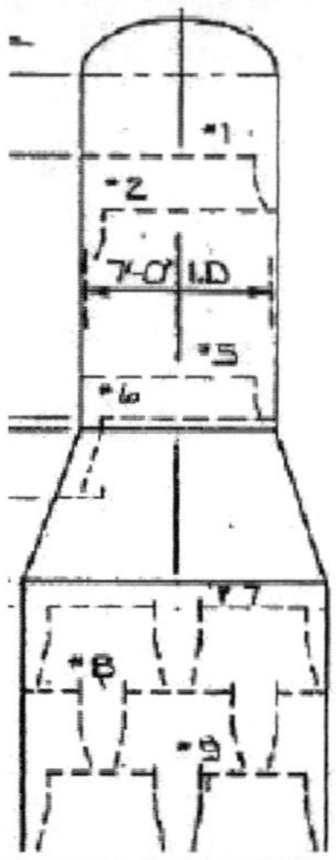

5

DHT Stripper

- The top section was severely flooded.
- Also, the refinery wanted to continue rate increases on the unit.
- With a system limit of 94% at Base+5% charge rate, high capacity trays would not stay un-flooded for long.
- **MPC had not installed any Beyond System Limit trays ever!**

	Original Design Ratings	Pre-revamp Tray Rating
Equation 13 (%)	73	127
Downcomer flood (%)	75	87
DC backup (in liq)	8.6	15.5
Dry tray dP (in liq)	3.1	10.4
DC backup (%TS+W)	32	57
System Limit (%)	50	94

6

Shell ConSep™ tray overview

- **<u>Con</u>**tact and **<u>Sep</u>**aration
- Combines a conventional contacting deck with a separation deck utilizing centrifugal force to separate entrained liquid from the vapor
- Allows operation above the jet flood point, but permits normal downcomer flow as well
- Tray arrangement can be a conventional single pass, multiple pass, or high capacity Shell HiFi tray
- Separation is performed using high capacity swirl tube separators
- Can be retrofitted to existing towers with tray spacing as low as 18 inches
- Operation reverts to that of a normal tray at turndown conditions

7

Tested at FRI

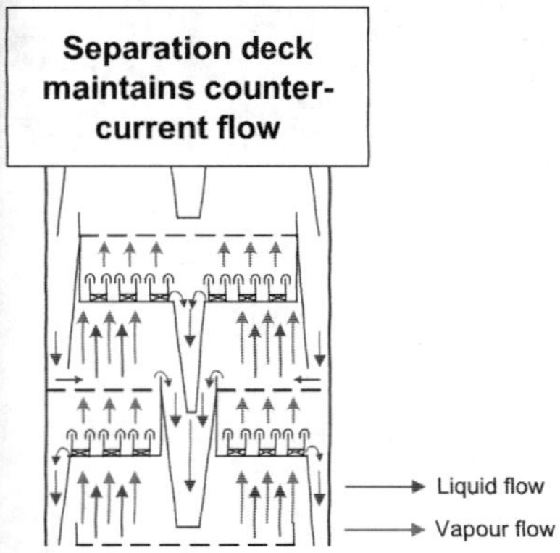

8

View showing HiFi tray downcomers & swirltubes

Tray Ratings

	Original Design Ratings	Pre-Revamp Tray Ratings	Shell ConSep™ Design Rates*
Equation 13 (%)	73	127	146
Downcomer flood (%)	75	87	100
DC backup (in liq)	8.6	15.5	20
Dry tray dP (in liq)	3.1	10.4	13.7
DC backup (%TS+W)	32	57	74
System Limit (%)	50	94	108

*Based on the original conventional tray design.

Shell ConSep™ Tray Ratings

	ConSep Design Ratings
Fraction Maximum useful capacity (%)	90
Downcomer Froth Backup (%)	59
Downcomer Clear Liquid (inch)	10.9
Tray Pressure drop (psi)	0.198

Revamped DHT Stripper Layout

- Five new Shell ConSep™ trays above the feed.
- The top tray is a false downcomer with a swirl tube deck to prevent liquid entrainment to the overhead.
- A seal pan collection system was installed to direct liquid into the feed tray false downcomers.
- Trays were supplied by Sulzer in 2013.
- Trays were installed on existing and expansion rings without welding to the column shell.

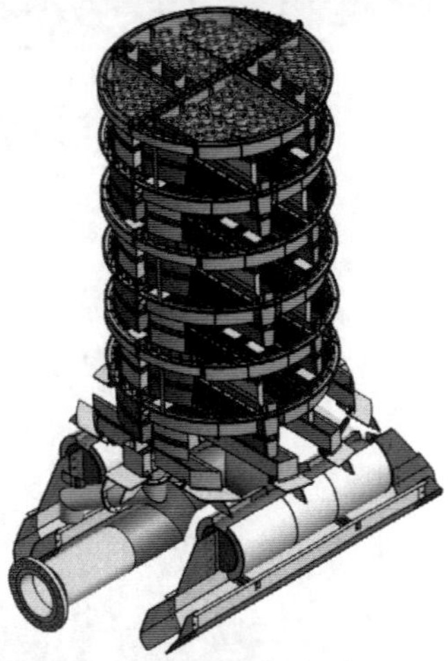

Shell ConSep™ trays installed in DHT Stripper

13

Shell ConSep™ Trays installed in DHT Stripper

14

15

ConSep Trays installed in DHT Stripper

16

Original Tray Ratings

	Original Design Ratings	Pre-revamp Tray Ratings*	Shell ConSep™ Design Rates*	Post-revamp Tray Ratings*
Equation 13 (%)	73	127	146	192
Downcomer flood (%)	75	87	100	150
DC backup (in liq)	8.6	15.5	20	33.8
Dry tray dP (in liq)	3.1	10.4	13.7	22.5
DC backup (%TS+W)	32	57	74	125
System Limit (%)	50	94	108	138

*Based on the original conventional tray design

17

Shell ConSep™ Tray Ratings

	ConSep™ Design Ratings	Post-revamp ConSep™ Ratings
Fraction Maximum useful capacity (%)	90	116
Downcomer Froth Backup (%)	59	90
Downcomer Clear Liquid (inch)	10.9	18.7
Tray Pressure drop (psi)	0.198	0.33

18

DHT Stripper – hydraulic analysis

- Rating calculations indicate that the ConSep trays have exceeded the typical capacity limits expected, including the maximum allowed swirl tube vapor capacity, downcomer backup and downcomer inlet velocity.
  - At moderate to high pressures, the maximum capacity of Shell ConSep™ trays is normally set by a downcomer liquid backup constraint. At lower pressure this limit is clearly conservative.
  - For this application, the liquid load to the swirl deck is still relatively low and therefore the swirl tubes have achieved higher vapor load factors than for what is typically designed.
  - As the liquid captured by the swirl deck is sufficiently degassed before being returned to the downcomers below, high downcomer velocities can be tolerated.

19

Results

- The overhead Naphtha stream is consistently below 400°F ASTM D3710 98% endpoint.
- The Diesel bottoms stream is consistently greater than 140°F flash point.
- Optimization of the unit after the Shell ConSep™ trays were installed resulted in ~40 vol% reduction of Naphtha recycle to the crude unit with ~120% increase in reflux.
- This change in the Naphtha distillation subsequently allowed the naphtha to be routed to the reformer, instead of back to the crude unit.
- The charge rate of the unit was increased ~5% right after the revamp. An additional ~10% unit charge rate increase test run was completed this year.

20

Questions

ExxonMobil

October 12th, 2017

Troubleshooting Tower Carryover

E. J. Grave, N.K. Yeh and J. C. Juarez
ExxonMobil Upstream Research Company
9th AIChE Southwest Process Technology Conference. Galveston, Texas.

Outline

- Introduction
- Facility overview / TEG Contactor original design
- Facility debottleneck
- The problem – TEG liquid carryover
- Troubleshooting – Results
- Conclusions
- Acknowledgments
- Questions

Introduction

- TEG carryover can dictate the capacity of the TEG contactor if the de-entrainment internals are not specified correctly or not modified during a revamp.
- This presentation summarizes the lessons learned in debottlenecking a "dirty" TEG contactor at our LaBarge Black Canyon Facility in Wyoming USA

ExxonMobil 3

Facility Overview

- Black Canyon Facility TEG System
  - Originally design 480 MMSCFD. Two 240 MMSCFD Trains (Process 1 and Process 2)
  - Dries gas containing 65% CO_2 and 5% H_2S
  - Operating pressure over 1300 psig
  - Very low TEG circulation rate (about 0.6 gal TEG / lb H_2O removed)
  - Gas density in the contactor is over 10 lb/ft3
  - Severe fouling over the entire TEG system
  - Increased operating cost in asset due to TEG carryover contamination of solvent at Shute Creek Facility and annual shutdown to clean the system and recharge TEG.

ExxonMobil 4

TEG Contactor Original Design

- Originally equipped with 12 single pass bubble cap trays.
- The feed gas was introduced into the contactor via a flush nozzle.
- The bottom of the nozzle only 4 inches above the HLL, potential liquid re-entrainment from the bottom liquid surface to the bottom tray.
- Vertically oriented (horizontal gas flow) vane pack mist eliminator with two windows.
  - Vane pack with 5/8" (16mm) blade spacing with no pockets.
  - The vanes drained via a downpipe behind the inlet weir of the first tray
  - The mist eliminator was undersized with a k-factor of 0.8 ft/second at 240MMSCFD.

ExxonMobil

5

TEG Contactor Original Design (cont.)

- Designed to process 240 MMSCFD.
- Could process 280 MMSCFD when the tower was clean (TEG losses ~ 0.6 gal/MMSCF).
- Capacity would drop by 50% within 9-12 months.
- Excessive pressure drop due to extreme fouling and foaming on the trays.

Fig. 1 - Bottom bubble cap tray after one year in service

Fig. 2 - Bottom tray outlet picket fence weir

Fig. 3 - Gas outlet vane pack mist eliminator at the top of the tower

ExxonMobil

6

Debottlenecking to 750 MMSCFD

- Future operations requirements:
  - Increase total throughput to 375 MMSCFD per process.
  - Eliminate annual planned shutdowns.
  - Reduce TEG carryover to Shute Creek Facility.
- August 2001 – Process 1 modifications
  - Replaced all bubble caps with Provalve® fixed valves.
  - Removed all downcomer inlet sumps.
  - Removed bottom seal pan and extended the downcomer from the bottom tray to the bottom liquid level to avoid trapping fouling material in the seal pan.
  - Installed Koch-Glitsch Inlet Vane Diffuser at the gas inlet nozzle.
- Process 1 tower performance after debottlenecking:
  - Tower operated an entire year at 375 MMSCFD
  - Liquid level at the bottom of the tower fairly steady
  - Tower pushed to 400 MMSCFD plant stability maintained with no visible hydraulic limitations

Fig. 4 - New Provalve®

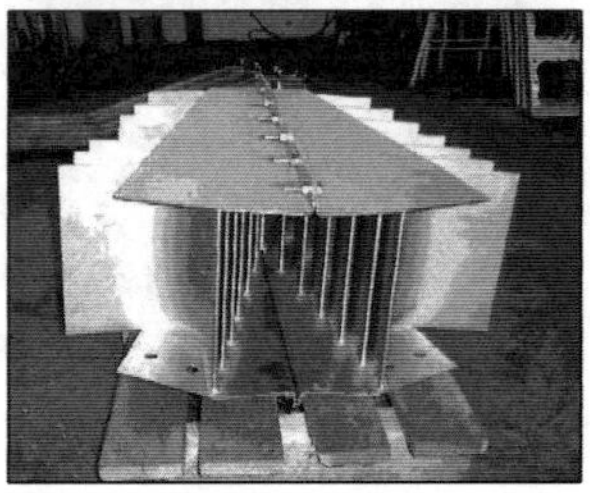

Fig. 5- Koch-Glitsch Inlet Vane Diffuser

ExxonMobil 7

Debottlenecking to 750 MMSCFD (cont.)

- Process 1 tower performance after debottlenecking (cont):
  - TEG carryover was high at 5.7 gal/MMSCF at 375 MMSCFD.
  - Due to the very low circulation rate and high gas velocities:
    - Trays were operating in the spray regime
    - Existing vane pack overwhelmed with liquid and small droplets (k-factor at 1.17 ft/second).
- Summer 2002 - Process 1 tower inspection
  - Amount of fouling on the Provalve® was considerably less than with bubble caps trays (~ 34% higher gas rates)
  - Bottom of the tower "dirtier" with more solids.
  - IVD also collected some solid material
- Summer 2002 - Process 2 modifications
  - Same modifications were completed in Process 2 tower.
  - One exception; IVD with fewer vanes (large vane spacing)

Fig. 6 - Bottom tray after one year of operation with Provalve®

ExxonMobil 8

The Problem – Liquid Carryover

- TEG carryover into the Shute Creek Facility higher than expected
  - Added considerable operating cost to its recovery
- High entrainment rates potential sources:
  - "Dirty" system contaminated the TEG, increasing foaming tendency and small droplets being entrained in greater amounts due to the high superficial vapor velocity at 375MMSCFD.
  - Undersized mist eliminators vane packs
- Carbon filtering system installed during scheduled shutdown in 2002 to treat the entire TEG circulation rate in Process 1.
  - Allowed a direct comparison between "dirty" and "clean" systems
  - Better stability, less to no foaming, and higher circulation rate.
  - TEG carryover dropped from 5.7 gal/MMSCF to 1.8 gal/MMSCF

ExxonMobil

The Problem – Liquid Carryover (cont.)

	Gas Rate	TEG Circ.	TEG Losses	
	MMSCFD	gpm	Gal/MMSCF	% of circ. TEG
Dirty Tests	319.4	10	1.0	2%
	369.4	12	5.7	12%
Clean Tests	319.7	12	0.8	1.5%
	375.1	12	1.8	3.9%
	374.9	18	2.5	3.6%

Table 1 - Process 1 Performance Test TEG Entrainment Summary with Carbon Filtration

- Two months after shutdown, process 2 was already seeing increased carryover and overall system instability.
- After installation of the Carbon filtration system TEG carryover was still unacceptable.
- Decision to replace the existing top mist eliminator vane pack with demisting cyclones.

ExxonMobil

Troubleshooting - First Fix Attempt

- 2003 - Demisting Cyclones installation
  - EM successful experience at high pressure and hydrocarbon service
  - EM first time application in TEG service
  - High CO2 concentration in the feed gas were outside of vendor available data
  - TEG losses expected to be less than 1.0 gal/MMSCF
- Demisting cyclone performance after installation
  - Similar to the vane packs at low gas rates 240-280 MMSCFD
  - As flow rates increased, performance deteriorated for both towers to an unacceptable level
  - Process 1 TEG losses climbed from 0.6 to 10.6 gal/MMSCF between 290 and 328 MMSCFD
  - Process 2 TEG losses were less than in Process 1 (> 4.0 gal/MMSCF) but still unacceptable
  - Observed pressure drop for both towers was 15-20% higher than calculated

Troubleshooting - First Fix Attempt (cont.)

Demisting Cyclones installation

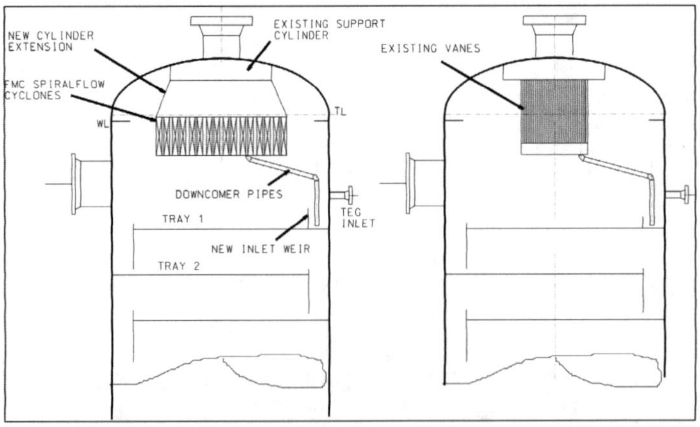

Fig. 7 - TEG Contactor Mist Eliminator Modifications

Troubleshooting - First Fix Attempt (cont.)

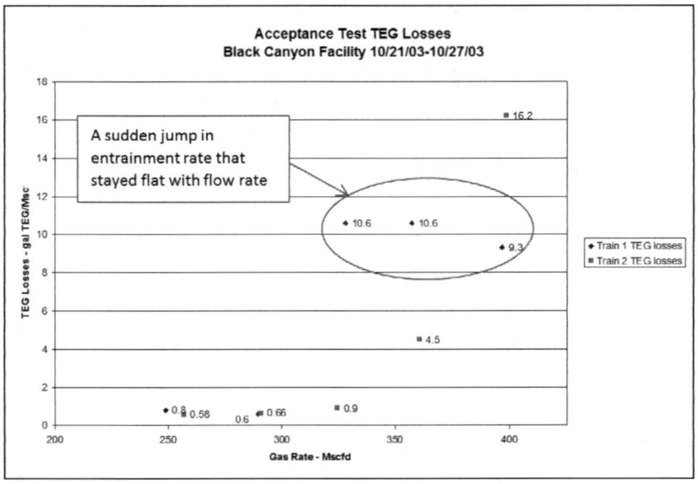

Fig. 8 - TEG Contactor Demisting Cyclones Performance Data

ExxonMobil 13

Troubleshooting - Development of hypothesis

1. Trays with increased flow rate were operating in the "blowing regime".
   - Design provisions: blocking the outlet weir with pickets and providing sufficient hole area.
   - Circulation rate increased to 18 gpm for the performance tests
   - Tower was performing as expected with acceptable gas dew point.
   - Therefore, this hypothesis was not very likely.

2. Trays were operating in a spray regime
   - Significant amount of liquid being entrained and overwhelming the cyclones.
   - High entrained liquid rate could explain the high pressure drop.

3. Foaming on the trays
   - History of foaming in the system
   - After the carbon filtration system was added, the TEG had less tendency to foam.
   - Cyclones should be able to break foam much better than the vane packs.
   - Therefore, this hypothesis was not very likely.

ExxonMobil 14

Troubleshooting - Development of hypothesis (cont.)

4. High viscosity liquid and high vapor density
   - Would cause film instability in the demisting cyclone resulting in surface re-entrainment.

5. Insufficient drainage head.
   - Due to the unexpected high liquid load to the cyclones the required drainage head was too high and the cyclones were unable to drain properly.

6. Syphoning.
   - High pressure drop was causing some liquid to be siphoned from the tray below.

7. Downpipes for drainage were not sealed properly and vapor was bypassing.
   - This can occur with leakage on the tray deck or around the inlet weir.

ExxonMobil 15

Troubleshooting - Second Fix Attempt

- November 2003 – Process 1 tower inspection:
  - No mechanical damage found
  - No fouling on the trays, cyclones or in the cyclones drain header.
  - Inlet weir not properly sealed against vessel wall.
  - Indication of some liquid bypassing the inlet weir.

- November 2003 – Tower modifications
  - Sealing of the downpipes were moved from behind inlet weir of tray 1 to the downcomer of tray 4.
  - Seal cup was added to the downpipes to assure sufficient static seal

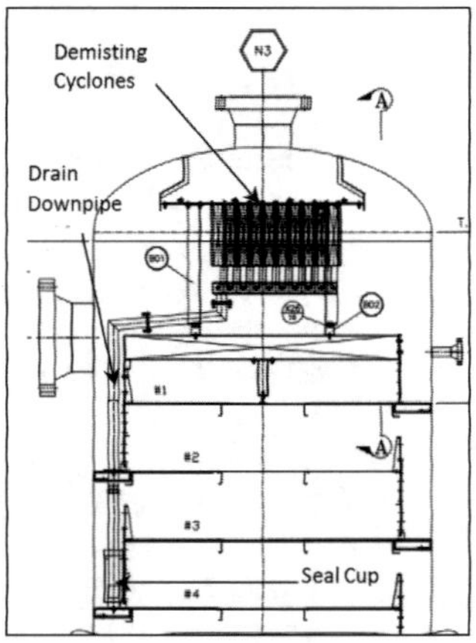

Fig. 9 - Downpipes Seal Cup Modification

ExxonMobil 16

Troubleshooting - Second Fix Attempt (cont.)

- Process 1 performance after installation
  - TEG losses were reduced to less than 6 gal/MMSCF. Still not acceptable.
  - TEG losses target per process guarantees and expected cyclone performance less than 1.0 gal/MMSCF.

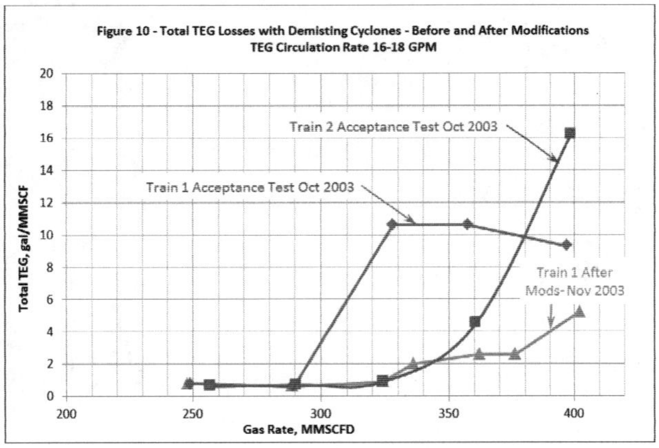

ExxonMobil 17

Troubleshooting – Back to Hypothesis

1. Insufficient number of cyclones
   - High viscosity and high vapor density causing film instability and re-entrainment.

2. Liquid flux rate hitting the cyclones from the trays is too large at the higher gas rates.
   - The high pressure drop is an indicator that the cyclones are seeing a larger liquid load than the original design assumptions.

ExxonMobil 18

Troubleshooting - Third Fix Attempt

- September 2004 – Process 1 tower modifications
  - More demisting cyclones were added
    - Reduce the momentum
    - Decrease the liquid load per cyclone
    - Reduced overall pressure drop and drainage head requirement
  - New vane pack system was installed
    - Covering the entire bubbling area (~ 15 in above the tray deck)
    - Reduced the liquid load to the cyclones.
    - The vanes have no pockets
    - Clever design (FMC Technologies) developed to integrate the vane pack into the tower with limited space

Fig. 11 -Third Fix Demising Cyclones deck

Fig. 12 - Third Fix New Vane Pack Deck

ExxonMobil

19

Troubleshooting - Results

- October 2004 - Performance after installation
  - The TEG loses decreased from about 6 gal/MMSCF to 0.21 gal/MMSCF at 375 MMSCFD.
  - To date the TEG losses are less than 0.6 gal/MMSCF at rates greater than 375 MMSCFD.

ExxonMobil

20

Conclusions

- A clean TEG system will minimize foaming and increase capacity as well as decrease maintenance, operating expense and downtime. As demonstrated, the full stream carbon filtration system reduced TEG liquid carryover by 65-70%.
- Although vendor data was available on demisting cyclones with glycol at elevated pressure and good experience was also available with cyclones in hydrocarbon service at high pressure, this application was outside the data range. The entrained liquid load is also an important parameter that often is ignored or is unknown.
- The combination of the high capacity trays and mist elimination equipment, along with the new solvent filtration system, allowed the gas capacity of the Black Canyon Facility to be almost doubled.

ExxonMobil 21

Acknowledgements

- The authors would like to thank our employer, **ExxonMobil**, for the opportunity and encouragement in the execution of this work. Also, we would like to recognize all of those who contributed, directly or indirectly, to this work, including: the operators and facility engineers at Black Canyon Facility and 3rd-parties such as **Koch-Glitsch** and **TechnipFMC Technologies**.

ExxonMobil 22

Thanks / Questions

ExxonMobil 23

Application of Air/Water Derived Mass Transfer Models to Hydrocarbon Distillation

Frank Seibert
Separations Research Program
The University of Texas at Austin

American Institute of Chemical Engineers
9th Southwest Process Technology Conference
Galveston, Texas
October 12, 2017

Outline

Background

Air/Water Mass Transfer Experimentation

Generalized Mass Transfer Models

Application to Hydrocarbon Distillation

SRP

Experimental Apparatus

16.8" ID

PVC

10 ft. Packed Height

L = 2.5 – 35 GPM/ft^2

U_G = 1.6 – 13 ft/s

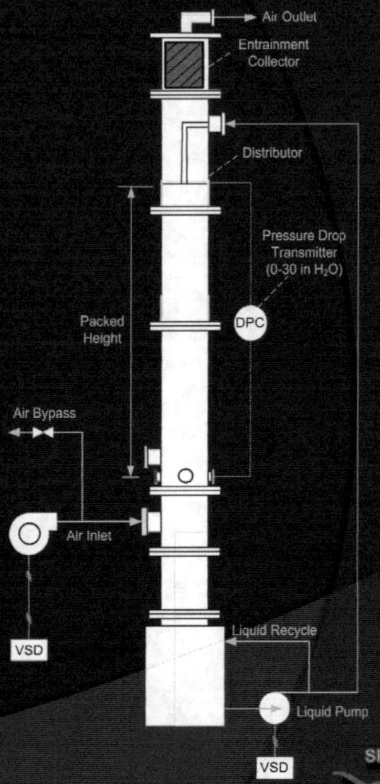

SRP

Background - Experimental Measurements

2002-2008	Contact Area (a_e)	Lewis, Wilson & Seibert
2008-2010	Contact Area (a_e)	Tsai
2010-2015	Contact Area (a_e) Liquid Film MTC (k_L) Gas Film MTC (k_G)	Perry, Wang & Seibert
2015-2017	Contact Area (a_e) Liquid Film MTC (k_L) Gas Film MTC (k_G)	Song

SRP

Chemical Systems

Lewis, Wilson & Seibert	Air/CO2/0.1N NaOH
Tsai	Air/CO2/0.1N NaOH Air/CO2/0.1N NaOH w Surf-→ σ Air/CO2/0.1N NaOH w polymer → μ_L
Perry, Wang & Seibert	Air/CO2/0.1 N NaOH Air/Toluene/Water Air/SO2/0.1 N NaOH
Song	Air/CO2/0.1 N NaOH Air/Toluene/0.1 N NaOH Air/SO2/0.1 N NaOH Air/CO2/0.1 N NaOH (w Glycerol) → μ_L Air/Toluene/Water (w Glycerol) → μ_L

Mass Transfer Modelling

2010	New Contact Area	Tsai (2010)
2015	Refinement of Contact Area New Liquid & Gas Film MTC	Wang (2015)
2017	Refinement of Contact Area Refinement of Gas Film MTC New Liquid Film MTC	Song (2017)

Air/Water Database - Example Packings

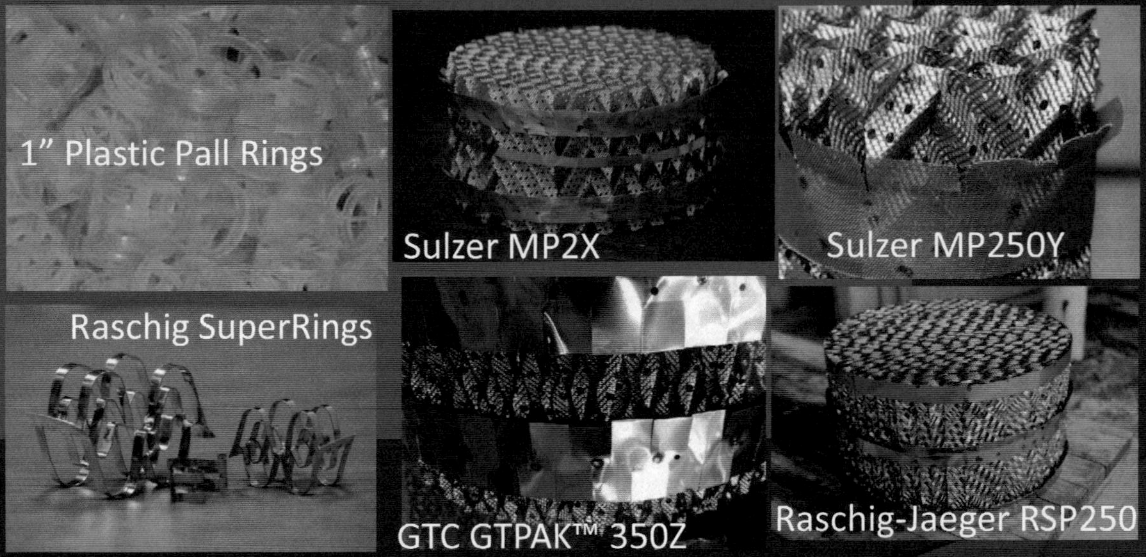

SRP Air/Water Mass Transfer Database

Contact Area (a_e) Only

31 Structured Packings
13 Random Packings

a_e, k_L and k_G

16 Structured Packings
5 Random Packings

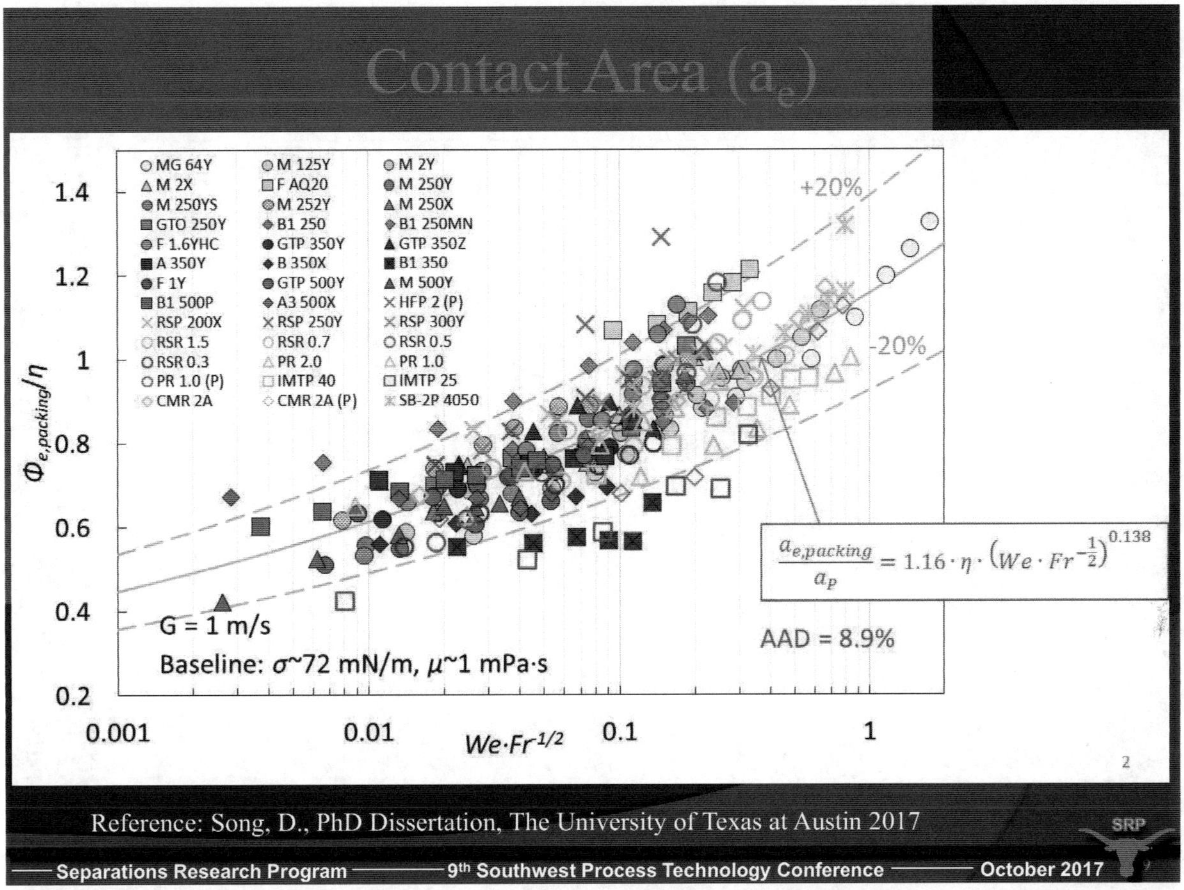
Contact Area (a_e)
MG 64Y
M 125Y
M 2Y
M 2X
F AQ20
M 250Y
M 250YS
M 252Y
M 250X
GTO 250Y
B1 250
B1 250MN
F 1.6YHC
GTP 350Y
GTP 350Z
A 350Y
B 350X
B1 350
F 1Y
GTP 500Y
M 500Y
B1 500P
A3 500X
HFP 2 (P)
RSP 200X
RSP 250Y
RSP 300Y
RSR 1.5
RSR 0.7
RSR 0.5
RSR 0.3
PR 2.0
PR 1.0
PR 1.0 (P)
IMTP 40
IMTP 25
CMR 2A
CMR 2A (P)
SB-2P 4050
+20%
-20%
$\Phi_{e,packing}/\eta$
$We \cdot Fr^{1/2}$
$\frac{a_{e,packing}}{a_p} = 1.16 \cdot \eta \cdot \left(We \cdot Fr^{-\frac{1}{2}}\right)^{0.138}$
G = 1 m/s
Baseline: σ~72 mN/m, μ~1 mPa·s
AAD = 8.9%
0.2
0.4
0.6
0.8
1
1.2
1.4
0.001
0.01
0.1
1
2
Reference: Song, D., PhD Dissertation, The University of Texas at Austin 2017
SRP
Separations Research Program
9th Southwest Process Technology Conference
October 2017

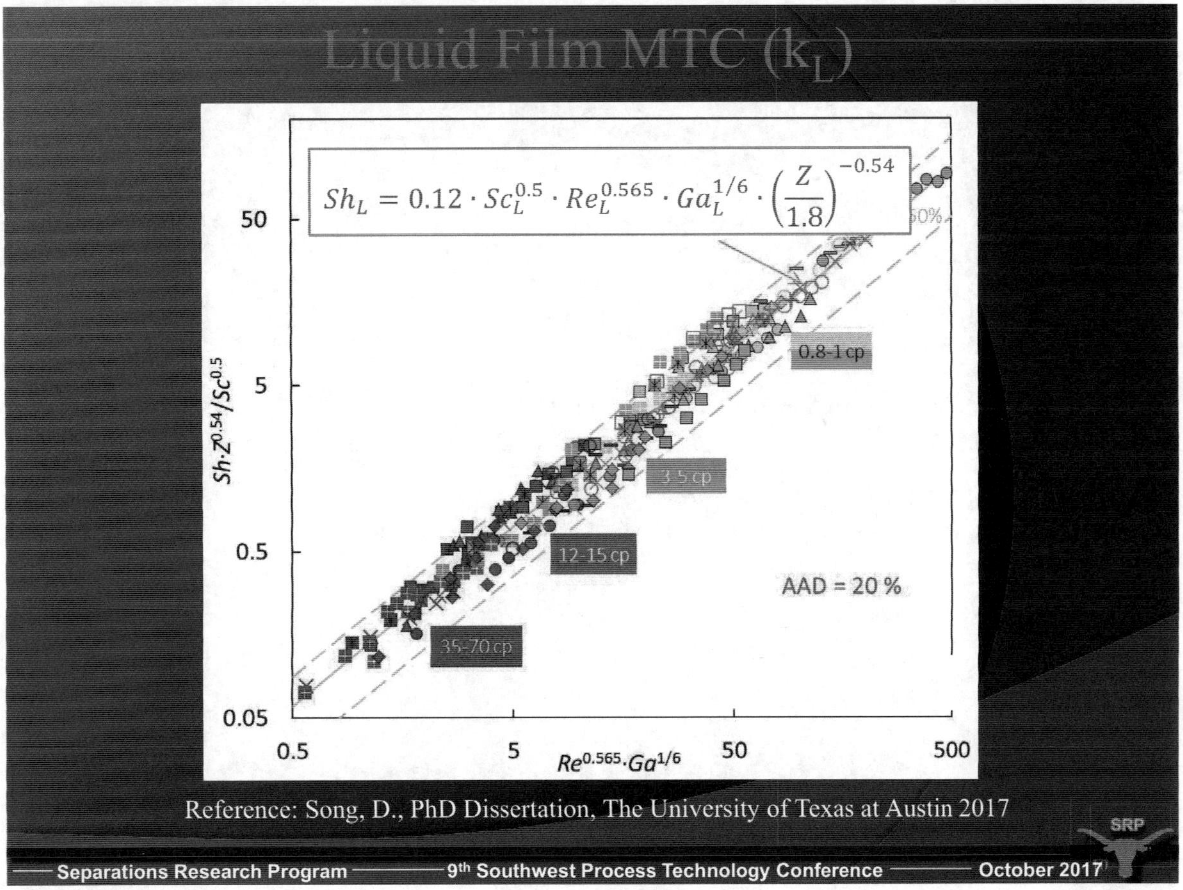
Liquid Film MTC (k_L)
$Sh_L = 0.12 \cdot Sc_L^{0.5} \cdot Re_L^{0.565} \cdot Ga_L^{1/6} \cdot \left(\frac{Z}{1.8}\right)^{-0.54}$
$Sh \cdot Z^{0.54}/Sc^{0.5}$
$Re^{0.565} \cdot Ga^{1/6}$
0.8-1 cp
3-5 cp
12-15 cp
35-70 cp
AAD = 20 %
0.05
0.5
5
50
0.5
5
50
500
Reference: Song, D., PhD Dissertation, The University of Texas at Austin 2017
SRP
Separations Research Program
9th Southwest Process Technology Conference
October 2017

Gas Film MTC (k_G)

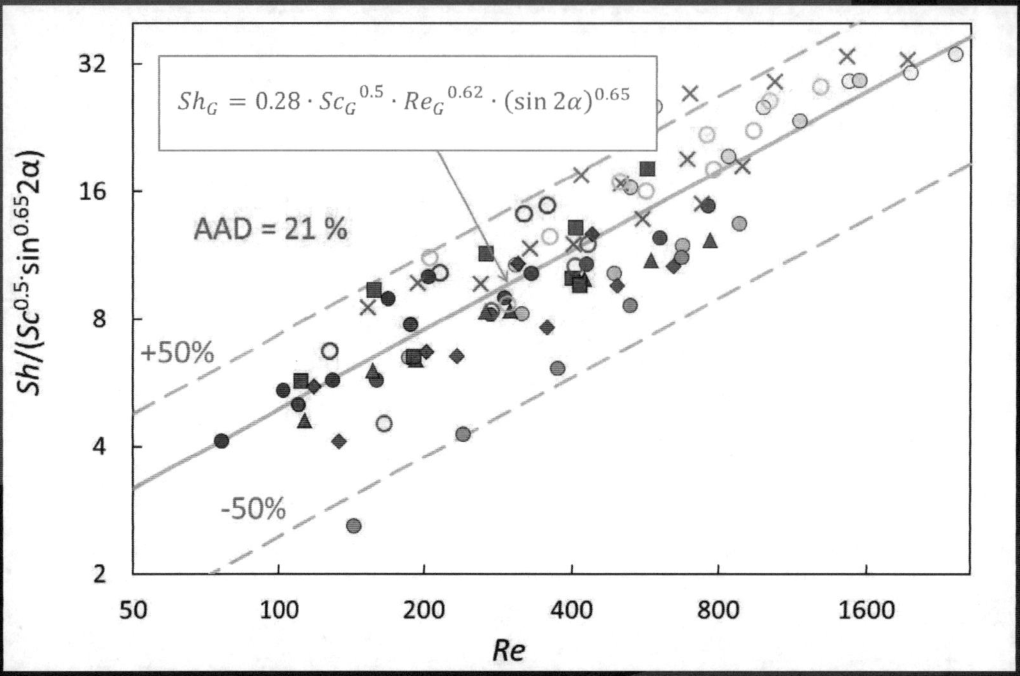

Reference: Song, D., PhD Dissertation, The University of Texas at Austin 2017

Generalized Mass Transfer Area Model

$$\frac{a_{e,packing}}{a_P} = 1.16 \cdot \eta \cdot \left(We \cdot Fr^{-\frac{1}{2}}\right)^{0.138} = 1.16 \cdot \eta \cdot \left[\left(\frac{\rho_L}{\sigma}\right) \cdot g^{1/2} \cdot u_L \cdot a_P^{-3/2}\right]^{0.138}$$

$$\eta = \eta_{type} \cdot \eta_{material} \cdot \eta_{loading}$$

η_{type}	Structured 1.0	Random/Hybrid $1.34 - 0.26\left(\frac{a_P}{250}\right)$
$\eta_{material}$	Stainless Steel 1.0	Plastic 0.62
$\eta_{loading}$	Pre-loading zone ($\Delta P < 400$ Pa/m) 1.0	Loading zone ($\Delta P \geq 400$ Pa/m) 1.15

Song, D., Ph.D. Dissertation, The University of Texas at Austin, 2017

Generalized Liquid Film MTC Model

$$Sh_L = 0.12 \cdot Sc_L^{0.5} \cdot Re_L^{0.565} \cdot Ga_L^{1/6} \left(\frac{Z_{P\{m\}}}{1.8}\right)^{-0.54}$$

$$k_L = 0.12 \cdot u_L^{0.565} \cdot \left(\frac{\mu_L}{\rho_L}\right)^{-0.4} \cdot D_L^{0.5} \cdot g^{1/6} \cdot a_P^{-0.065} \left(\frac{Z_{P\{m\}}}{1.8}\right)^{-0.54}$$

Song, D., Ph.D. Dissertation, The University of Texas at Austin, 2017

Generalized Gas Film MTC Model

$$Sh_G = 0.28 \cdot {Sc_G}^{0.5} \cdot {Re_G}^{0.62} \cdot \left(\frac{\sin 2\alpha}{\sin(2 \times 45°)}\right)^{0.65}$$

$$k_G = 0.28 \cdot u_G^{0.62} \cdot \left(\frac{\mu_G}{\rho_G}\right)^{-0.12} \cdot D_G^{0.5} \cdot a_P^{0.38} \cdot (\sin 2\alpha)^{0.65}$$

$a_{effective}$ = 45° (for random and hybrid packings)

Song, D., Ph.D. Dissertation, The University of Texas at Austin, 2017

Air/Aqueous Physical Properties

Temperature	60 ˚F	80 ˚F	100 ˚F
Liquid Density, kg/m^3	1,000	1,000	1,000
Liquid Viscosity, cP	1.29	1.01	0.89
Diffusivity of Toluene in Water, cm^2/s	6.89E-6	9.14E-6	9.48E-6
Vapor Density, kg/m^3	1.22	1.18	1.14
Vapor Viscosity, cP	0.0182	0.0187	.0193
Diffusivity of SO_2 in Air, cm^2/s	1.19E-1	1.27E-1	1.35E-1

Cyclohexane/n-Heptane Physical Properties

- ChemCad 7.1
  - Densities, Viscosities, Surface Tension
  - Avg MW
  - Top & Bottom
  - Checked with Hand Calculations
- Vapor Diff Coefficient - Fuller, Schettler and Giddings (I&EC 1956)
- Liquid Diff Coefficient - Scheibel (*I&EC* 1954)

Average Physical Properties

System	Air/Aqueous	Cyclohexane/n-Heptane			
		Pressure, kPa			
Parameter	101	16.5	33.3	165	414
Avg. Temperature, C	26	39.5	55.9	105.9	143.4
Liq. Density, kg/m^3	1000	713	699	649	608
Liq. Viscosity, kg/m-s	1.01E-03	5.39E-04	4.38E-04	2.87E-04	2.43E-04
Liq. Diffusion Coeff, m^2/s	9.14E-10	3.94E-09	4.73E-09	7.56E-09	9.55E-09
Vap. Density, kg/m^3	1.18	0.618	1.14	5.07	12.19
Vap. Viscosity, kg/m-s	1.82E-05	6.75E-06	7.15E-06	8.30E-06	9.25E-06
Vap. Diffusion Coeff, m^2/s	1.19E-05	2.11E-05	1.16E-05	2.94E-06	1.39E-06
Relative Volatility		1.97	1.84	1.57	1.42
Slope of equilibrium line		0.89	0.91	0.95	0.97
Surface Tension, N/m	0.07	0.0206	0.0188	0.0132	0.0098

C6C7 Distillation Physical Properties Relative to Air/Aqueous System

Liquid Density 0.713 → 0.608
Liquid Viscosity 0.539 → 0.243
Liquid Diff Coefficient 4.31 → 10.44

Vapor Density 0.524 → 10.33
Vapor Viscosity 0.371 → 0.508
Vapor Diff Coefficient 1.77 → 0.117

Surface Tension 0.294 → 0.14

Top and Bottom C6C7 Physical Properties

Liquid

Parameter	Top	Bottom	AVG
Temperature, C	48.3	63.5	
MW	85.02	99.11	
Density, kg/m3	745.3	652.4	698.9
Viscosity, kg/m-s	5.97E-4	2.8E-4	4.38E-4
Diffusion Coefficient, m2/s	2.26E-9	7.20E-9	4.73E-9

Vapor

Parameter	Top	Bottom	AVG
Temperature, C	48.3	63.5	
MW	85.02	98.16	
Density, kg/m3	1.07	1.21	1.14
Viscosity, kg/m-s	7.6E-6	6.7E-6	7.15E-6
Diffusion Coefficient, m2/s	1.2E-5	1.11E-5	1.16E-5

Parameter	Top	Bottom	AVG
$X_{mol\%,C6}$	94	6	0.50
Slope of equilibrium line	0.58	1.66	0.91
Surface Tension, dynes/cm	0.0215	0.0161	0.0188

SRP Distillation Database

- Cyclohexane/n-Heptane Test Mixture
- Total Reflux
- Column Diameter (i.d.) = 16.8 inches
- Packing Height ~ 10 ft
- Four Operating Pressures, 2.4-60 psia
- Four Temperatures, 87 -310 F
- # of Available Packings = 60

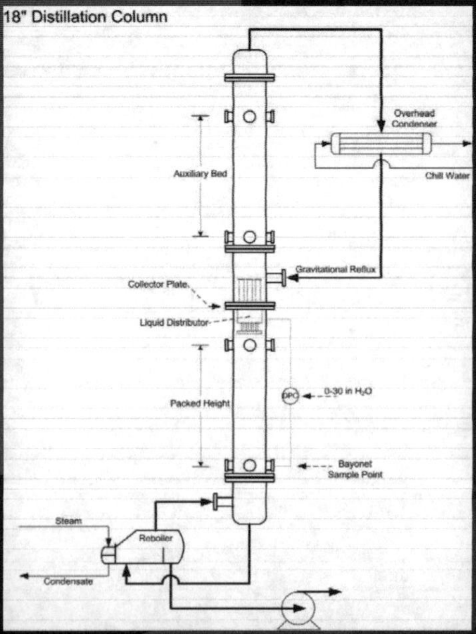

Application to Distillation

$$HTU_V = \frac{U_V}{k_G\, a_e} \qquad HTU_L = \frac{U_L}{k_L\, a_e}$$

@ Total Reflux $\quad HTU_{OV} = HTU_V + mHTU_L$

$$m = \frac{\alpha}{[1 + (\alpha - 1)X]^2}$$

$$HETP = \frac{\ln m}{m - 1} HTU_{OV}$$

SRP

Separations Research Program — 9th Southwest Process Technology Conference — October 2017

Distillation Packings Evaluated

Structured Packings	MOC	a_P (m^2/m^3)	Corrugation Angle
MP500Y	Sheet Metal	500	45°
MP250Y	Sheet Metal	250	45°
RSP250	Sheet Metal	250	"45°"
A3-500	Gauze	500	60°
B1-250	Sheet Metal	250	45°
B1-250	Sheet Metal	250	60°
MaxPak.90	Sheet Metal	250	90°

SRP

Separations Research Program — 9th Southwest Process Technology Conference — October 2017

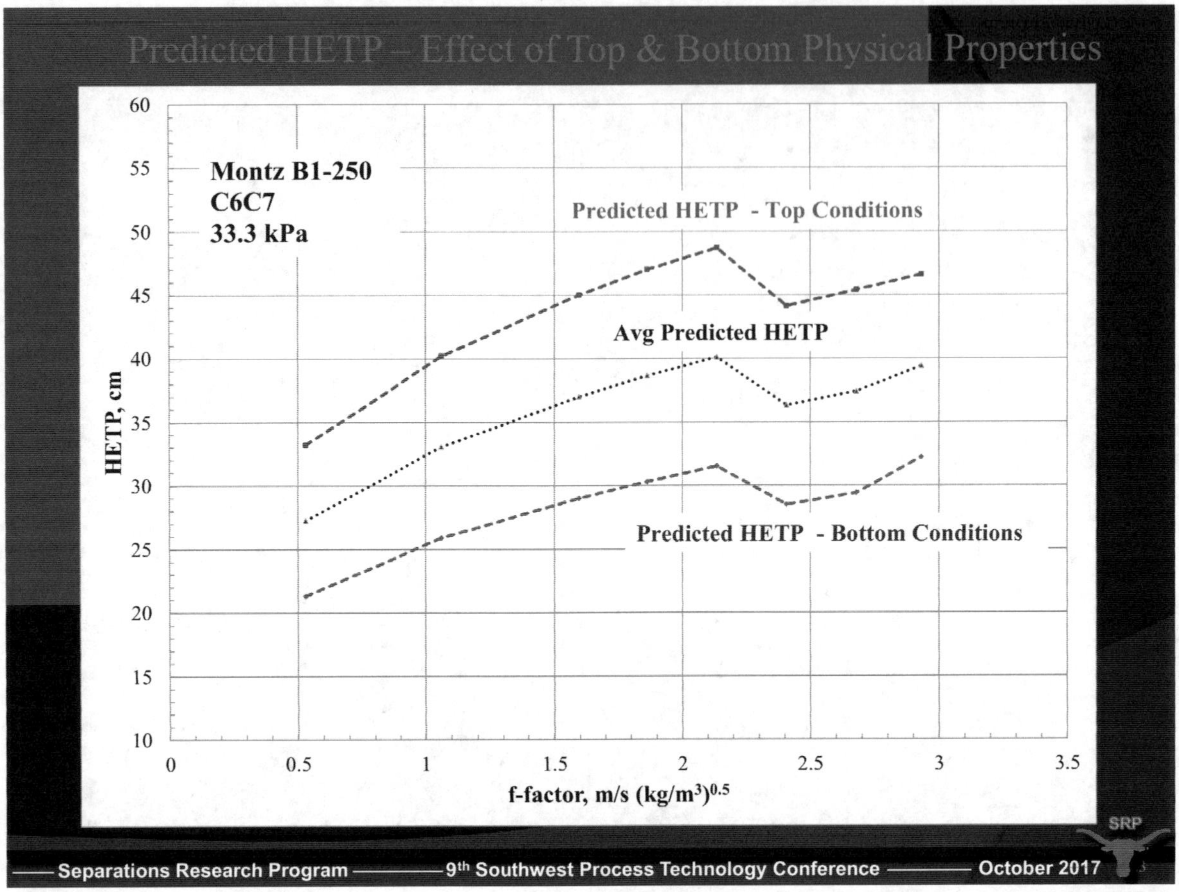
Predicted HETP – Effect of Top & Bottom Physical Properties
Montz B1-250
C6C7
33.3 kPa
Predicted HETP - Top Conditions
Avg Predicted HETP
Predicted HETP - Bottom Conditions
HETP, cm
f-factor, m/s (kg/m³)^0.5
SRP
Separations Research Program
9th Southwest Process Technology Conference
October 2017

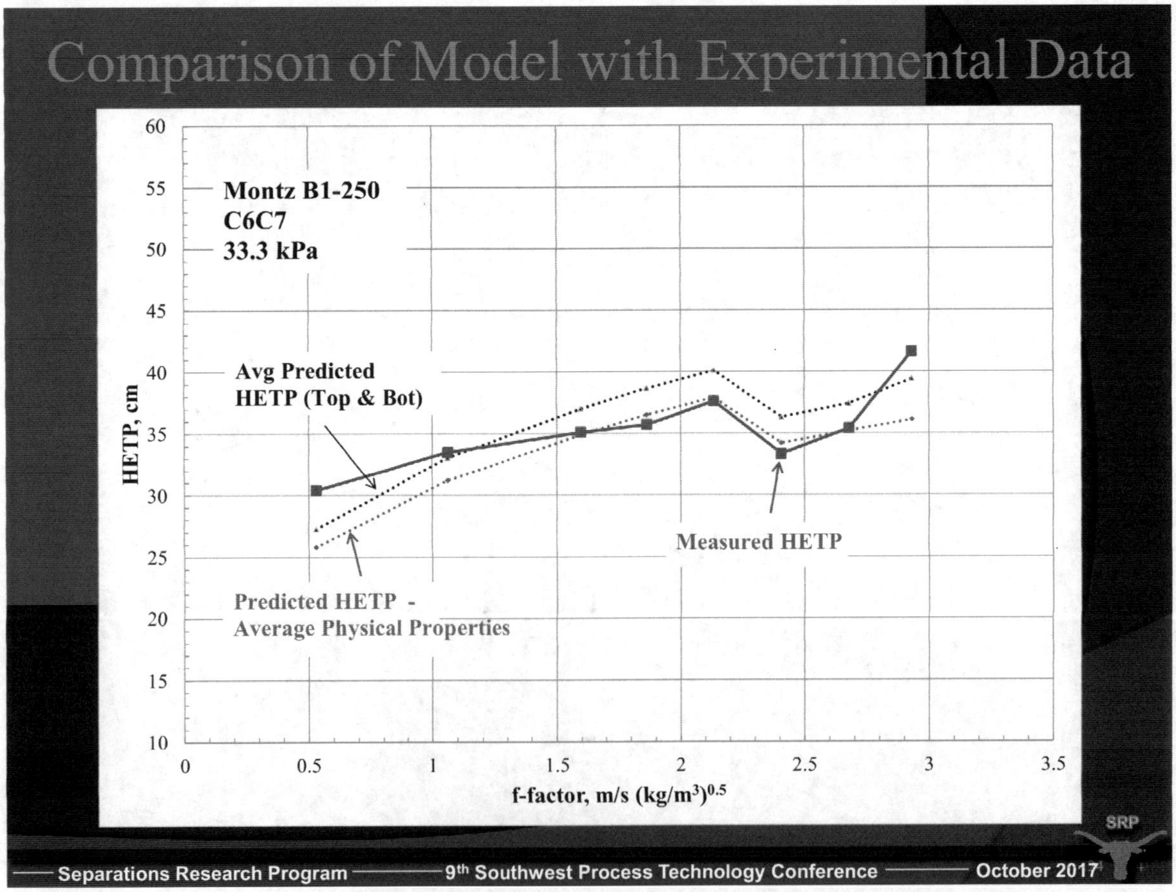
Comparison of Model with Experimental Data
Montz B1-250
C6C7
33.3 kPa
Avg Predicted
HETP (Top & Bot)
Measured HETP
Predicted HETP -
Average Physical Properties
HETP, cm
f-factor, m/s (kg/m³)^0.5
SRP
Separations Research Program
9th Southwest Process Technology Conference
October 2017

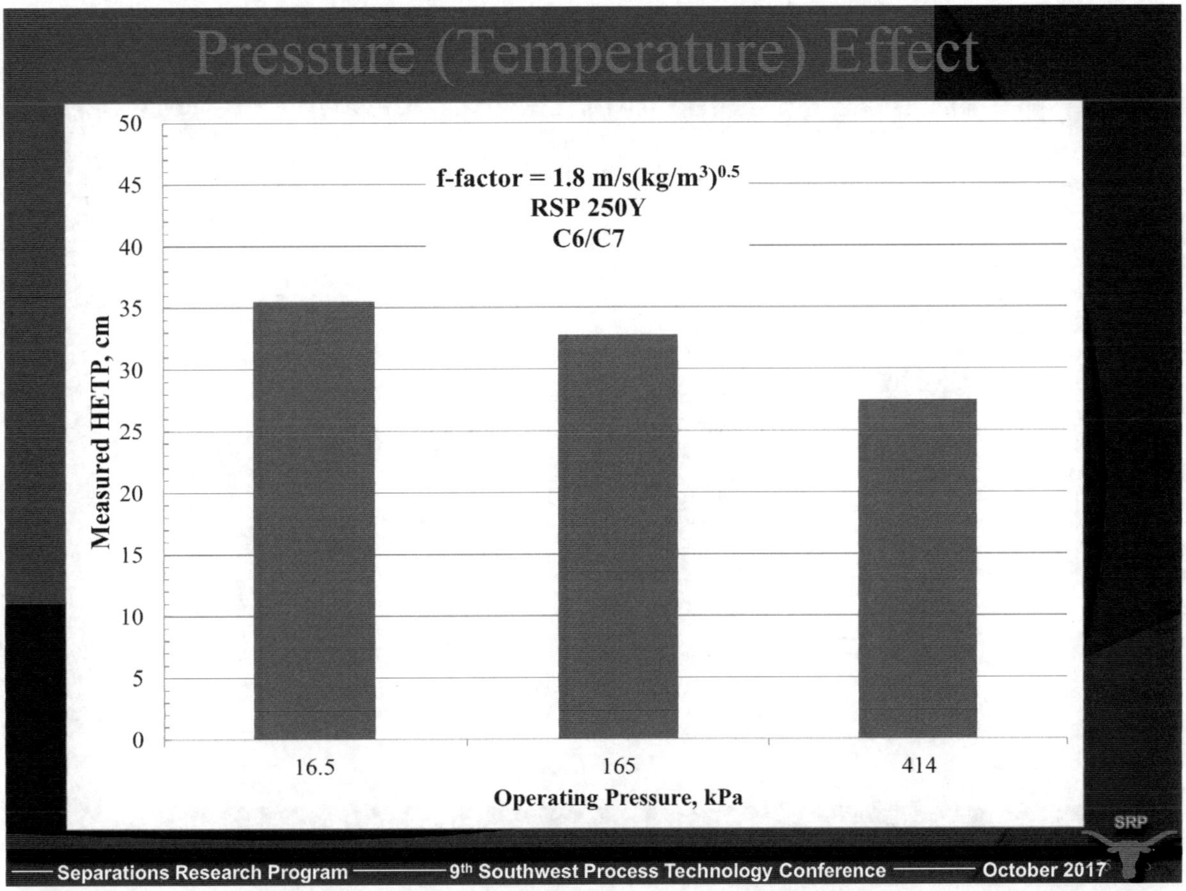
Pressure (Temperature) Effect
f-factor = 1.8 m/s(kg/m³)^0.5
RSP 250Y
C6/C7
Measured HETP, cm
50
45
40
35
30
25
20
15
10
5
0
16.5
165
414
Operating Pressure, kPa
SRP
Separations Research Program
9th Southwest Process Technology Conference
October 2017

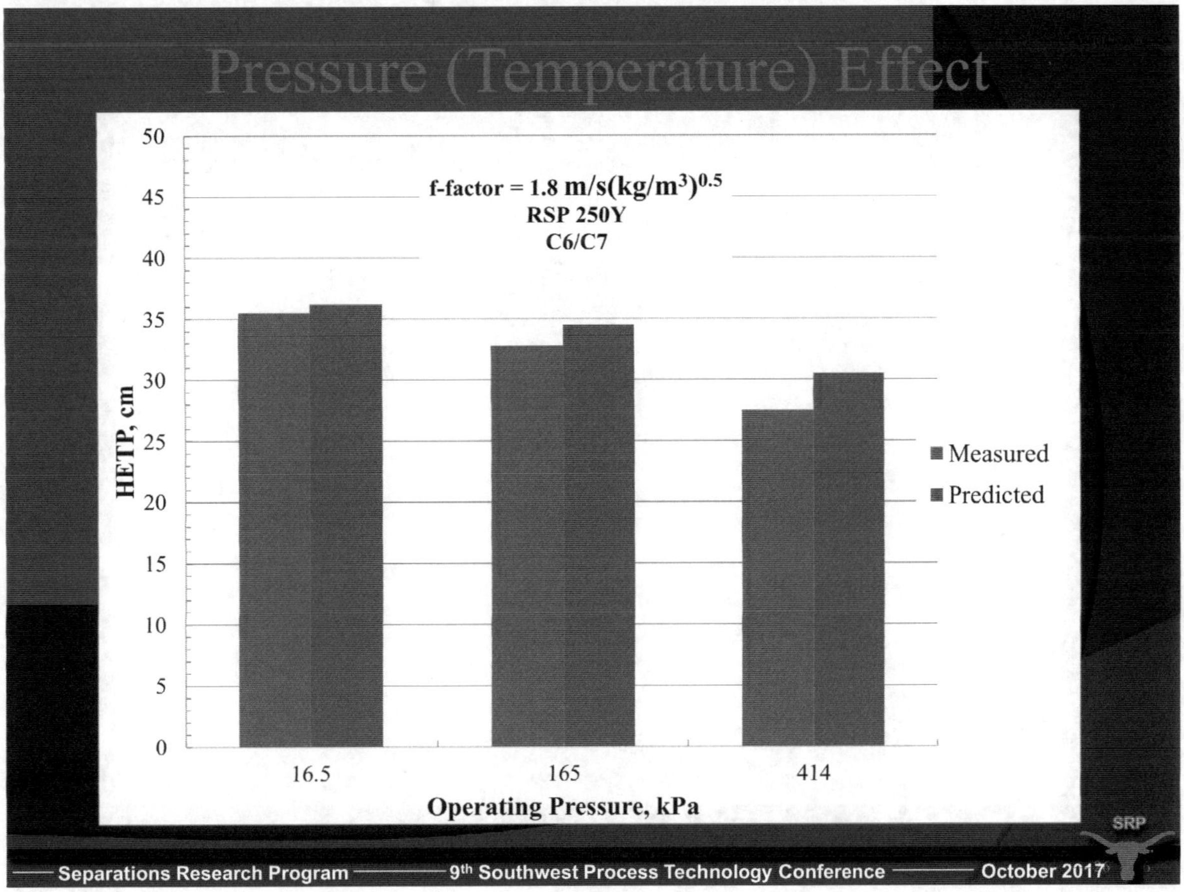
Pressure (Temperature) Effect
f-factor = 1.8 m/s(kg/m³)^0.5
RSP 250Y
C6/C7
HETP, cm
50
45
40
35
30
25
20
15
10
5
0
16.5
165
414
Operating Pressure, kPa
Measured
Predicted
SRP
Separations Research Program
9th Southwest Process Technology Conference
October 2017

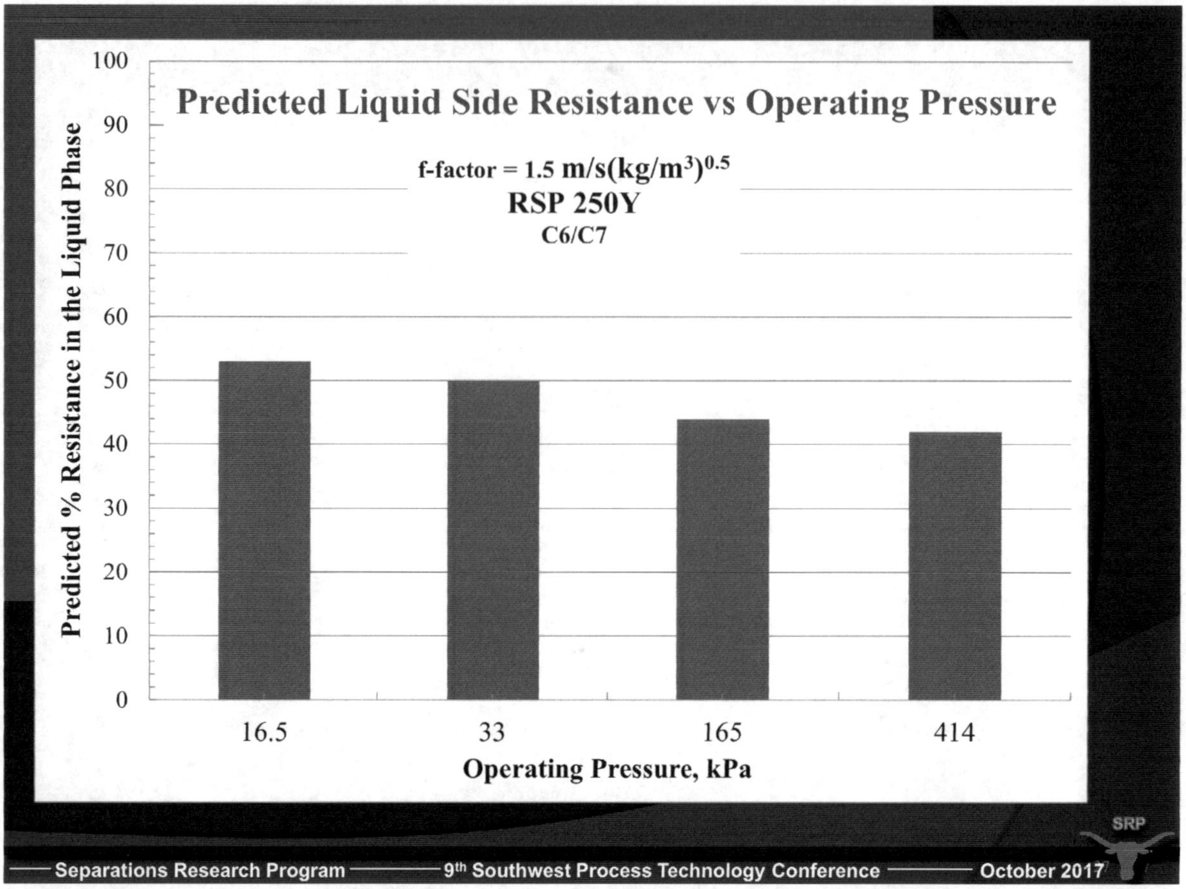
Predicted Liquid Side Resistance vs Operating Pressure
f-factor = 1.5 m/s(kg/m³)^0.5
RSP 250Y
C6/C7
Predicted % Resistance in the Liquid Phase
100
90
80
70
60
50
40
30
20
10
0
16.5
33
165
414
Operating Pressure, kPa
SRP
Separations Research Program
9th Southwest Process Technology Conference
October 2017

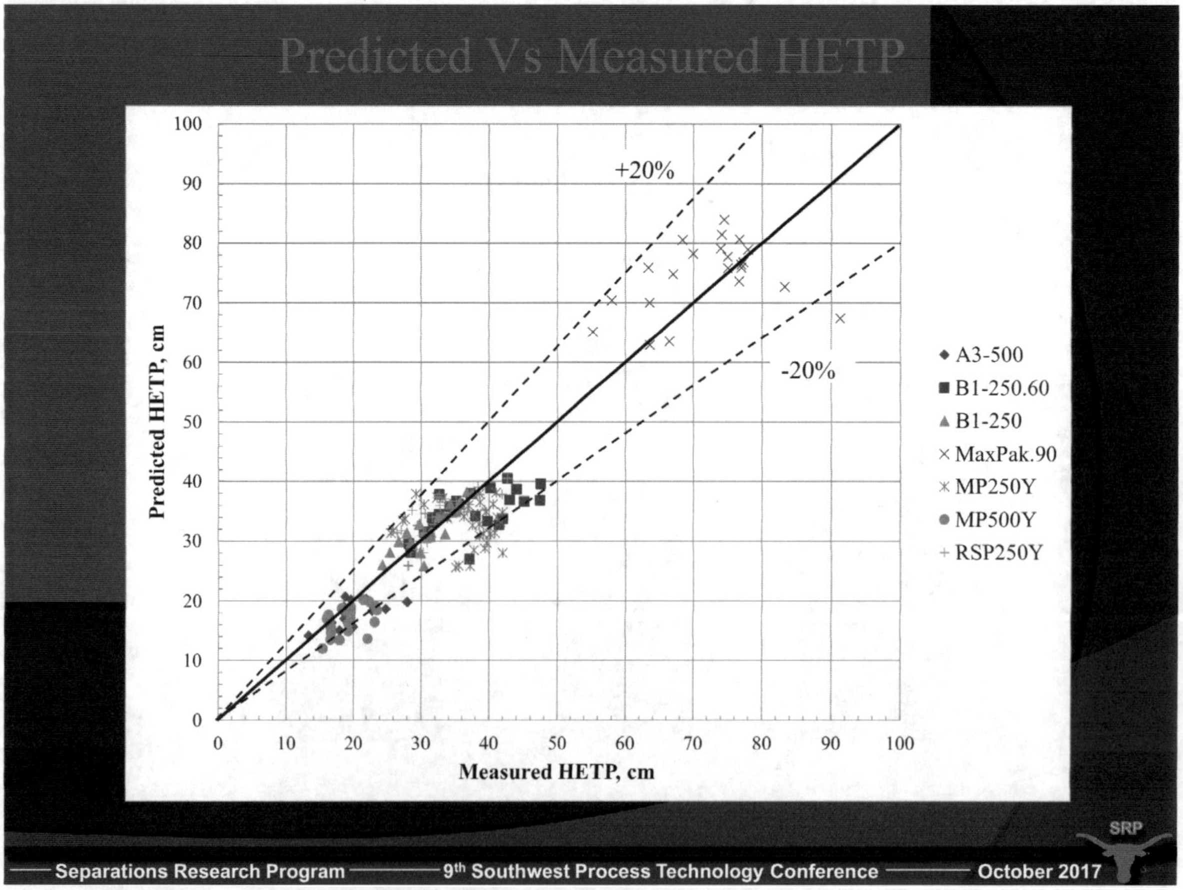
Predicted Vs Measured HETP
Predicted HETP, cm
Measured HETP, cm
+20%
-20%
0
10
20
30
40
50
60
70
80
90
100
A3-500
B1-250.60
B1-250
MaxPak.90
MP250Y
MP500Y
RSP250Y
SRP
Separations Research Program
9th Southwest Process Technology Conference
October 2017

Model – Data Agreement

Structured Packings	Avg. Relative Error, %
MP500Y	13.9
MP250Y	15.8
MaxPak.90	9.2
RSP250	7.1
B1-250.60	10.2
B1-250	5.0
A3-500	10.9

Average Relative Error =10.3%

Summary

- Packing HETP Varies with Physical Properties
- Air/Aqueous Derived Models Match C6/C7 Distillation Data
  - Decreasing f-factor → Decreasing HETP
  - Increasing P → Decreasing HETP
  - Decreasing P → Increasing Predicted % Liquid Side Resistance

Acknowledgements
Di Song
Chao Wang
Micah Perry
Bob Tsai
Ian Wilson
Chris Lewis
Gary Rochelle
Chemstations
SRP
Separations Research Program
9th Southwest Process Technology Conference
October 2017

Thank You
32

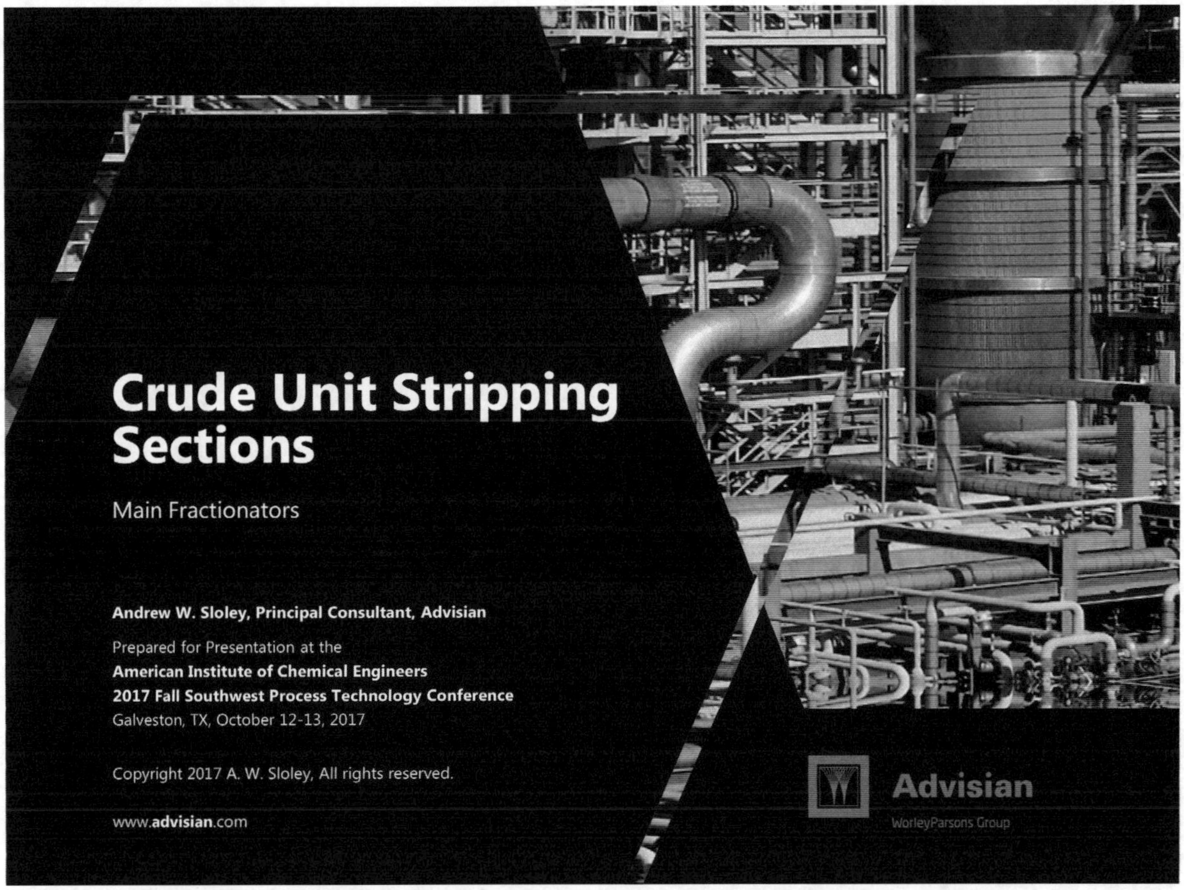
Crude Unit Stripping Sections
Main Fractionators
Andrew W. Sloley, Principal Consultant, Advisian
Prepared for Presentation at the
American Institute of Chemical Engineers
2017 Fall Southwest Process Technology Conference
Galveston, TX, October 12-13, 2017
Copyright 2017 A. W. Sloley, All rights reserved.
www.advisian.com
Advisian
WorleyParsons Group

But First, What We Do
Consulting and Engineering in:
Refinery and Petrochemicals
• Molecule management
• Concept selection
• FEED
• Economics and planning
• Technology selection
• Refinery/petrochemical integration
Operational Excellence
• Benchmarking
• Gap analysis
• Safety and environment
Sulfur and Gas Treating
• Oxygen enrichment
• Gas treating solvents
Infrastructure and Environmental
• Water
• Materials handling
Management, Economics, and Business Operations
EcoNomics
Advisian / 2

Overview

Stripping

- Main fractionator services:
  - Nearly all atmospheric towers
  - Roughly half of vacuum towers
- Inert gas (steam) stripping
- Flash point control for storage and products (asphalt)
- Bulk recovery of material

Stripping Section (Atmospheric Tower)

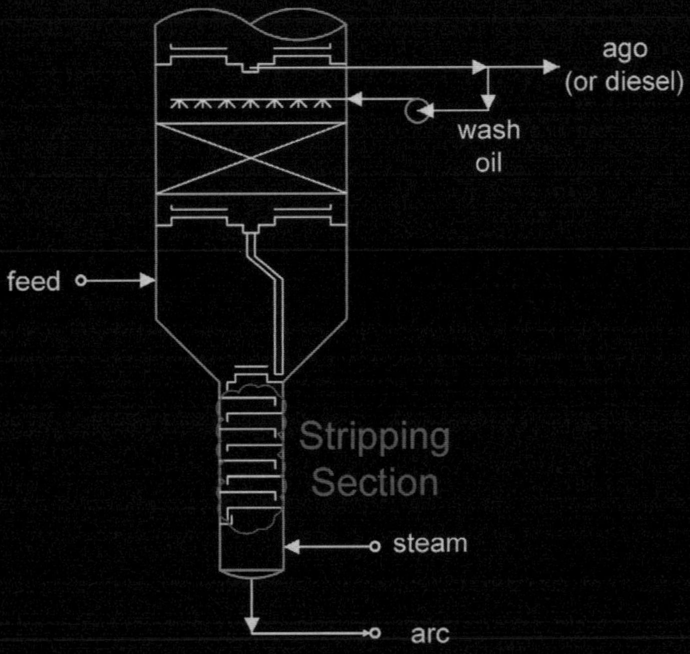

Advisian / 5

How Does Stripping Work?

- Reduces partial pressure
- Bubble point liquid from flash zone vaporizes
- Stages use lighter material to enhance partial pressure effects

$$pressure_{total} = partial\ pressure_{hydrocarbon} + partial\ pressure_{steam}$$

Advisian / 6

Objectives

Flash Point Control

- Flash point specification on bottoms:
  - Safer storage in tanks
  - Specification on asphalt streams for safe handling
- Straightforward
- Relatively easy
- Only a small amount of material removal

Bulk Recovery

- Atmospheric preflash stripping: recover light material as naphtha or kerosene:
  - Improve downstream operation: higher capacity, improved operations
- Atmospheric tower stripping: recover light material as diesel or AGO:
  - Improve upstream operation: lower heater outlet temperature
  - Improve downstream operation: vacuum heater, vacuum tower, vacuum system
- Vacuum tower stripping: recover light material as HVGO:
  - Improve upstream operation: lower heater outlet temperature
  - Reduce lowest value product (vrc) volume

Performance

Performance

- Stripping steam rate
- Number of stripping stages

Atmospheric Stripping

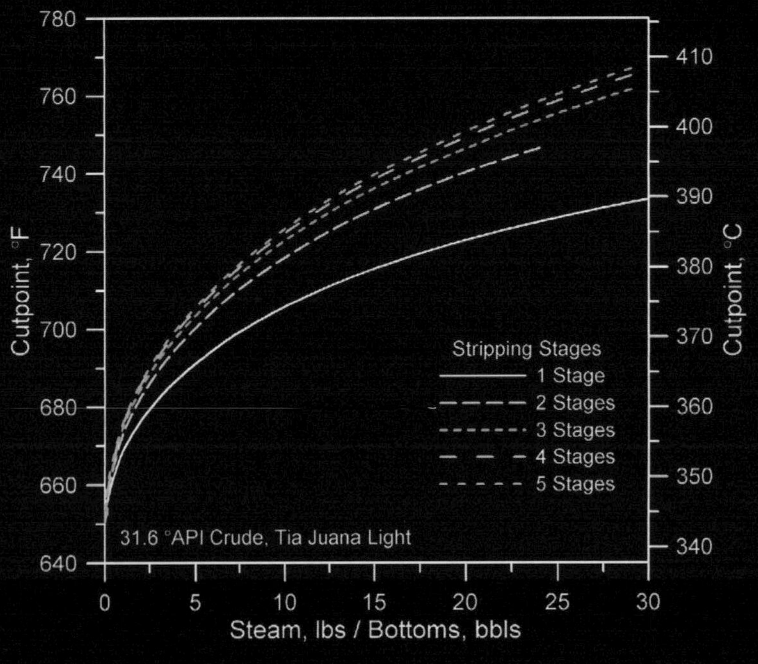

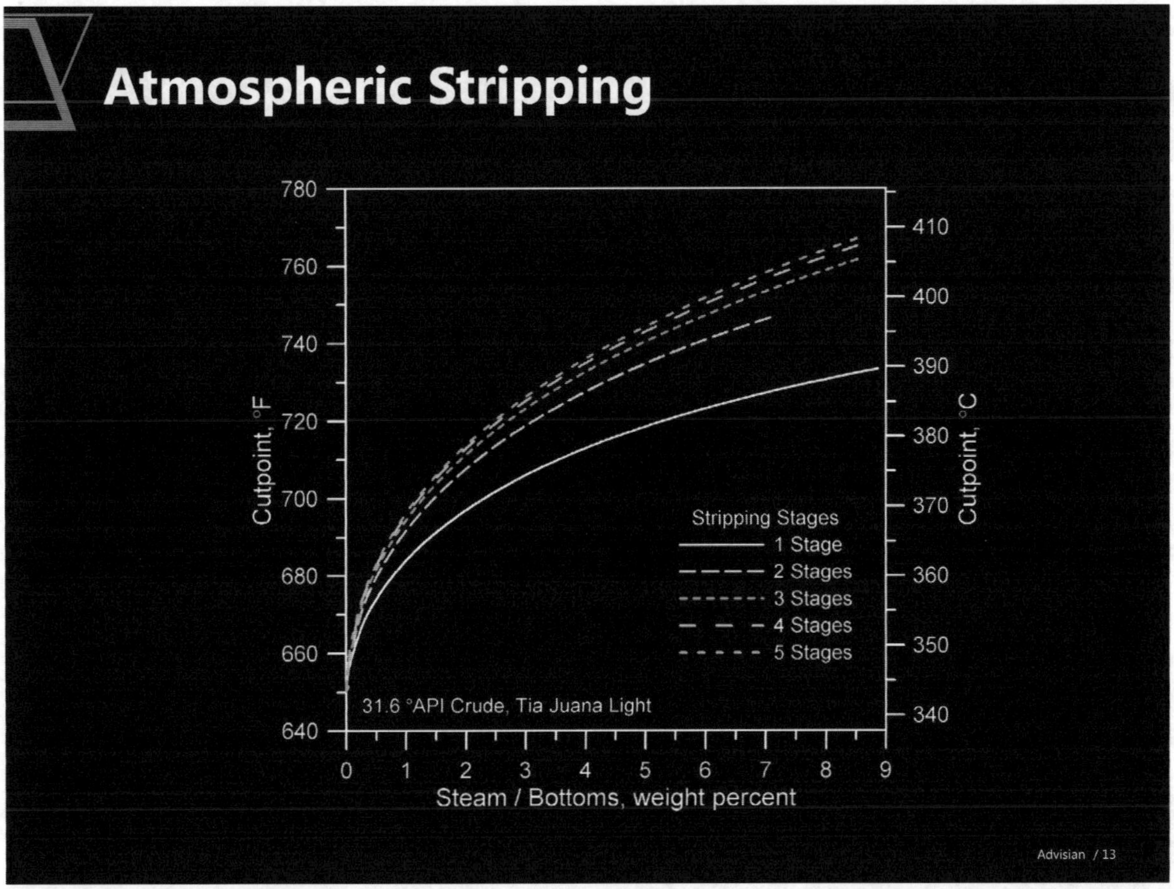
Atmospheric Stripping
Cutpoint, °F
Cutpoint, °C
Steam / Bottoms, weight percent
Stripping Stages
1 Stage
2 Stages
3 Stages
4 Stages
5 Stages
31.6 °API Crude, Tia Juana Light
780
760
740
720
700
680
660
640
410
400
390
380
370
360
350
340
0
1
2
3
4
5
6
7
8
9
Advisian / 13

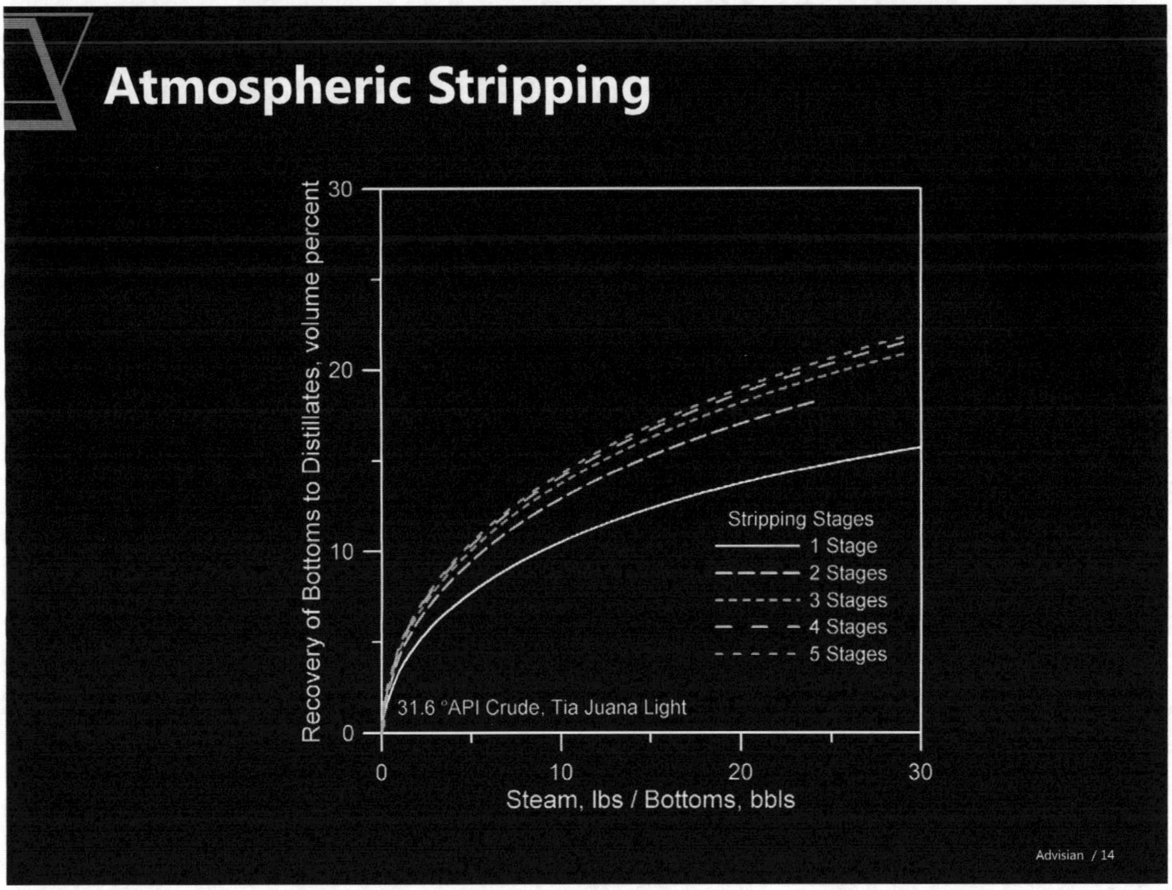
Atmospheric Stripping
Recovery of Bottoms to Distillates, volume percent
Steam, lbs / Bottoms, bbls
Stripping Stages
1 Stage
2 Stages
3 Stages
4 Stages
5 Stages
31.6 °API Crude, Tia Juana Light
30
20
10
0
0
10
20
30
Advisian / 14

Vacuum Stripping

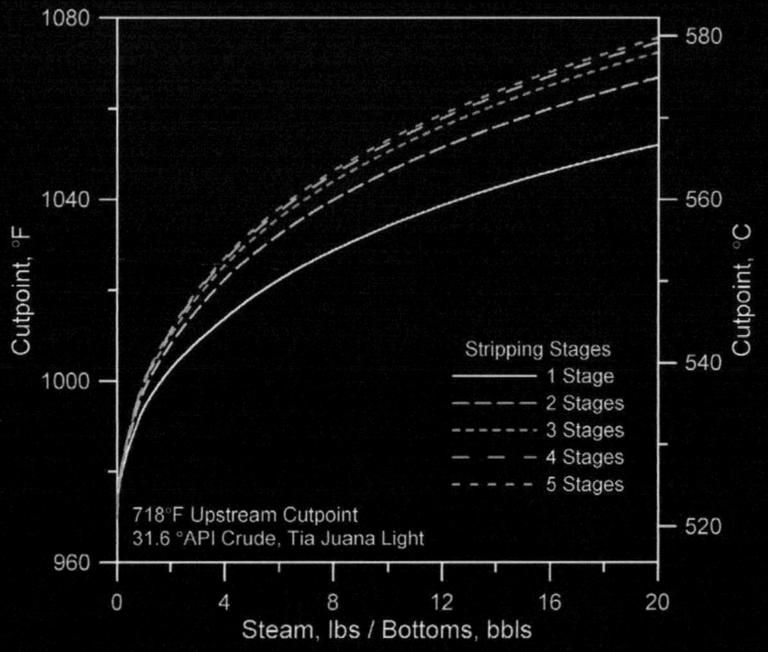

Advisian / 15

Vacuum Stripping

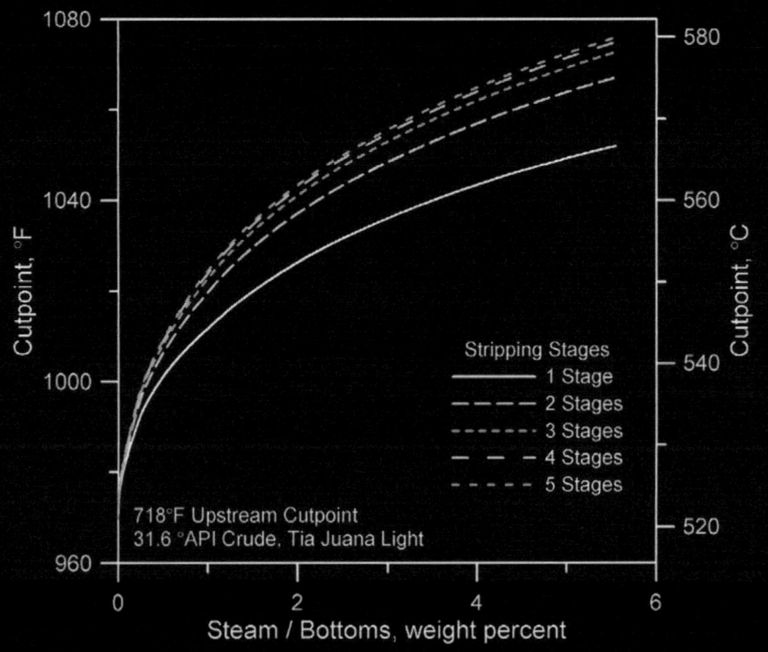

Advisian / 16

Vacuum Stripping

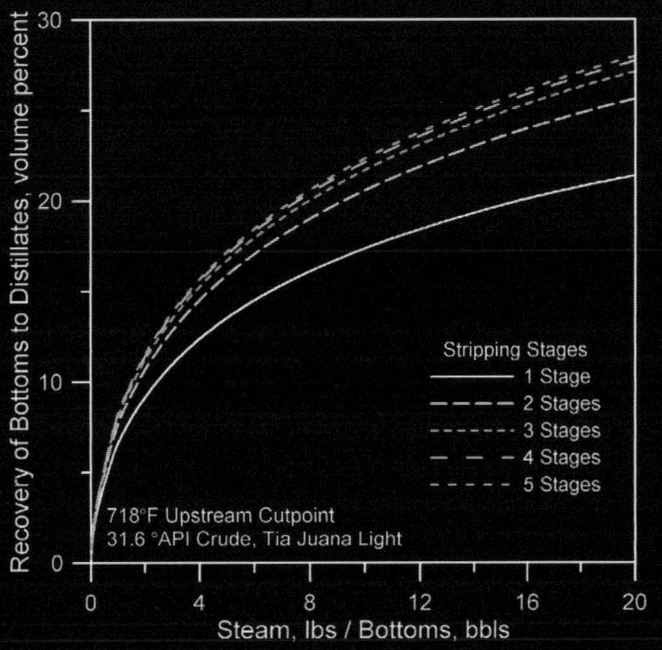

Most Benefit Captured After Two Stages

- One stage has a large benefit
- Two stages capture nearly all the benefit
- Curves shown at constant pressure:
  - Too much stripping steam can cause towers to pressure up
  - Pressure rise can outweigh additional stripping steam

Most Stripping Sections < = One Stage

- Atmospheric Tower, one stage:
  - Cutpoint increase up to 50°F (28°C)
  - Equivalent of up to 10% reduction in bottoms rate
- Vacuum Tower, one stage:
  - Cutpoint increase up to 40°F (22°C)
  - Equivalent of up to 15% reduction in bottoms rate
- Performance varies with unit and crude type

Heater Tradeoffs

Stripping and Flash Zone HMB

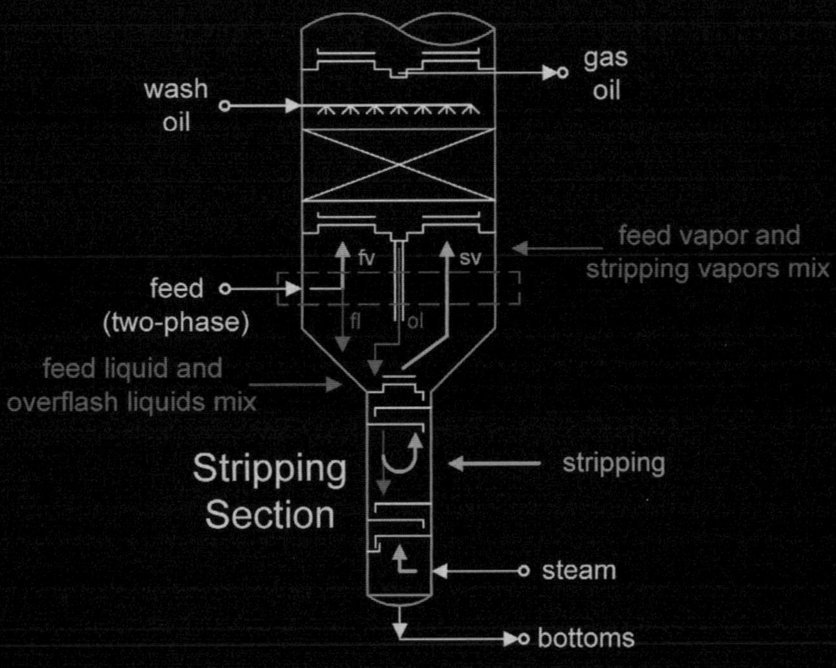

Material Balance Equations

$$\textit{total distillate} = \textit{flash zone}_{hcv} - \textit{overflash}_{l}$$

$$\textit{flash zone}_{hcv} = \textit{total distillate} + o_{l}$$

$$\textit{feed}_{hcv} + \textit{stripping}_{hcv} = \textit{flash zone}_{hcv}$$

- Constant yield has constant flash zone vapor
- Vapor can come from heater outlet or stripping
- Improved stripping can allow for lower heater outlet temperatures
- Roughly 18°F (10°C) temperature drop halves coking rate

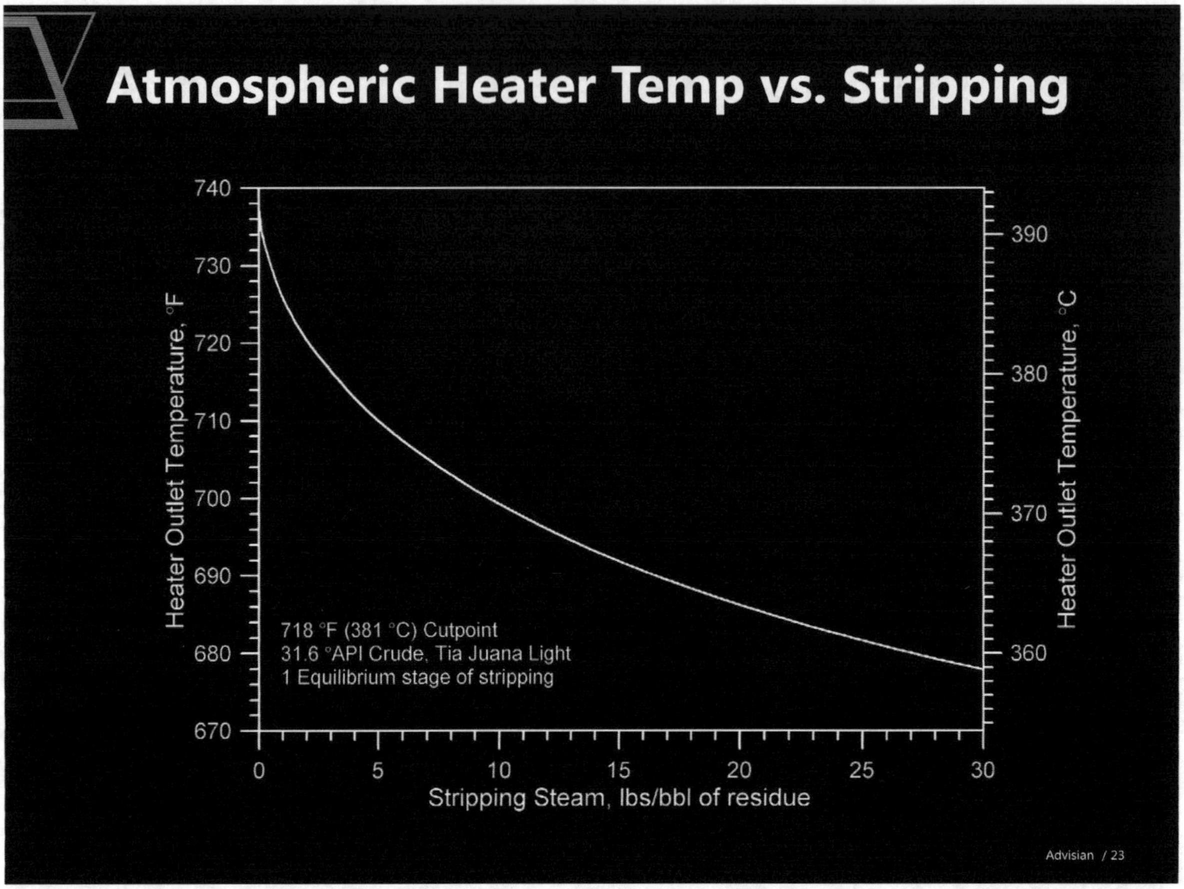
Atmospheric Heater Temp vs. Stripping
740
730
720
710
700
690
680
670
Heater Outlet Temperature, °F
390
380
370
360
Heater Outlet Temperature, °C
718 °F (381 °C) Cutpoint
31.6 °API Crude, Tia Juana Light
1 Equilibrium stage of stripping
0
5
10
15
20
25
30
Stripping Steam, lbs/bbl of residue
Advisian / 23

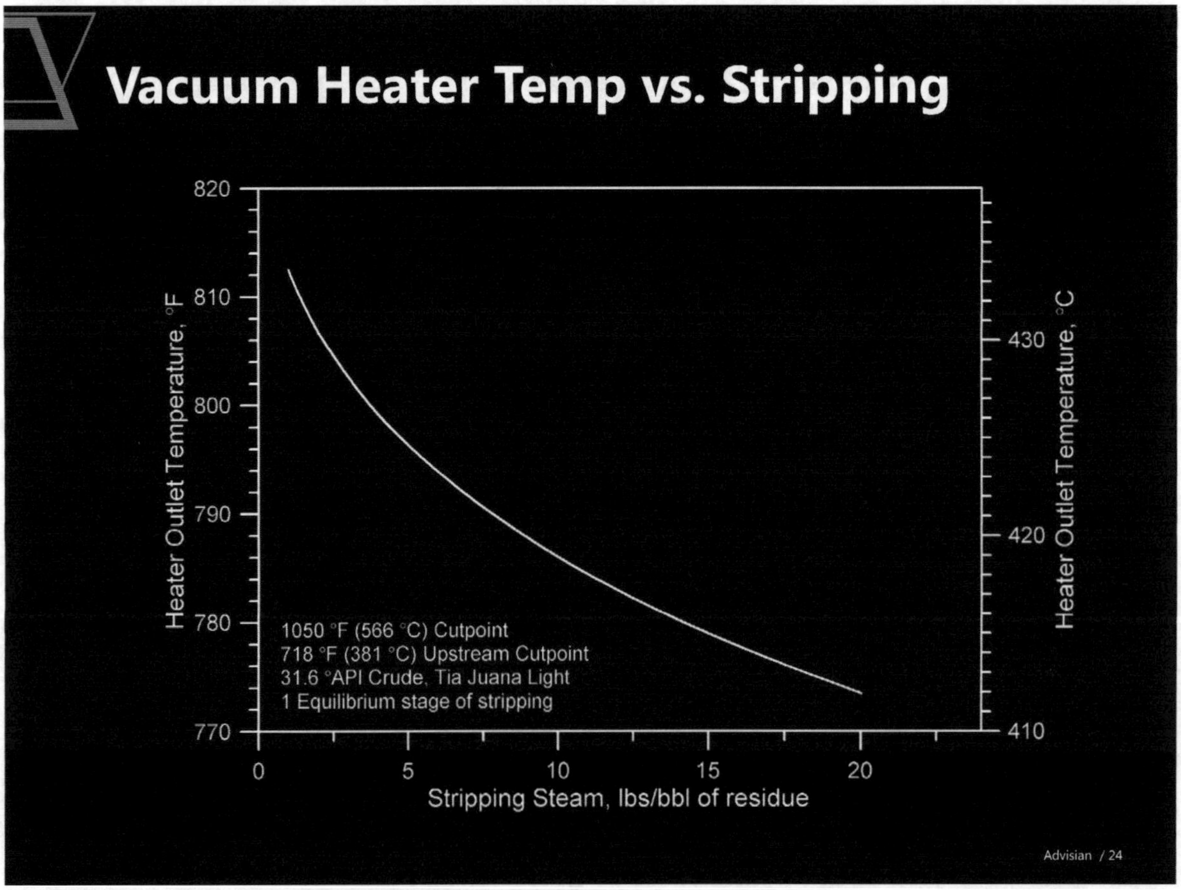
Vacuum Heater Temp vs. Stripping
820
810
800
790
780
770
Heater Outlet Temperature, °F
430
420
410
Heater Outlet Temperature, °C
1050 °F (566 °C) Cutpoint
718 °F (381 °C) Upstream Cutpoint
31.6 °API Crude, Tia Juana Light
1 Equilibrium stage of stripping
0
5
10
15
20
Stripping Steam, lbs/bbl of residue
Advisian / 24

Monitoring Performance

Performance Monitoring

- Sampling can be difficult
- Initial parts of distillation curves can have significant errors
- Where does vaporization energy come from?
  - The hydrocarbon stream
  - Very little mass of steam
- Temperature drops in stripping sections
- Use temperature to monitor performance

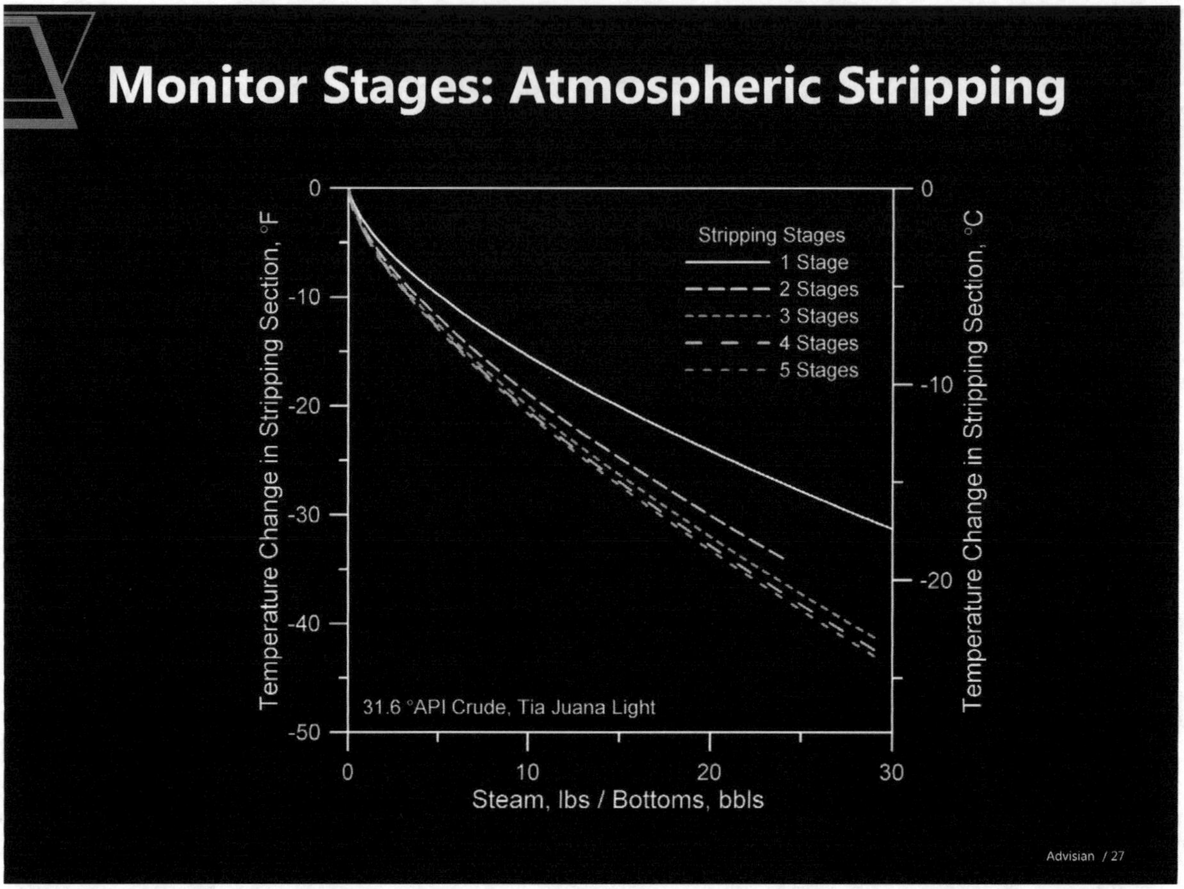
Monitor Stages: Atmospheric Stripping
Temperature Change in Stripping Section, °F
Temperature Change in Stripping Section, °C
Stripping Stages
1 Stage
2 Stages
3 Stages
4 Stages
5 Stages
31.6 °API Crude, Tia Juana Light
0
-10
-20
-30
-40
-50
0
-10
-20
0
10
20
30
Steam, lbs / Bottoms, bbls
Advisian / 27

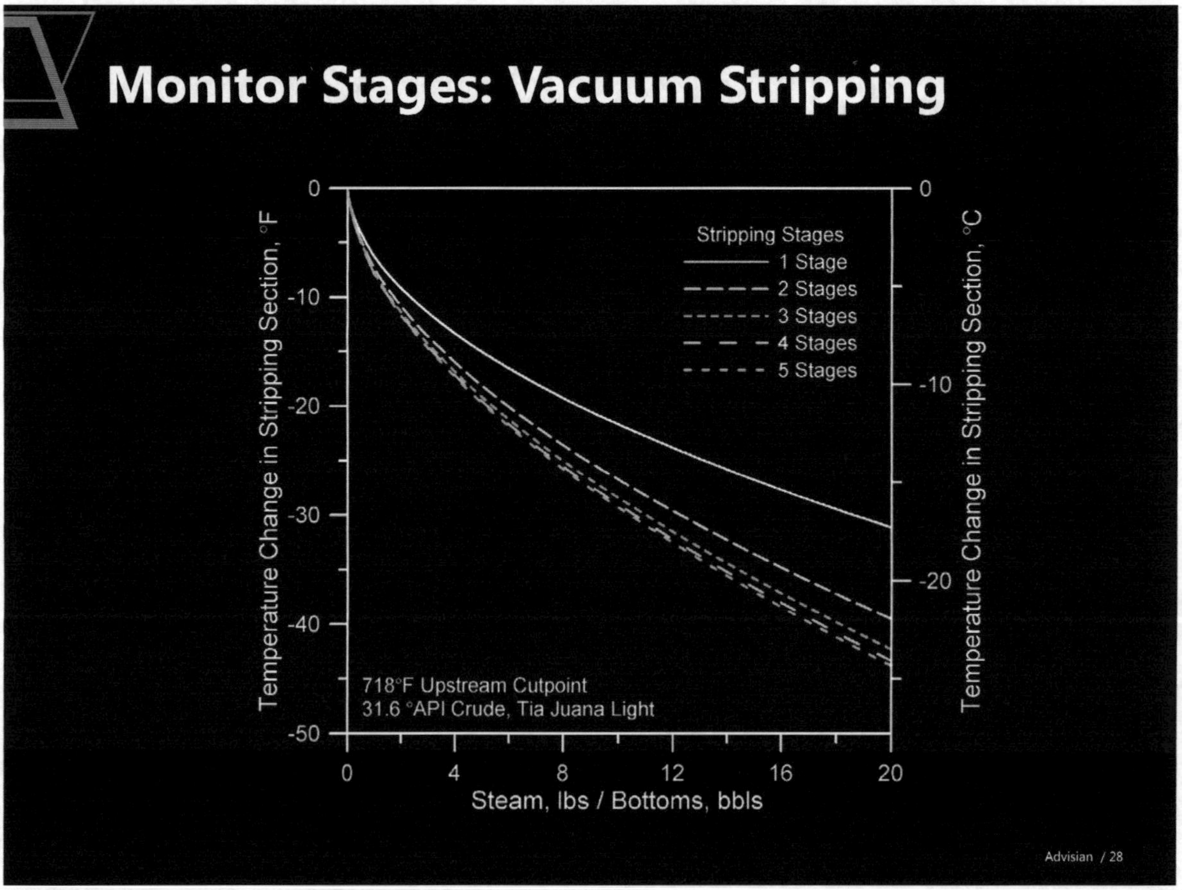
Monitor Stages: Vacuum Stripping
Temperature Change in Stripping Section, °F
Temperature Change in Stripping Section, °C
Stripping Stages
1 Stage
2 Stages
3 Stages
4 Stages
5 Stages
718°F Upstream Cutpoint
31.6 °API Crude, Tia Juana Light
0
-10
-20
-30
-40
-50
0
-10
-20
0
4
8
12
16
20
Steam, lbs / Bottoms, bbls
Advisian / 28

Vacuum Stripping has Larger DT

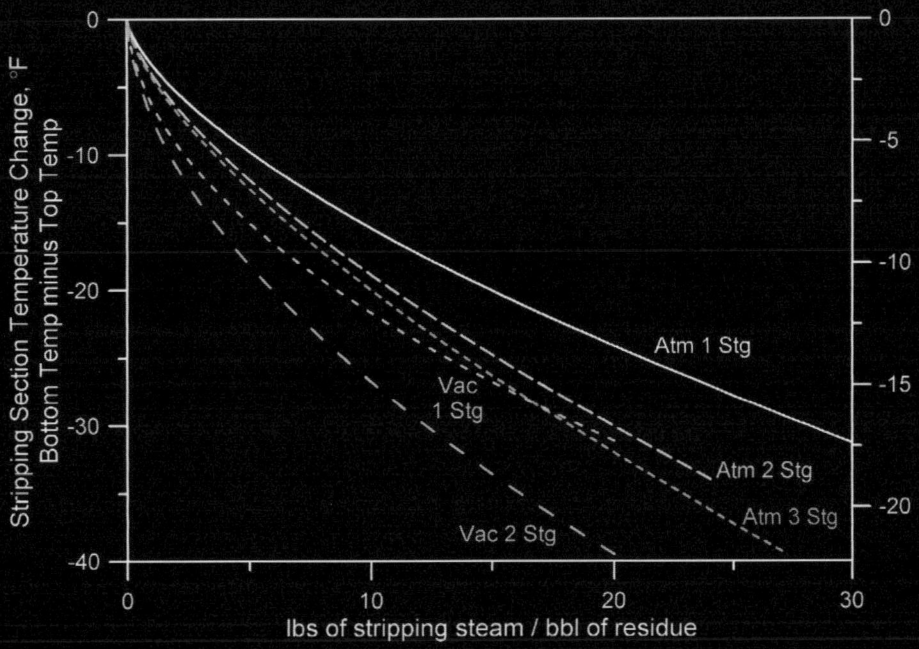

Equipment Limitations

Equipment Limits

- Minimum pressure drop required:
  - Vapor rate low, holes far apart, efficiency low
  - Pressure drop required for mixing
- Too high a pressure drop floods trays
- 2:1 vapor handling range:
  - Sets minimum and maximum steam rates
- Vapor rates change over trays, multiple designs required
- Efficiency low, 25% - 30% at best

Vapor Load Proxy: Fixed Tray Layout

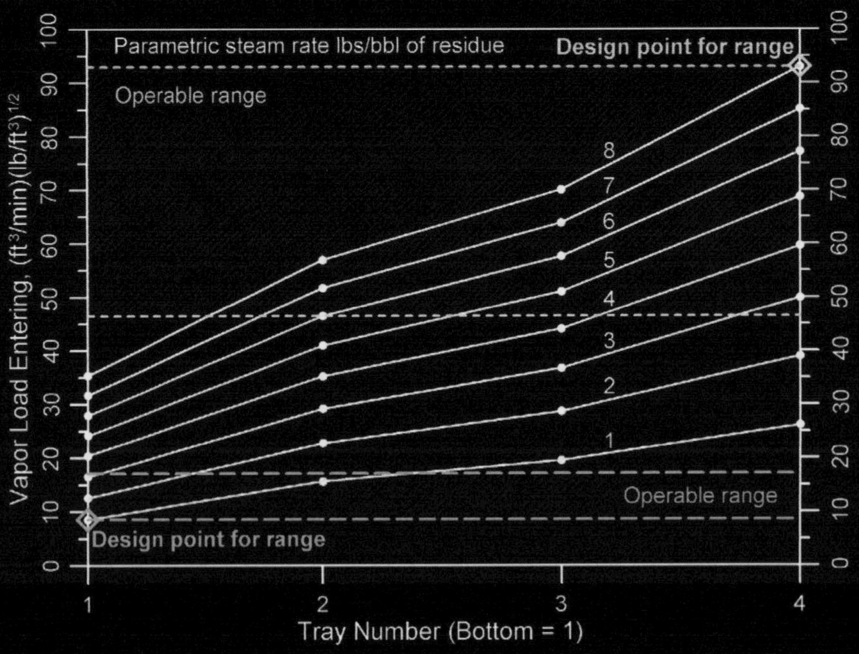

Summary

Stripping

- Critical for overall unit operation
- Sensitive to stages and steam rates
- Upstream tradeoffs possible
- Monitor with temperature profiles

Disclaimers

DISCLAIMER

This presentation has been prepared by a representative of Advisian.

The presentation contains the professional and personal opinions of the presenter, which are given in good faith. As such, opinions presented herein may not always necessarily reflect the position of Advisian as a whole, its officers or executive.

Any forward-looking statements included in this presentation will involve subjective judgment and analysis and are subject to uncertainties, risks and contingencies—many of which are outside the control of, and may be unknown to, Advisian.

Advisian and all associated entities and representatives make no representation or warranty as to the accuracy, reliability or completeness of information in this document and do not take responsibility for updating any information or correcting any error or omission that may become apparent after this document has been issued.

To the extent permitted by law, Advisian and its officers, employees, related bodies and agents disclaim all liability—direct, indirect or consequential (and whether or not arising out of the negligence, default or lack of care of Advisian and/or any of its agents)—for any loss or damage suffered by a recipient or other persons arising out of, or in connection with, any use or reliance on this presentation or information.

Advisian / 36

Advisian

WorleyParsons Group

About Us

Consulting and Engineering in

Refinery and Petrochemicals

- Molecule Management
- Economics and Planning
- Technology Selection
- Refinery-Petrochemical Integration

Logistics

- Hydrocarbon Logistics
- Bulk liquid terminals
- Railway corridors
- Market analysis
- Risk Analysis
- Safety and Environment

Advisian

WorleyParsons Group

Sulfur and Gas Treating

- Oxygen enrichment
- Gas treating solvents

Infrastructure and Environmental

- Water
- Materials handling

EcoNomics™

Advisian / 38

Process Safety
Risk Management

Leveraging Learnings from Incidents and Near Misses Using Process Safety Networks

The Dow Chemical Company

Sheila F. Van Geffen
Process Safety Technology Leader
Houston, TX

Process Safety
Risk Management

Biography – Sheila Van Geffen

- Process Safety Technology Leader, The Dow Chemical Company
- Located in Houston, Texas
- 27 years industry experience, over 12 years in Process Safety
- Have worked at Dow sites in Louisiana, Minnesota, Michigan, and Texas
- B.S. degree in Chemical Engineering, University of Alabama

Process Safety
Risk Management

Leveraging Learnings from Incidents and Near Misses Using Process Safety Networks

The Dow Chemical Company

Sheila F. Van Geffen

Process Safety Technology Leader

Houston, TX

Process Safety
Risk Management

The Dow Chemical Company Process Safety Incidents by Year

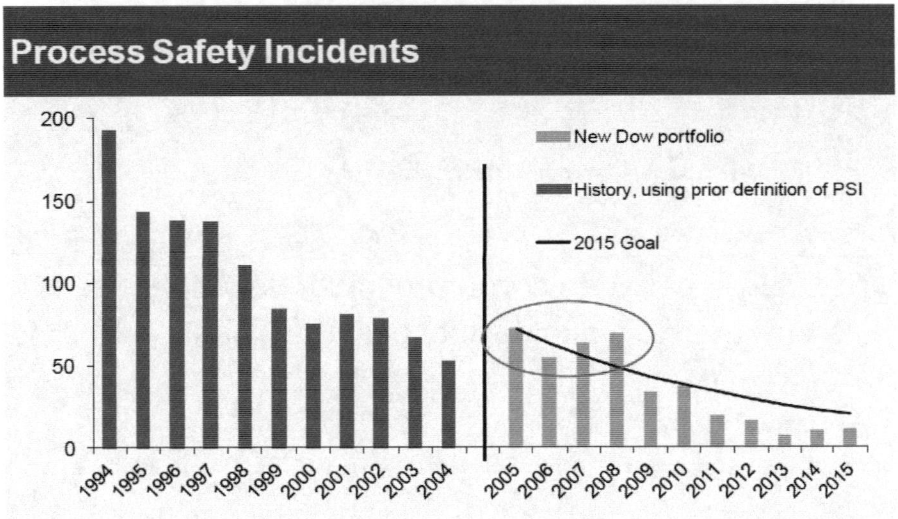

Process Safety
Risk Management

DOW

Process Safety Operational Discipline

- Process Safety Focal Point and Networks
- Process Safety Near Misses
- Root Cause Investigations and Sharing of Learnings

Enables success on a day-to-day basis

Process Safety
Risk Management

DOW

Process Safety Focal Point Role

Process Safety Focal Point

- "Boots on the ground" for each facility
- Experienced engineer or leader
- Liaison between corporate initiatives/expertise and the individual facility activities

Process Safety
Risk Management

Process Safety Focal Point Responsibilities

- Implementation of Process Safety Requirements
- Process Safety reviewer on management of change
- Root Cause Investigations with Process Safety impacts.
  - Process Safety Near Misses (PSNMs) and
  - Process Safety and Containment Events (PSCEs)

Process Safety
Risk Management

Process Safety Networks

- Network of PS Focal Points aligned to a geography or technology
- Includes experts and specialists
- Sharing best practices and innovative approaches to managing hazards.

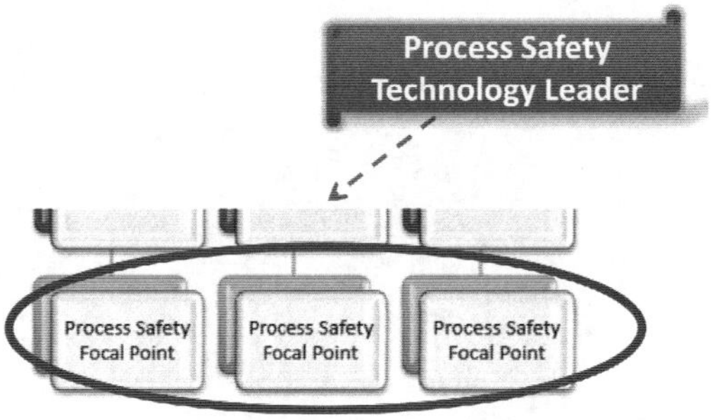

Process Safety
Risk Management

Process Safety Network Key Expectations

- Every facility represented
- Everybody shares & engaged in team
- Everybody shares learnings and discusses within their facility
- Follow-up discussion the next meeting

11/14/2017 9

Process Safety
Risk Management

Process Safety Near Miss Program

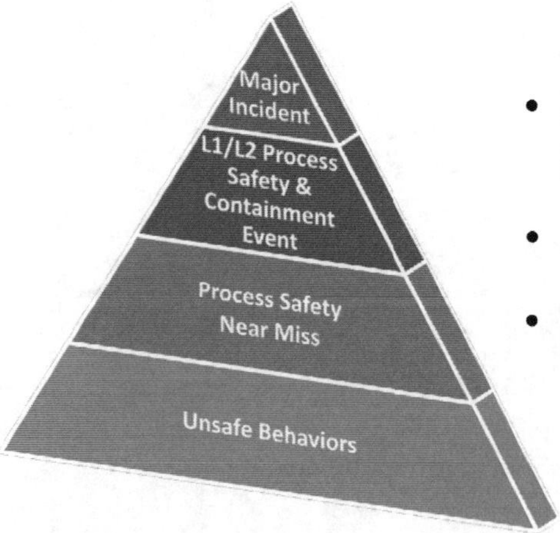

- A PSNM can only occur if something fails to work as intended
- ...but does not meet the criteria for a L1 or L2 PSCE.
- Dow recorded ~3000 PSNMs in 2016

11/14/2017 10

Relief Device Failure or Activation
Uncontrolled Reaction
Mechanical Overpressure
DOW
Uncontrolled Mobile Equipment
Unplanned Mixing
TECHNOLOGY CENTERS
PROCESS SAFETY NEAR MISS
Fire or Explosion
Process Safety Cardinal Rules
SIS Activation (with high learning value)
Personnel Exposure
Actual or Potential Chemical Release
Failure of a Protection Layer

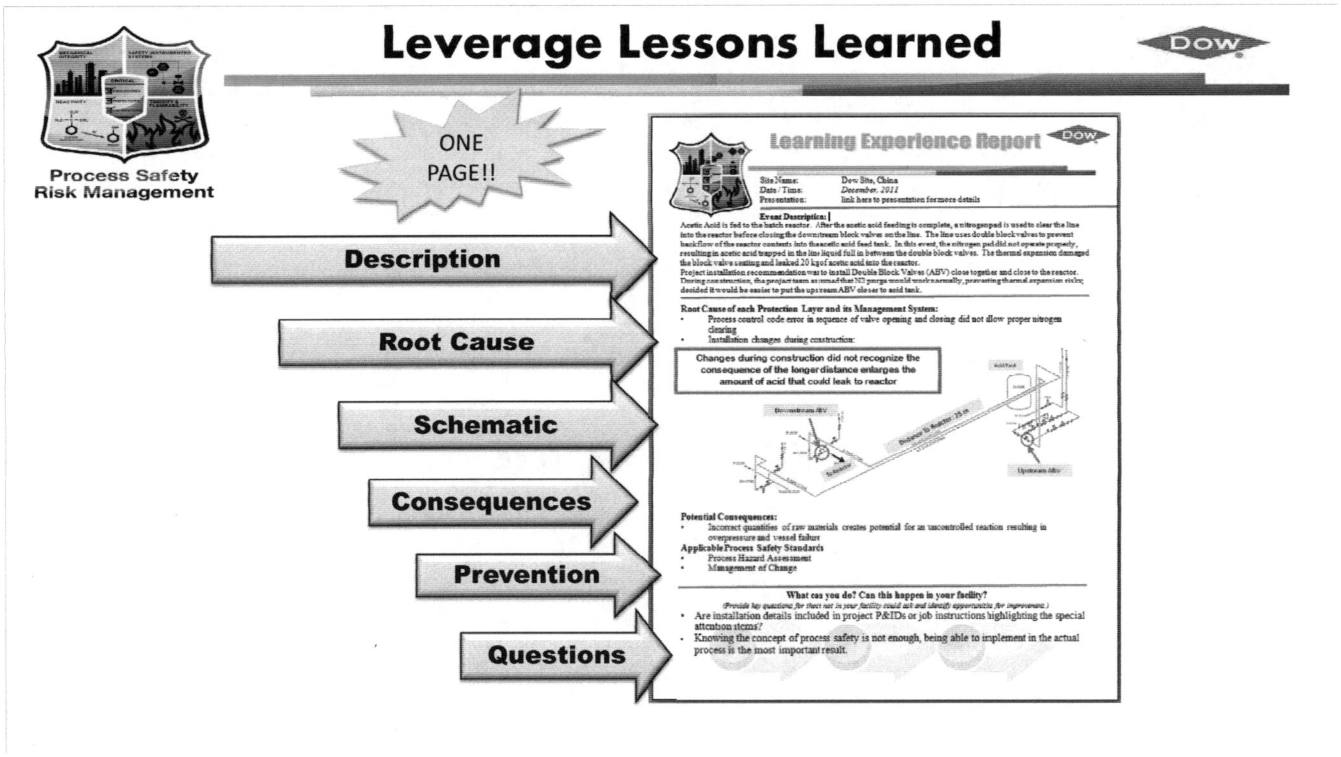

Leverage Lessons Learned
DOW
Process Safety Risk Management
ONE PAGE!!
Description
Root Cause
Schematic
Consequences
Prevention
Questions
Learning Experience Report
Site Name: Dow Site, China
Date / Time: December, 2011
Presentation: link here to presentation for more details
Event Description:
Acetic Acid is fed to the batch reactor. After the acetic acid feeding is complete, a nitrogen pad is used to clear the line into the reactor before closing the downstream block valves on the line. The line uses double block valves to prevent backflow of the reactor contents into the acetic acid feed tank. In this event, the nitrogen pad did not operate properly, resulting in acetic acid trapped in the line liquid full in between the double block valves. The thermal expansion damaged the block valve seating and leaked 20 kg of acetic acid into the reactor.
Project installation recommendation was to install Double Block Valves (ABV) close together and close to the reactor. During construction, the project team assumed that N2 purge would work normally, preventing thermal expansion risks; decided it would be easier to put the upstream ABV closer to acid tank.
Root Cause of each Protection Layer and its Management System:
Process control code error in sequence of valve opening and closing did not allow proper nitrogen clearing
Installation changes during construction:
Changes during construction did not recognize the consequence of the longer distance enlarges the amount of acid that could leak to reactor
Downstream ABV
Distance to Reactor: 25 m
Upstream ABV
Potential Consequences:
Incorrect quantities of raw materials creates potential for an uncontrolled reaction resulting in overpressure and vessel failure
Applicable Process Safety Standards
Process Hazard Assessment
Management of Change
What can you do? Can this happen in your facility?
Are installation details included in project P&IDs or job instructions highlighting the special attention items?
Knowing the concept of process safety is not enough, being able to implement in the actual process is the most important result.

Process Safety
Risk Management

Process Safety Operational Discipline

- The face of Process Safety embedded in a plant's organization
- Recognition of potential consequences and driving prevention
- Easy format for communication and continued learning

11/14/2017 13

Process Safety
Risk Management

Thank you

11/14/2017 14

Process Safety
Risk Management

References

- *Are Today's Engineering Designs Preventing Tomorrow's Process Safety Incidents?*
  - **2013 Mary Kay O'Connor Process Safety Center International Symposium**
- *Investigating Process Safety Near Misses To Improve Performance*,
  - **Process Safety Progress, Volume 32, Issue 2, pages 170-174, June 2013.**
- Next Generation Root Cause Investigation and Analysis - Elimination of Repetitive Incidents through Strengthening Management Systems
  - **2015 Global Congress on Process Safety**
- Dow Learnings And Actions From The Deepwater Horizon Accident
  - **2015 Global Congress on Process Safety**
- Beyond HAZOP and LOPA – Four Different Company Approaches
  - **2016 Global Congress on Process Safety**
- Reducing Process Safety Events: An Approach Proven By Sustainable Results
  - **2017 Global Congress on Process Safety**

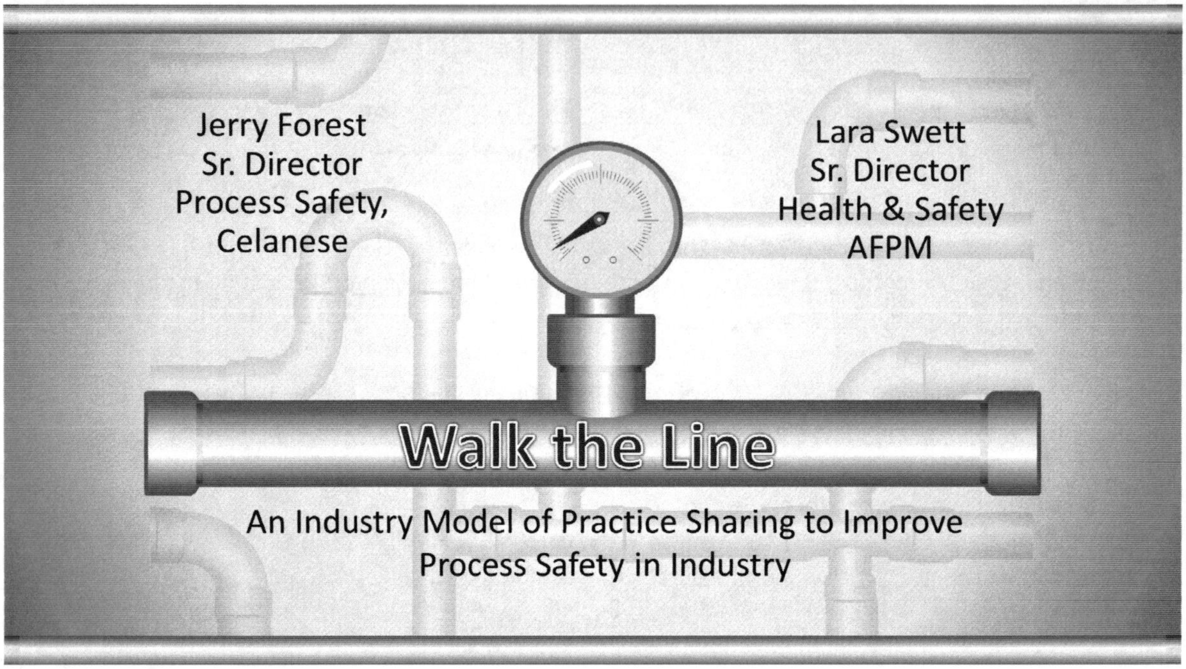

Jerry Forest

- Senior Director of Process Safety at Celanese
- Over 30 yrs petrochemical and refining experience
- Manufacturing background in isocyanates, hydrazine at Celanese and LyondellBasell and predecessor companies
- Mission: apply 20 years of manufacturing experience to PSM to save lives, protect the environment and prevent incidents
- Adjunct Lecturer in Chemical Engineering on PSM at LSU

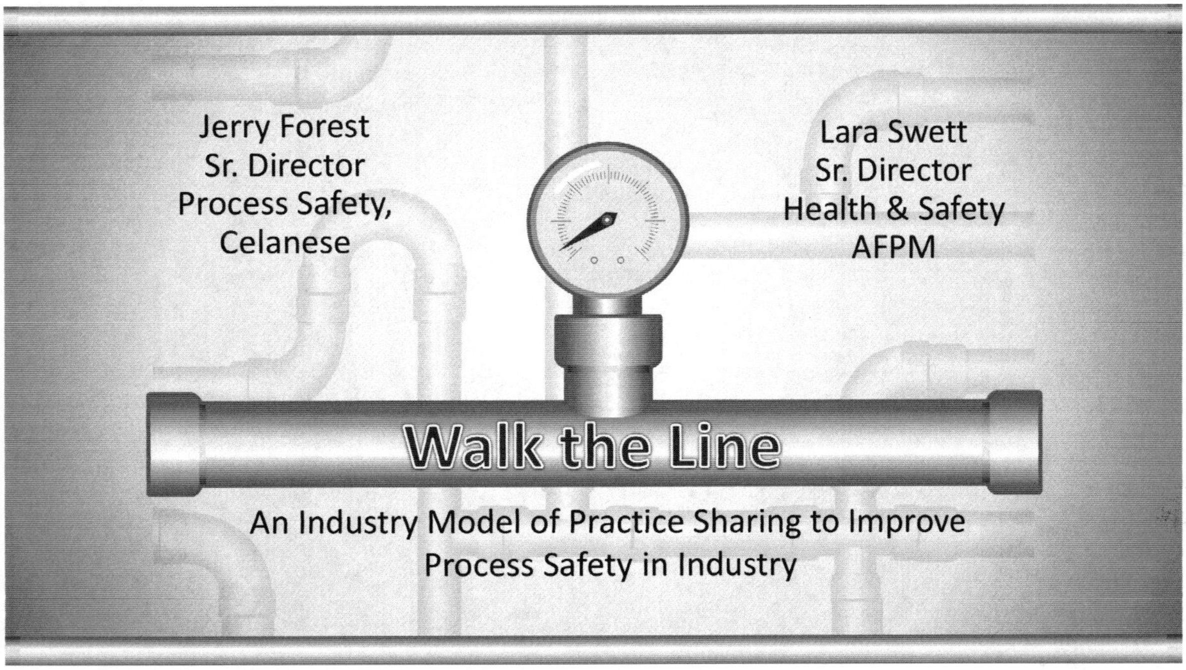
Jerry Forest
Sr. Director
Process Safety,
Celanese
Lara Swett
Sr. Director
Health & Safety
AFPM
Walk the Line
An Industry Model of Practice Sharing to Improve Process Safety in Industry

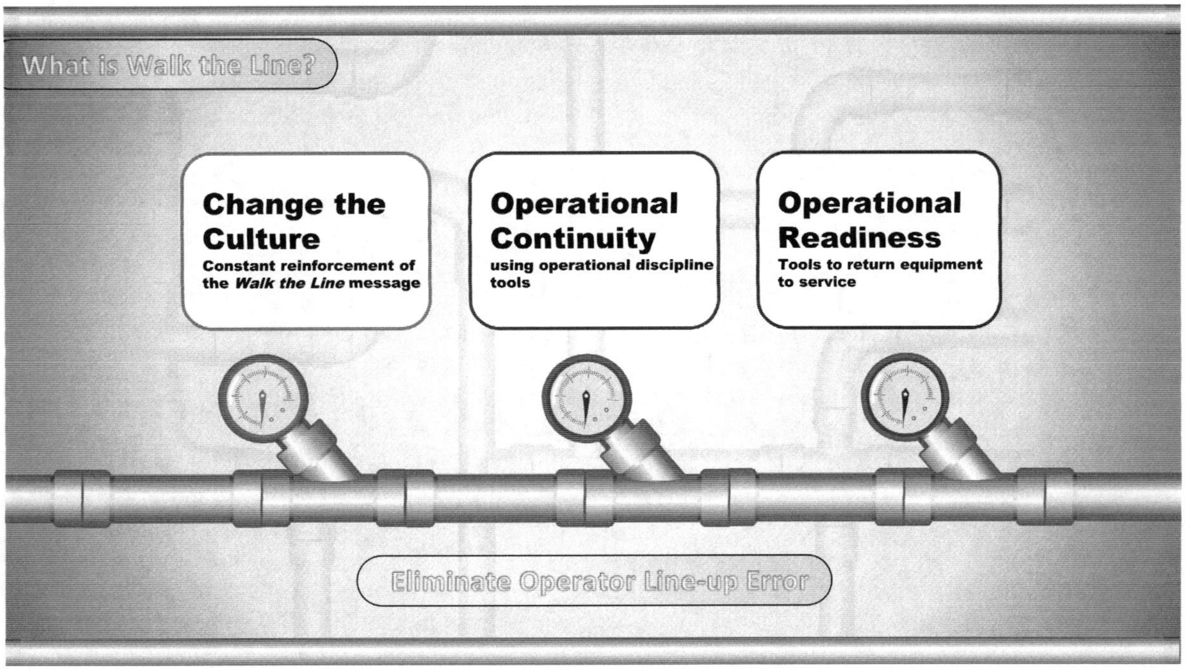
What is Walk the Line?
Change the Culture
Constant reinforcement of the *Walk the Line* message
Operational Continuity
using operational discipline tools
Operational Readiness
Tools to return equipment to service
Eliminate Operator Line-up Error

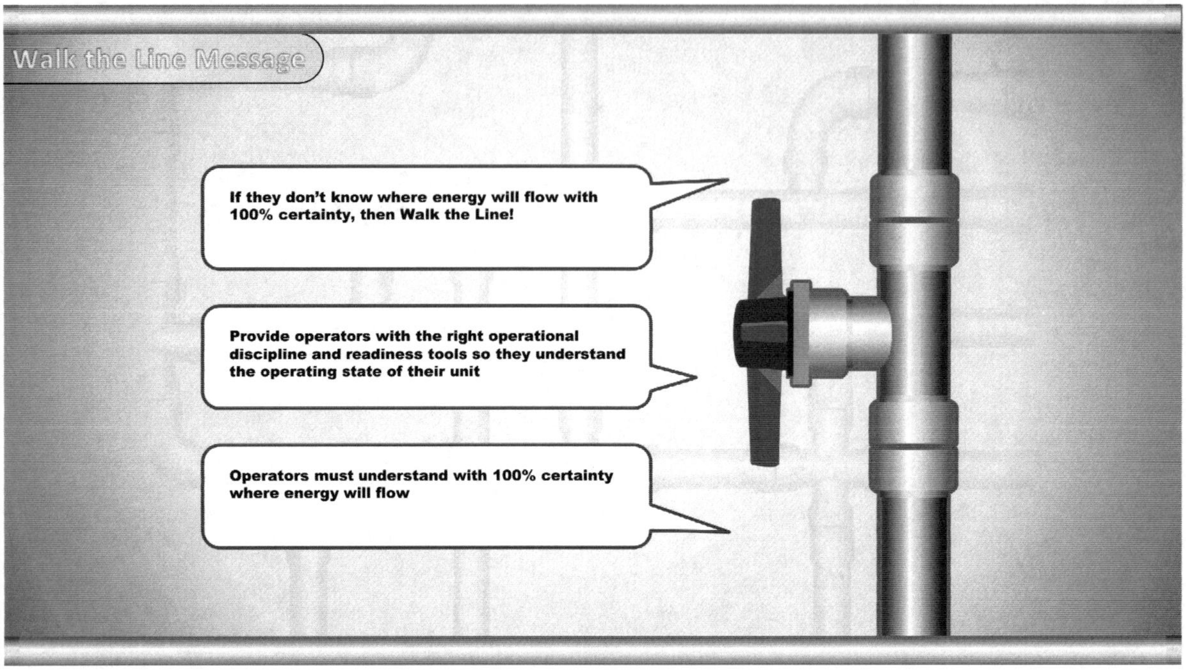
Walk the Line Message
If they don't know where energy will flow with 100% certainty, then Walk the Line!
Provide operators with the right operational discipline and readiness tools so they understand the operating state of their unit
Operators must understand with 100% certainty where energy will flow

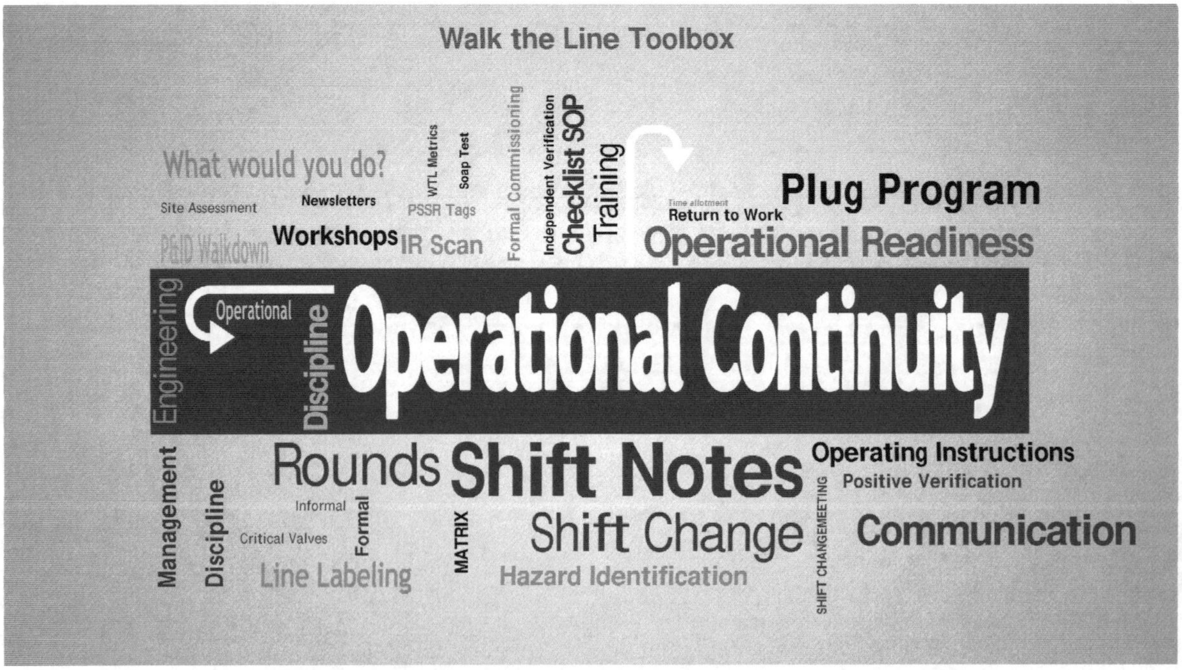
Walk the Line Toolbox
What would you do?
Site Assessment
Newsletters
P&ID Walkdown
Workshops
WTL Metrics
Soap Test
PSSR Tags
IR Scan
Formal Commissioning
Independent Verification
Checklist SOP
Training
Return to Work
Plug Program
Operational Readiness
Engineering
Operational
Discipline
Operational Continuity
Management
Discipline
Rounds
Informal
Formal
Critical Valves
Line Labeling
MATRIX
Shift Notes
Shift Change
Hazard Identification
SHIFT CHANGEMEETING
Operating Instructions
Positive Verification
Communication

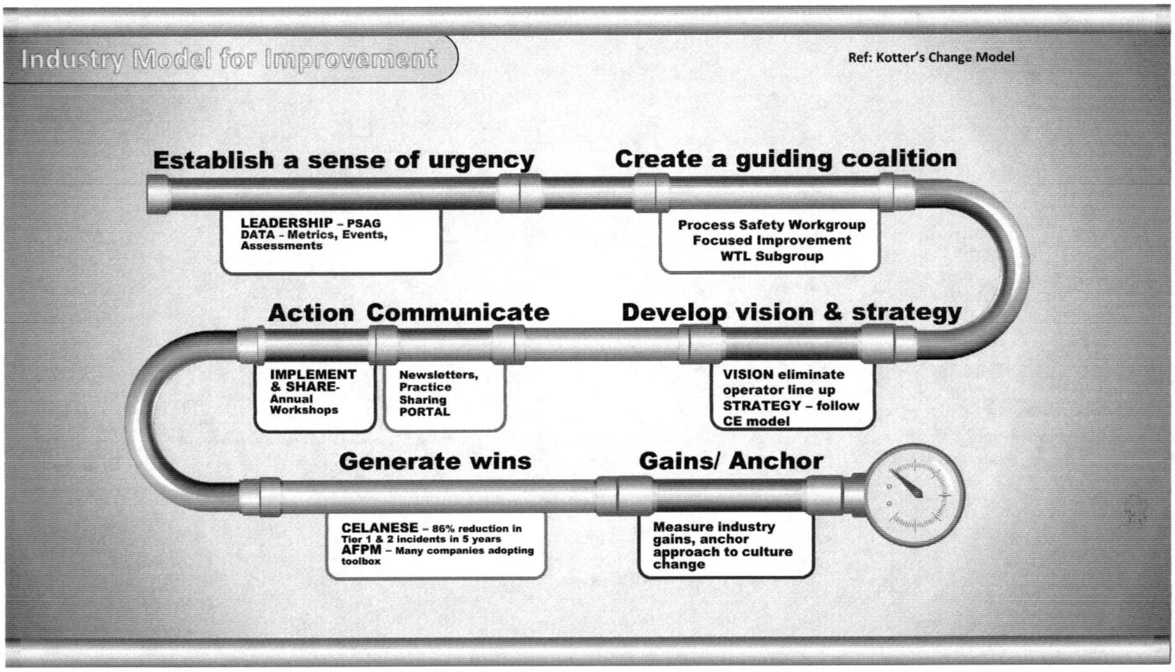
Industry Model for Improvement
Ref: Kotter's Change Model
Establish a sense of urgency
LEADERSHIP - PSAG
DATA - Metrics, Events, Assessments
Create a guiding coalition
Process Safety Workgroup
Focused Improvement
WTL Subgroup
Develop vision & strategy
VISION eliminate operator line up
STRATEGY - follow CE model
Action Communicate
IMPLEMENT & SHARE- Annual Workshops
Newsletters, Practice Sharing PORTAL
Generate wins
CELANESE - 86% reduction in Tier 1 & 2 incidents in 5 years
AFPM - Many companies adopting toolbox
Gains/ Anchor
Measure industry gains, anchor approach to culture change

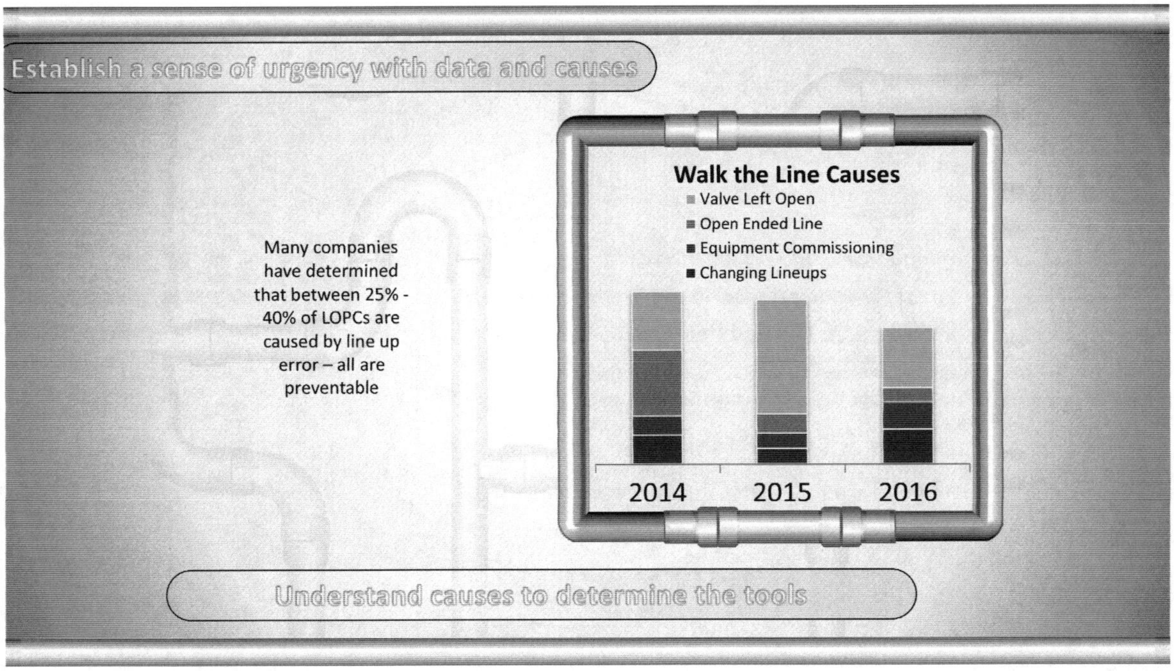
Establish a sense of urgency with data and causes
Many companies have determined that between 25% - 40% of LOPCs are caused by line up error – all are preventable
Walk the Line Causes
Valve Left Open
Open Ended Line
Equipment Commissioning
Changing Lineups
2014
2015
2016
Understand causes to determine the tools

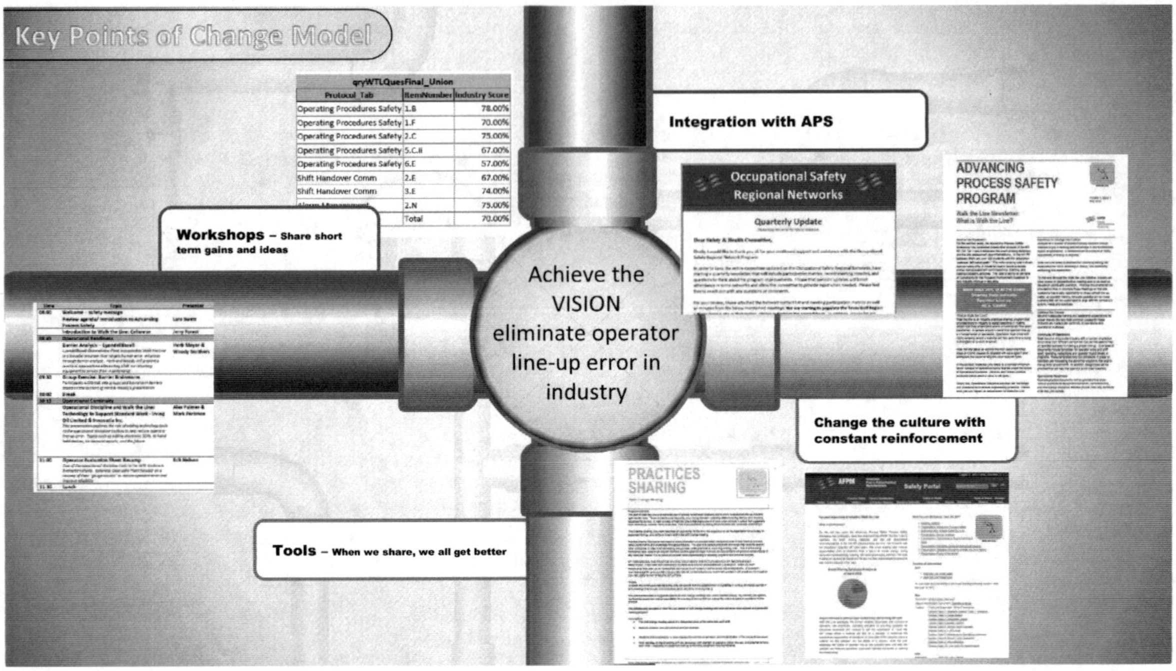
Key Points of Change Model
qryWTLQuesFinal_Union
Protocol Tab | ItemNumber | Industry Score
Operating Procedures Safety | 1.B | 78.00%
Operating Procedures Safety | 1.F | 70.00%
Operating Procedures Safety | 2.C | 75.00%
Operating Procedures Safety | 5.C.ii | 67.00%
Operating Procedures Safety | 6.E | 57.00%
Shift Handover Comm | 2.E | 67.00%
Shift Handover Comm | 3.E | 74.00%
2.N | 75.00%
Total | 70.00%
Integration with APS
Occupational Safety Regional Networks
Quarterly Update
ADVANCING PROCESS SAFETY PROGRAM
Workshops – Share short term gains and ideas
Achieve the VISION eliminate operator line-up error in industry
Change the culture with constant reinforcement
PRACTICES SHARING
Safety Portal
Tools – When we share, we all get better

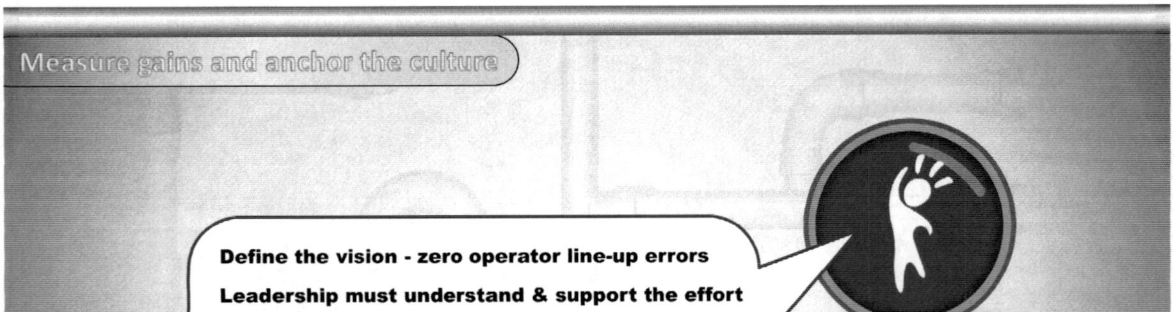
Measure gains and anchor the culture
Define the vision - zero operator line-up errors
Leadership must understand & support the effort
Analyze WTL incidents for causes
Use the tools to address the causes
• Constantly reinforce the message
Measure and celebrate success
Share learnings with other sites & companies

Backup

http://www.afpm.org/sv1/

Next Generation Process Hazard Analysis (PHA)

Why?

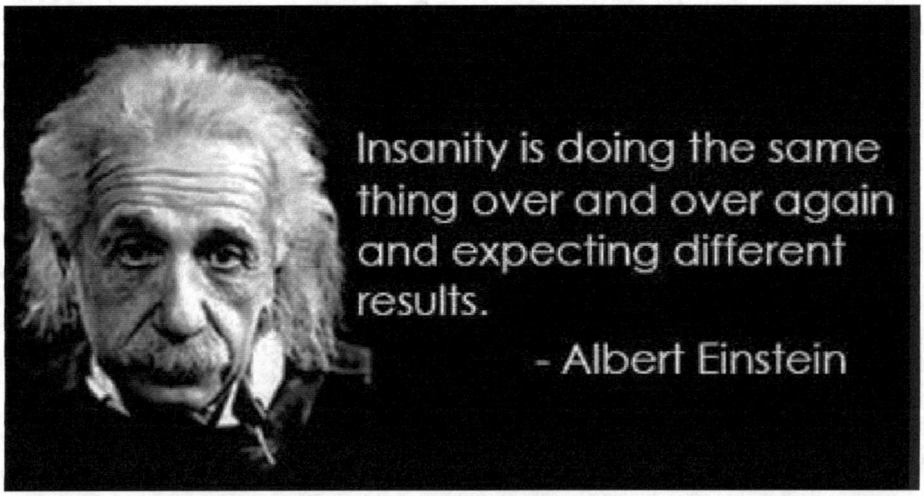

2

Current PHA Challenges

- High-level risks and significant events are not aligned - priorities and utilization of resources are misaligned.
- Reliance on risk ranking alone does not work, we lose sight of the big picture.
- PHAs do not produce the results, or deliver the product, that we desire and value.
- We expect more from our PHAs than what they are designed to do.
- What does the data say? Are our efforts appropriately aligned with what our incidents are telling us?
- What is the plan to get out of this loop?

3

Current PHA Challenges

Where Would We Prefer To Focus Our Efforts?

- High-Level, Well-Managed Risks
- Lessons from Recent Incidents

Our Challenge

- Limited Resources
- Outdated Mindset; ***and***
- Satisfy OSHA and Company Requirements

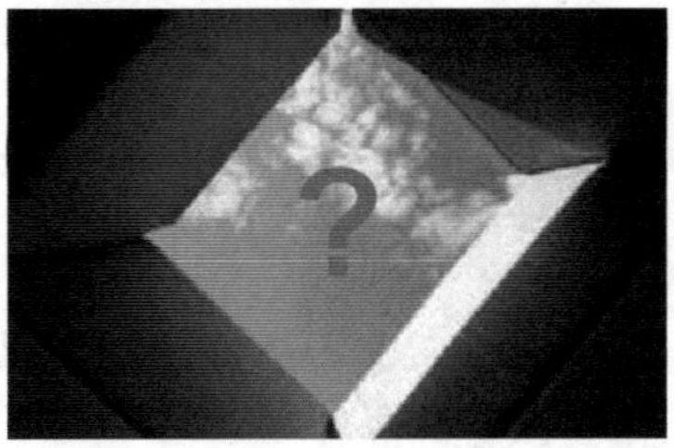

4

NextGen Purpose

- Ensure that our control systems are designed, installed, and maintained fail-safe:
  - Critically evaluate the control logic and design of every loop in order to validate it has been "designed fail-safe"
  - Verify the proper and adequate field installation of each component of the final control element to ensure it has been "installed fail-safe"
  - ..... been "maintained fail-safe"
- Bonus!
  - Train and educate our process engineers in reading P&ID's, understand control/interlock logic, and lay the foundation to critically evaluate their design.

5

NextGen Anticipated Benefits

- Ensure that the:
  - Original *design is sound*
  - Equipment that is installed is correct *per the design*
  - Equipment is *maintained and available* when called upon
- Confirm the accuracy and completeness, and correct as necessary, P&ID's, ladder logic, loop diagrams, control narratives, etc.

Designed Fail-Safe – Installed Fail-Safe – Maintained Fail-Safe

6

NextGen Anticipated Benefits

- Performed in a group setting
- Knowledge transfer:
  - Allow engineers early in their career to have a solid foundation in process and controls engineering
    - Read, edit and develop P&ID's, and other drawings
    - Critically evaluate control function, logic, and design
    - Identify by sight the function of components in the field
    - Understand how control systems are installed in the field
  - Capture knowledge from seasoned veterans

Designed Fail-Safe – Installed Fail-Safe – Maintained Fail-Safe

7

NextGen Anticipated Benefits

- Take the PHA from the conference room to the field!

8

Streamlined Revalidation Process

- Definitions
  - Deficiency – errors in applying PHA methodology
  - Gap – omissions; failure to address established requirements
  - Revalidate – to declare the previous PHA valid
- Depending on quality of PHA
  - Update and Revalidate
  - Retrofit, Update and Revalidate
  - Redo
- Revalidation Process
  - Baseline regulatory requirements
  - Verification of past PHA quality and completeness
  - Changes since last PHA
    - MOC's
    - Top events
    - Incidents
    - Process parameters

9

The NextGen Process

2 Key Components

- Training
- Loop Evaluation and Function Check

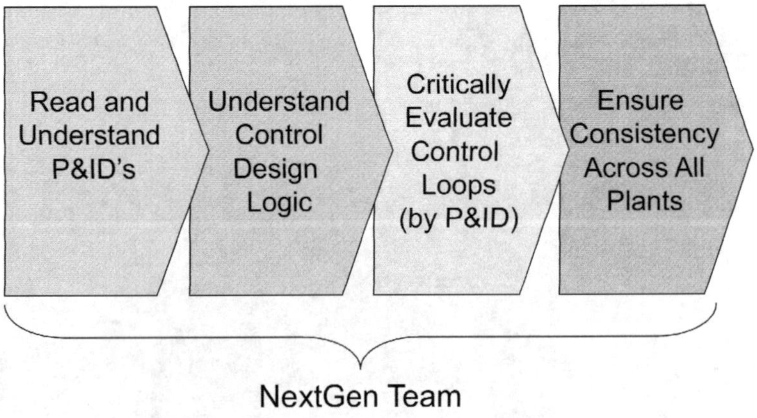

10

The NextGen Process

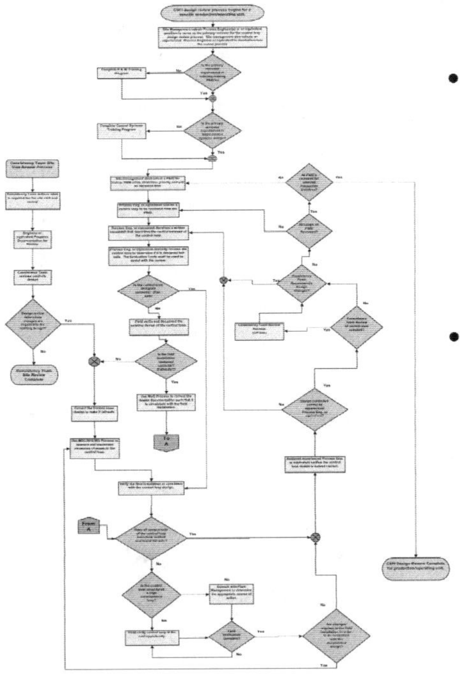

- Detailed NextGen flow chart developed

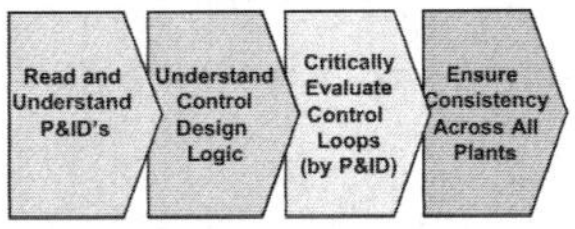

- Four Critical Elements
  - P&ID Training
  - Controls Training
  - Loop Evaluation Tool
    - Paperwork, Logic, and Field
  - Consistency Review
    - At Plant and by External Team

11

P&ID Training

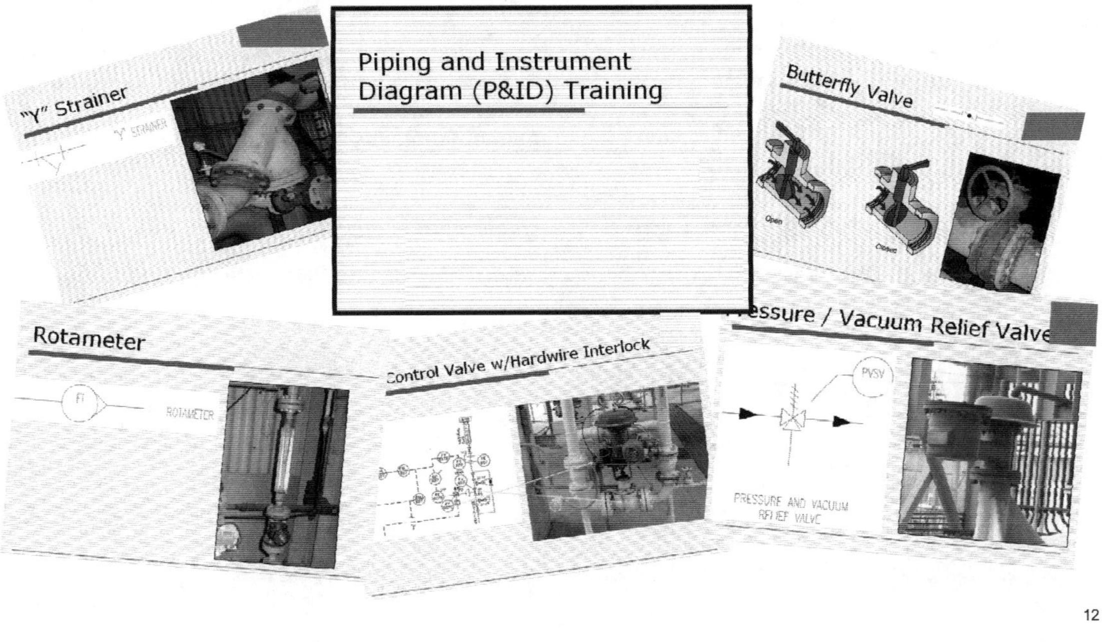

12

Controls Training

Understand Control Design Logic

- Eight Lessons
  1. Process Measurement Types and Final Control Elements
  2. Basic Process Control
  3. Control System Architecture
  4. Alarms
  5. Interlocks
  6. Fail Safe Design
  7. Good Engineering Practices for Instrument Systems
  8. How to Use the Control Loop Evaluation Tool
- Training done with instructor

13

Control Loop Evaluation Tool

Critically Evaluate Control Loops (by P&ID)

- The evaluation tool is an Excel spreadsheet with three specific sheets:
  - Consequence of Failure
  - Detailed Loop Review
  - Field Review
- 5 evaluations completed as part of pilot testing
  - Findings include: trip on bad pv; alarm set point adjustments; P&ID and other paperwork inconsistencies
- Plant manages follow-up action items

Designed Fail-Safe – Installed Fail-Safe – Maintained Fail-Safe

14

Consistency Review

Ensure Consistency Across All Plants

- Plant review done by experienced process engineer or equivalent
- Upload into SharePoint site for review by external team
  - Feedback to engineer to ensure same thorough review
    - Specified number of loop evaluations per engineer
  - Review action items for consistency

Designed Fail-Safe – Installed Fail-Safe – Maintained Fail-Safe

15

NextGen Timeline

- Kickoff Meeting
- Streamline Revalidation Process In Motion
- P&ID Training Piloted
- P&ID Training Complete
- Goals Finalized ("x" P&ID's per engineer)
- Controls Training and Evaluations Tool Piloted
- Controls Training Started
- Complete Loop Evaluations
- Consistency Review per Engineer Complete
- Team Meeting to Review Program and Next Year Goals

16

Lessons learned from a crude oil train derailment

Monir Ahammad, Ph.D.
Lecturer and Research Associate
Mary Kay O'Connor Process Safety Center
Texas A&M University
College Station, TX 77843

MARY KAY O'CONNOR PROCESS SAFETY CENTER
TEXAS A&M ENGINEERING EXPERIMENT STATION

Monir Ahammad

MARY KAY O'CONNOR PROCESS SAFETY CENTER
TEXAS A&M ENGINEERING EXPERIMENT STATION

Lecturer and Research Associate
Mary Kay O'Connor Process Safety Center

Ph.D. in Chemical Engineering
Texas A&M University

M.S. and B.S. in Chemical Engineering
Bangladesh University of Engineering and Technology

- Extensive lab experience in LNG
- CFD modeling, Root Cause Analysis, PHAs, PSM, SOP development, Consequence Analysis and Risk Assessment
- Interned with Huntsman Chemical Co.

2

Lessons learned from a crude oil train derailment

Monir Ahammad, Ph.D.
Lecturer and Research Associate
Mary Kay O'Connor Process Safety Center
Texas A&M University
College Station, TX 77843

TEES | MARY KAY O'CONNOR PROCESS SAFETY CENTER
TEXAS A&M ENGINEERING EXPERIMENT STATION

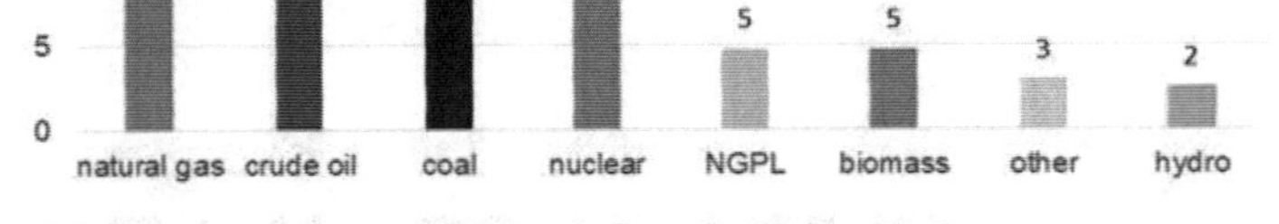

Note: Natural gas is dry gas, NGPL is natural gas plant liquids, other is geothermal, solar, and wind, and hydro is conventional hydroelectric.
Source: U.S. Energy Information Administration, *Monthly Energy Review*, April 2017, preliminary data

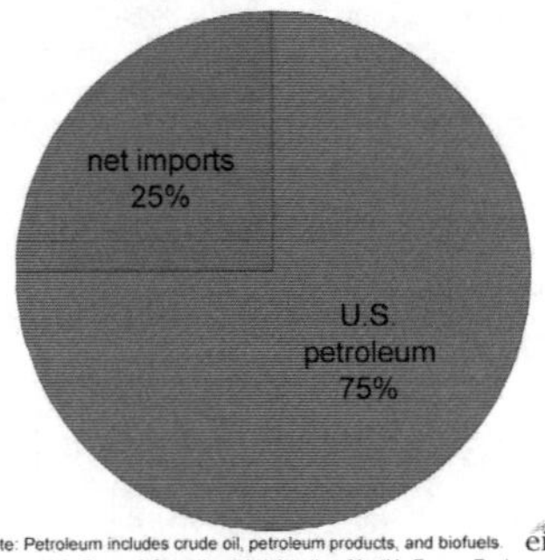

Note: Petroleum includes crude oil, petroleum products, and biofuels.
Source: U.S. Energy Information Administration, *Monthly Energy Review*, Table 3.3a, March 2017, preliminary data

MARY KAY O'CONNOR PROCESS SAFETY CENTER
TEXAS A&M ENGINEERING EXPERIMENT STATION

CCrude-by-rail movements (2014)

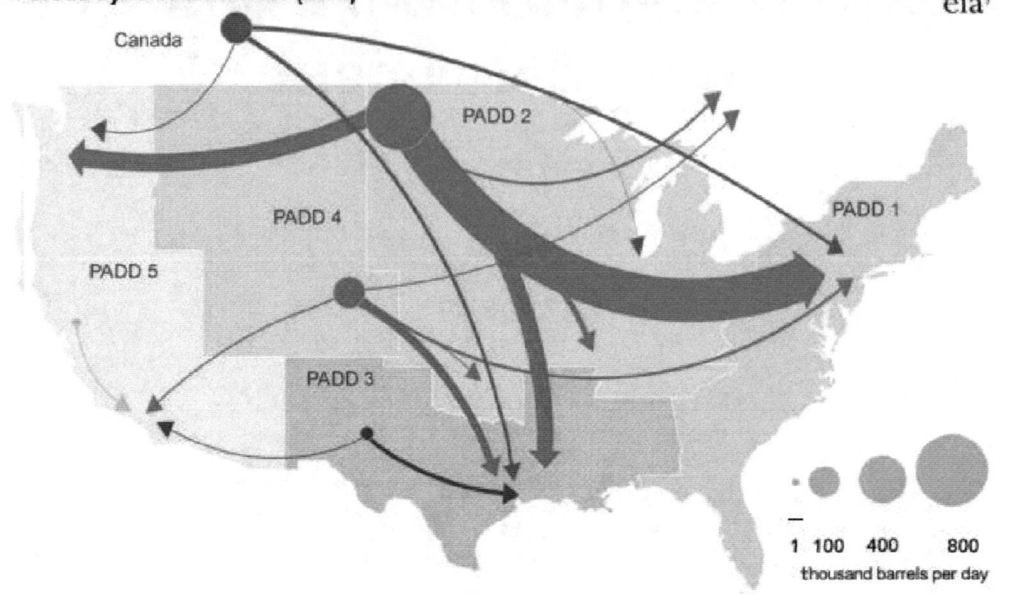

www.eia.gov/petroleum/drilling/

Oil Train Disasters

MARY KAY O'CONNOR PROCESS SAFETY CENTER
TEXAS A&M ENGINEERING EXPERIMENT STATION

Apr 2017, MI: A train collision causes 10-12 cars to derail. Crude oil tank cars caught fire.
Jun 3, 2016 OR: derailment in Columbia river gorge
Nov. 8, 2015 WI: 13 train cars flip over; dozens of homes evacuated
Jul 17, 2015 MT: 35,000 gallons of crude spill; half-mile radius evacuated
May 6, 2015 ND: town evacuated after crude oil train derails, explodes
Mar. 7, 2015 ON: river system contaminated
Mar. 5, 2015 IL: train derails; fires burn for days
Feb. 16, 2015 WV: house destroyed; drinking water intake pumps closed
Feb. 14, 2015 AL: train carrying Alberta tar sands derails
Feb. 14, 2015 ON: 29 cars derail; fires burn for days
Jan. 29, 2015 CA: lawsuit filed over expansion of Bakersfield crude terminal
Aug. 14, 2014 WA: train carrying Bakken crude derails
Jun 20, 2014 MB: Dot-111s derail in rail yard
Jun 16, 2014 AL: oil tankers derail; mandatory evacuation of surrounding area
Jun 7, 2014 PA: derailment over Youghiogheny river
May 12, 2014 NY: tankers derail in rail yard
May 8, 2014 SK: tankers derail
Apr 30, 2014 VA: 30,000 gallons spilled; fireballs erupted; 300 people evacuated
Apr 30, 2014 NY: derailment in rail yard
Feb. 13, 2014 PA: train crashes into building; 3,000 gallons spilled
Feb 3, 2014 MN: 12,000 gallons leaked along 68 miles of track
Jan 31, 2014 MI: 90,000 gallons spilled; dozens of homes evacuated
Jan 20, 2014 PA: train derails; nearby expressway closed
Jan 7, 2014 NB: derailment sets off large fire; 150 people evacuated
Dec 30, 2013 ND: trains collide; massive explosion set off; 1,400 residents evacuated
Dec 10, 2013 NY: train derails
Dec 3, 2013 NY: oil tanker collides with truck; driver seriously injured

Nov 8, 2013 AL: train derails; fires burn for more than a day
Oct 19, 2013 AB: train derails; 100 residents evacuated
Jul 6, 2013 QC: runaway train derails; 47 people killed
Jun 27, 2013 AB: train derails; nearly falls into river
May 21, 2013 SK: trail derails; 24,000 gallons spilled
Apr 3, 2013 ON: train derails; 16,500 gallons spilled
Mar 27, 2013 MN: train derails; 20,000 gallons spilled
Jan 24, 2013 SK: train collides with road grader; one person dies
Jan 19, 2013 AB: freight train collides with oil tanker; fire engulfs

Oil Train Disasters

MARY KAY O'CONNOR PROCESS SAFETY CENTER
TEXAS A&M ENGINEERING EXPERIMENT STATION

https://earthjustice.org/features/map-crude-by-rail

7

MARY KAY O'CONNOR PROCESS SAFETY CENTER
TEXAS A&M ENGINEERING EXPERIMENT STATION

- Is DOT 111, a general purpose tank car suitable to transport crude oil safely?
- Does the physical properties (*e.g.,* flammability, vapor pressure) of the crude can dramatically change the consequence after derailment?

8

1. Objectives of this study

- Increase understanding of the physical/combustion processes describing the consequences of the crude oil train derailment
- Identify and perform a comprehensive analysis of the phenomena that may have significant public safety concern
- Mount Carbon, West Virginia, crude oil train derailment incident is selected as a representative case

2. Mount Carbon Incident Summary

MARY KAY O'CONNOR PROCESS SAFETY CENTER
TEXAS A&M ENGINEERING EXPERIMENT STATION

- February 12, 2015
- 362,300 gal crude oil release
- 26 train cars were derailed
- Train speed: 33 mph
- Fire reported almost immediately

Description	Count
No Breaches/Thermal Tears	7
Minimum Damage / Leakage	3
Fire damage	4
Fire Damage/Thermal Tear	13
Total	27

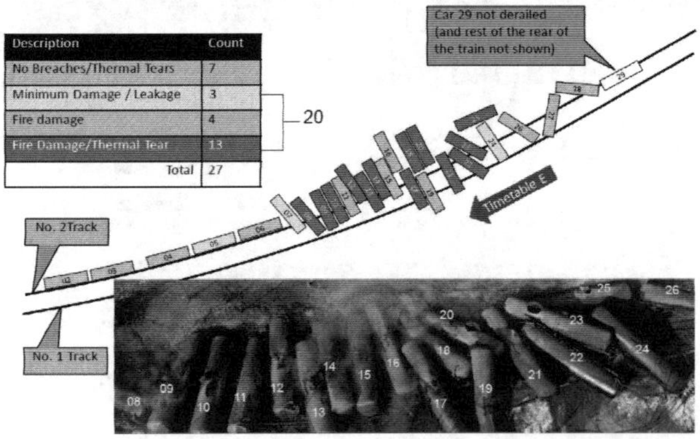

- First fireball after 24 minutes. The 13th fireball after 22 hours of derailment.

Incident propagation

MARY KAY O'CONNOR PROCESS SAFETY CENTER
TEXAS A&M ENGINEERING EXPERIMENT STATION

- Vertical Split Head in the rail caused 27 railcars to derail
- Train car 16 and 19 were punctured releasing over 50,000 gal crude oil
- Crude oil vapor ignited
- Remaining tank cars were "cooked" in the flame, the temperature and vapor composition of crude oil increased
- When the pressure in the tank car was above the strength of tank car super heated crude oil released and ignited as "fireballs"
- 1st fire ball about 24 minutes after the initial pool fire and last (13th) fireball on the next day

11

3. Mass boil-off resulting from speed

MARY KAY O'CONNOR PROCESS SAFETY CENTER
TEXAS A&M ENGINEERING EXPERIMENT STATION

- Assumption
  - 100% of kinetic energy converted to thermal energy
  - 33mph

Conclusion

- The conversion of kinetic energy did not significantly contribute to the vaporization of crude oil in the tank car

12

4. Conditions inside a tank car

MARY KAY O'CONNOR PROCESS SAFETY CENTER
TEXAS A&M ENGINEERING EXPERIMENT STATION

Assumptions

1. Vapor-liquid equilibrium calculations using Peng-Robinson equation-of-state
2. 1 atm pressure in tank
3. 20% vapor space by volume
4. Non-blanketed tank car

Conclusions

- At transportation conditions it is impossible for a flammable condition to exist in the vapor space
- Only in very unlikely conditions, *i.e.,* tank cars with large vapor spaces (>20%), low volatility crude oil, and very cold temperatures, a flammable atmosphere is possible
- A fire in the vapor space within the tank would be impossible during transportation

13

6. Phenomena leading to hazardous releases and consequences

MARY KAY O'CONNOR PROCESS SAFETY CENTER
TEXAS A&M ENGINEERING EXPERIMENT STATION

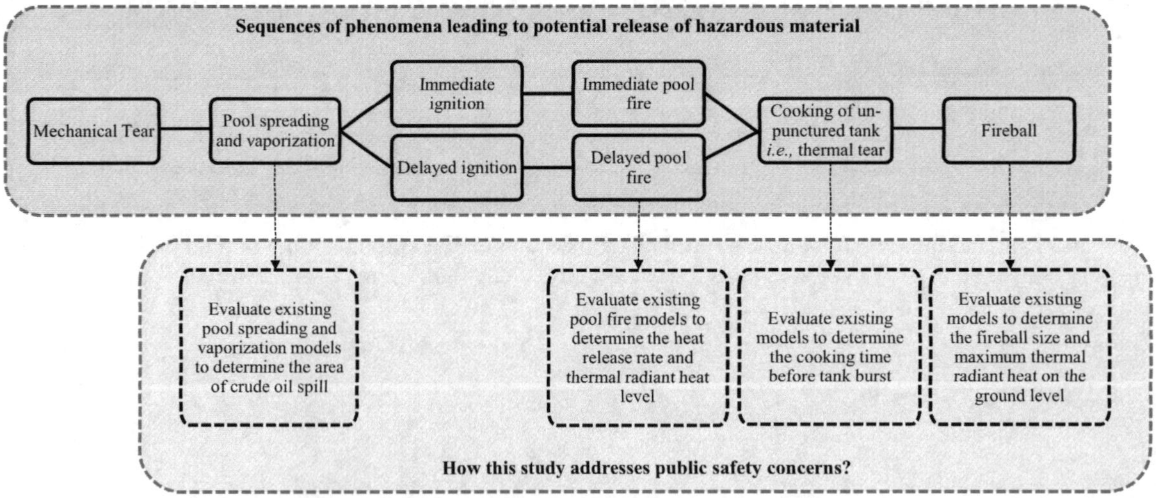

14

7. Study of pool spreading, vaporization, and pool fire

Pool Fire Modeling

MARY KAY O'CONNOR PROCESS SAFETY CENTER
TEXAS A&M ENGINEERING EXPERIMENT STATION

Case-1: Catastrophic rupture of tank car
Case-2: Liquid leak through 1 inch hole
Case-3: Liquid leak through 6 inch hole
Case-4: Liquid leak through 12 inch hole

Assumptions

- Inventory: 100 m^3 crude oil
- Substrate type: wet soil. Note that the actual soil had 6 inches of snow.
- Release direction for cases 2,3, and 4: Liquid impinging on the ground
- Atmospheric temperature 15°F
- Wind speed = 1.5 m/s, Stability F
- Spill temperature = 15°F
- Plain landscape

Pool Fire

MARY KAY O'CONNOR PROCESS SAFETY CENTER
TEXAS A&M ENGINEERING EXPERIMENT STATION

Early Pool Fire

Late Pool Fire

17

Effect of Crude Composition

MARY KAY O'CONNOR PROCESS SAFETY CENTER
TEXAS A&M ENGINEERING EXPERIMENT STATION

thermal radiant heat contour at ground level

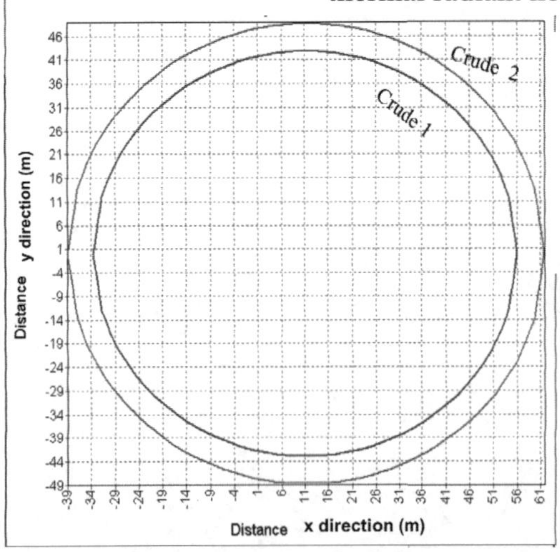

(a) Early Pool fire

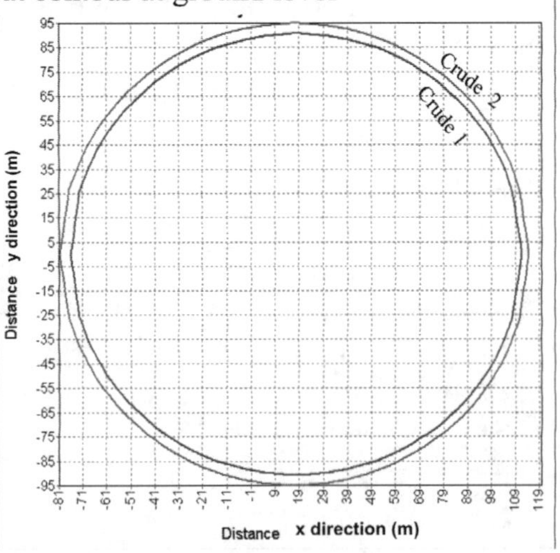

(b) Late pool fire;

18

MARY KAY O'CONNOR PROCESS SAFETY CENTER
TEXAS A&M ENGINEERING EXPERIMENT STATION

Conclusion

- Crude oil composition (*i.e.,* volatile or heavy) has little impact on the extent of the overall affected area

Research Gaps

- Existing models for pool spread, vaporization, and fire do not handle multicomponent mixtures.
- Soil types (*e.g.,* 6 inch of snow) and un-validated terrain type make the modeling more challenging.

19

MARY KAY O'CONNOR PROCESS SAFETY CENTER
TEXAS A&M ENGINEERING EXPERIMENT STATION

8. Thermal tear and fluid release

20

Estimated rupture time for different type of crude oil

MARY KAY O'CONNOR PROCESS SAFETY CENTER
TEXAS A&M ENGINEERING EXPERIMENT STATION

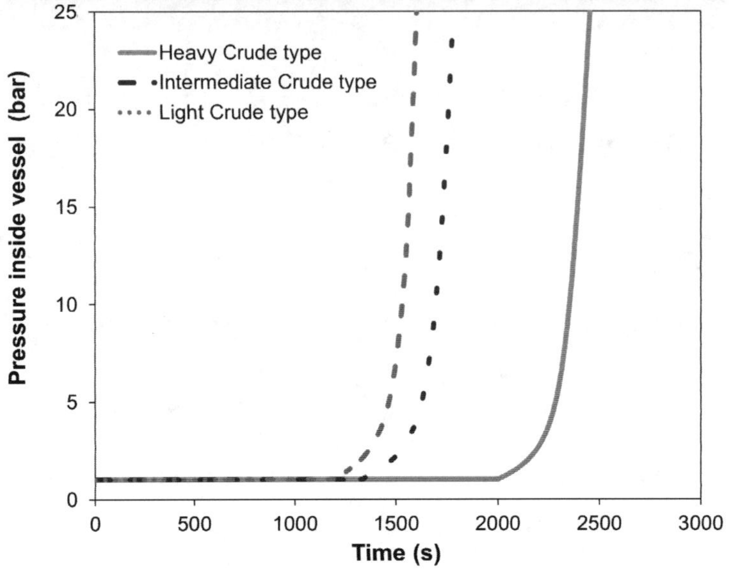

Variation of vessel internal pressure since the start of the pool fire for three different types of crudes

MARY KAY O'CONNOR PROCESS SAFETY CENTER
TEXAS A&M ENGINEERING EXPERIMENT STATION

Assumptions

1. Heat input from pool fire is calculated based on wetted surface of vessel
2. No relief device was functional
3. Thermal expansion of the vessel metal was not considered
4. API 521 heat flux calculation equation is used to calculate the heat flux from pool fire.

Conclusions

- The time to reach the burst pressure is a function of the volatility of the crude oil.
- However, the difference in time to reach burst pressure (for the different types of crudes) has no significant impact on magnitude of the consequences.

MARY KAY O'CONNOR PROCESS SAFETY CENTER
TEXAS A&M ENGINEERING EXPERIMENT STATION

9. Study of Fireball

23

Estimation of Fireball Mass: Part 1

MARY KAY O'CONNOR PROCESS SAFETY CENTER
TEXAS A&M ENGINEERING EXPERIMENT STATION

- Fireball occurred 65 minutes after the incident
- Calculated mass of fuel from fireball diameter
- Uncertainties in measurement of fireball radius

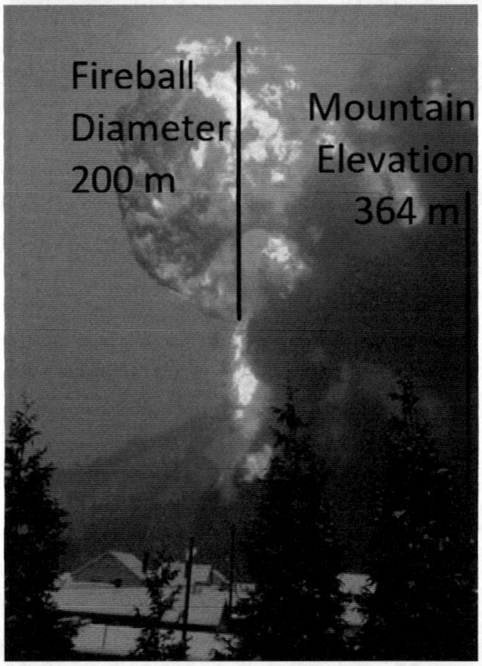

24

Estimation of Fireball Mass: Part 2

MARY KAY O'CONNOR PROCESS SAFETY CENTER
TEXAS A&M ENGINEERING EXPERIMENT STATION

- The viewer is standing close to the river on the side opposite to the fireball

Time = 0 s

25

Estimation of Fireball Mass: Part 2

MARY KAY O'CONNOR PROCESS SAFETY CENTER
TEXAS A&M ENGINEERING EXPERIMENT STATION

- Fireball elevation is roughly the same as the mountain

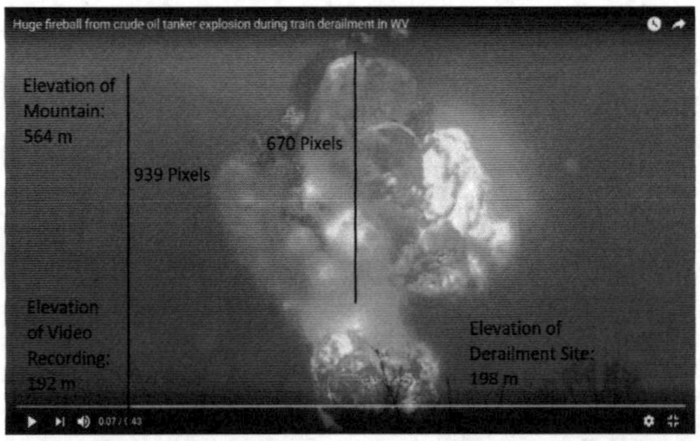

Time = 7 s

26

Estimation of Fireball Mass: Part 2

MARY KAY O'CONNOR PROCESS SAFETY CENTER
TEXAS A&M ENGINEERING EXPERIMENT STATION

- Google Earth Pro was used

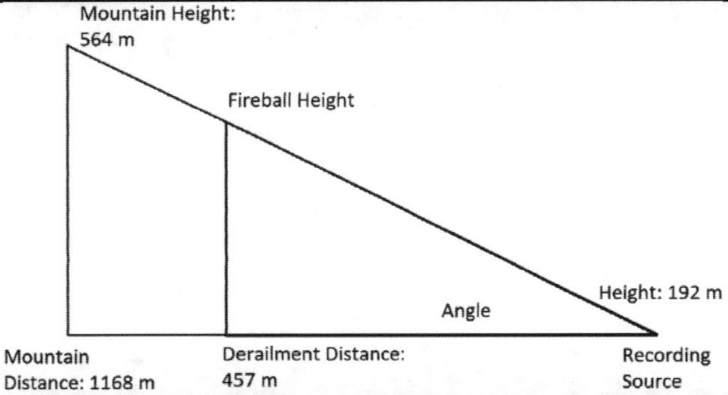

27

Estimation of Fireball Mass: Part 2

MARY KAY O'CONNOR PROCESS SAFETY CENTER
TEXAS A&M ENGINEERING EXPERIMENT STATION

- Maximum fireball diameter was 141 m
- Calculated mass of fireball from fireball diameter

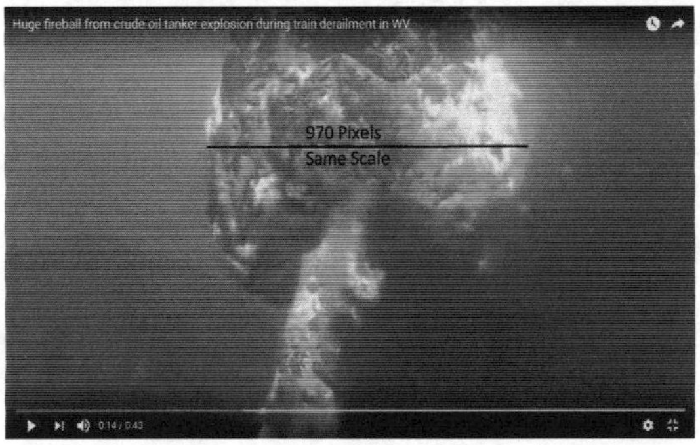

Time = 14 s

28

Estimation of mass of fuel in the fireball

MARY KAY O'CONNOR PROCESS SAFETY CENTER
TEXAS A&M ENGINEERING EXPERIMENT STATION

Models	Mass of fuel (kg) Part 1	Mass of fuel (kg) Part 2
TNO Model	38,000	13,000
HSE Model	46,000	16,000

Models	Mass of fuel (kg) Part 1	Mass of fuel (kg) Part 2
Hardee and Lee Model (1973)	47,000	16,000
Marshall Model (1987)	48,000	17,000
CCPS QRA Model	55,000	19,000

29

Radiant heat from fireballs: A sensitivity analysis

MARY KAY O'CONNOR PROCESS SAFETY CENTER
TEXAS A&M ENGINEERING EXPERIMENT STATION

- Existing models vary significantly in predicting fireball mass and radiation level
- A systematic analysis is needed to indicate the radiant heat from different fireball sizes as guidance to public safety concern

Assumption

- Burst pressure was assumed to be 240 psig

30

Effect of model selection

MARY KAY O'CONNOR PROCESS SAFETY CENTER
TEXAS A&M ENGINEERING EXPERIMENT STATION

- DNV and TNO model produces the same level of radiant heat at different distances whereas the HSE model over-predicts the radiant heat.

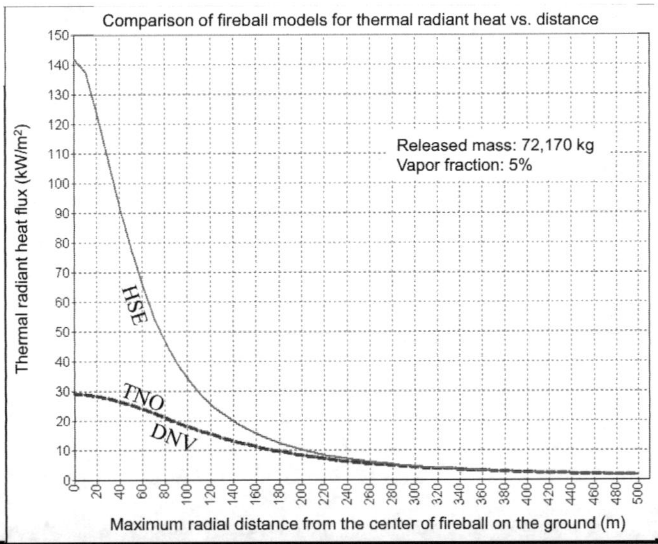

31

Effect of released mass in fireball calculations

MARY KAY O'CONNOR PROCESS SAFETY CENTER
TEXAS A&M ENGINEERING EXPERIMENT STATION

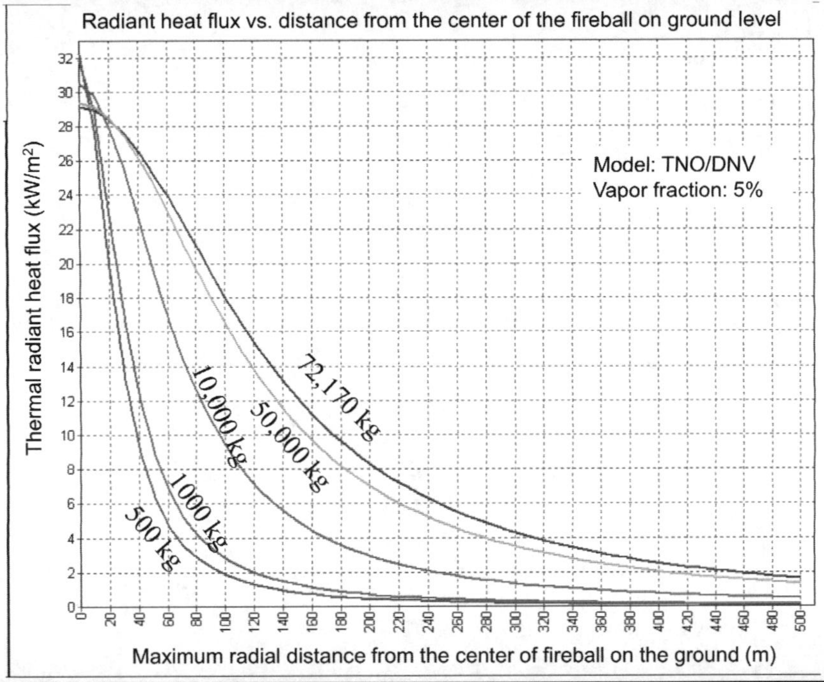

32

Effect of vapor fraction of the released mass

MARY KAY O'CONNOR PROCESS SAFETY CENTER
TEXAS A&M ENGINEERING EXPERIMENT STATION

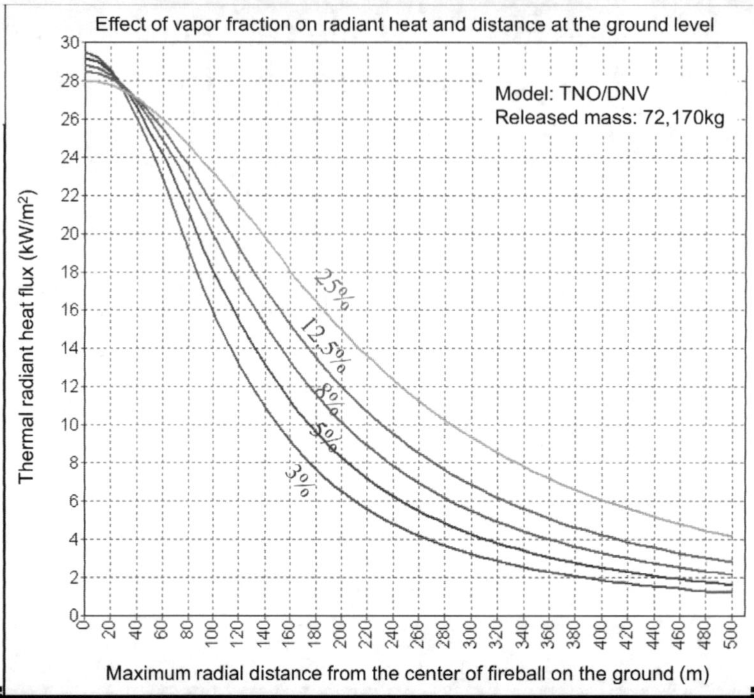

MARY KAY O'CONNOR PROCESS SAFETY CENTER
TEXAS A&M ENGINEERING EXPERIMENT STATION

Conclusions

- Photographic analysis of a fireball in the Mount Carbon incident shows significant variation in the estimated mass of the fireball.

Research gap

- Significant variability is present among the predictions from fireball models. Experimental validation is necessary in selecting the right model for crude oil fireball.

MARY KAY O'CONNOR PROCESS SAFETY CENTER
TEXAS A&M ENGINEERING EXPERIMENT STATION

10. Lessons Learned

35

Summary of conclusions

MARY KAY O'CONNOR PROCESS SAFETY CENTER
TEXAS A&M ENGINEERING EXPERIMENT STATION

- Simulated the consequences reported in the incident within certain engineering uncertainties.
- The effect of train speed in mass vaporization of crude oil is insignificant.
- The possibility of a flammable vapor mixture inside an un-punctured tank car is very unlikely to occur during normal transportation.
- The major physical and combustion phenomena that has the potential to cause fire and fireball scenario were sequentially identified.
- Existing models are limited for capturing the consequences multi-component system such as crude oil and result in conservative safety distances.

36

Summary of conclusions (2)

MARY KAY O'CONNOR PROCESS SAFETY CENTER
TEXAS A&M ENGINEERING EXPERIMENT STATION

- Existing tools, and models are evaluated to study the cooking of un-punctured tank cars.
- A systemic sensitivity analysis has been performed to study the effectiveness of fireball models in predicting the safety distance
- The crude oil composition (*i.e.,* volatility or vapor pressure) has little or no impact on the overall consequences
- Bursting time of a cooked tank car depends on the time to reach its IBP. Once IBP reached, the failure time is almost same for light and heavy crude
- The vapor fraction of the flammable mass in fireball, affects the radiative heat distances. Experimental studies are needed to validate the model

37

Acknowledgement

MARY KAY O'CONNOR PROCESS SAFETY CENTER
TEXAS A&M ENGINEERING EXPERIMENT STATION

DEPARTMENT OF TRANSPORTATION UNITED STATES OF AMERICA

Federal Railroad Authority

38

MARY KAY O'CONNOR
PROCESS SAFETY CENTER
TEXAS A&M ENGINEERING EXPERIMENT STATION

Thank You

Questions?

monir@tamu.edu

39

Next Generation of Thermodynamic Models and their Impact on Process Modeling and Simulation

Southwest Process Technology Conference

Conference: Southwest Process Technology Conference

Year: 2017

Proceeding: 9th AIChE Southwest Process Technology Conference

Group: Southwest Process Technology Conference

Session: Thermodynamics & Process Simulation

Time: Thursday, October 12, 2017 - 4:00pm-4:30pm

Authors:

Georgis, D., Process Systems Enterprise Inc.
Natarajan, S., Process Systems Enterprise Lts
Papaioannou, V., Imperial College London

Process industries are always looking for ways to improve design and operation mainly in terms of profitability and safety while maintaining certain constraints which are typically process specific. Rigorous process optimization requires high-fidelity predictive mathematical models based on first-principles. Accurate thermodynamic predictions are a key requirement for high-fidelity predictive mathematical models. In additional to this, high degree of predictability and robustness, perceived in this case as the ability to perform reliable calculations over a range of thermodynamic conditions, is imperative for process safety analysis.

Recent advances in thermodynamic property prediction have led to the development of methodologies that can be reliably applied to the modeling of complex, highly non-ideal systems and over a wide range of conditions. As an example of such a methodology, in this presentation we focus on the implementation of the SAFT-ï§ Mie group contribution approach within an advanced process modeling platform. The physical basis of this methodology lends SAFT-ï§ Mie with the ability to accurately describe properties of highly non-ideal pure components and mixtures, and the formulation within the scope of group-contribution allows for property prediction where little, or no, experimental data is available. In this presentation, the fundamentals of this thermodynamic methodology and its implementation within an advanced process modeling environment will be briefly discussed. The performance of the theory in describing the properties of several examples of complex molecules will be demonstrated. Finally, the benefits of integrating this approach within a process simulation environment will be illustrated through different case studies.

Performance Enhancement & Profitability Improvements of Refinery Operations using CFD

Southwest Process Technology Conference

Conference: Southwest Process Technology Conference

Year: 2017

Proceeding: 9th AIChE Southwest Process Technology Conference

Group: Southwest Process Technology Conference

Session: Thermodynamics & Process Simulation

Time: Thursday, October 12, 2017 - 3:30pm-4:00pm

Authors:

Kumar, P., Tridiagonal Solutions Inc.
Vedapuri, D., Tridiagonal Solutions Inc.

Efficient operations of Refinery processes rely on optimal flow distribution, fluid mixing and phase separation. Poor flow, mixing or separation bottlenecks a process, reduces its efficiency and can result in frequent shut downs for maintenance. Optimizing flow can debottleneck a process, lead to an enhancement of the process performance and reduce the frequency of shut down. This then improves the profitability of the refinery operations. Advancement in simulation technology has enabled building reliable computational models of refinery process units. This presentation discusses the use of computational fluid dynamics (CFD) techniques to improve refinery operations. The presentation focuses on three case studies:

- Analysis & improvement of crude blending operations
- Design improvements to vapor horns for efficient removal of entrained liquids in a vapor fractionator
- Retrofit of a Desalter to enhance performance and increase capacity

The case studies illustrate the opportunity to improve performance and avoid need to purchase new equipment through use of flow modeling techniques.

Relief System Design Breakthrough Enhances Steady-State Rigor; Seamlessly Integrates Dynamics; Makes Best-Practice Safety and Operability Analysis Cost-Efficient

Southwest Process Technology Conference

Conference: Southwest Process Technology Conference

Year: 2017

Proceeding: 9th AIChE Southwest Process Technology Conference

Group: Southwest Process Technology Conference

Session: Thermodynamics & Process Simulation

Time: Thursday, October 12, 2017 - 1:30pm-2:00pm

Authors:

Anderson, D.

A new, higher level of workflow integration is introduced for steady-state and dynamic sizing and performance validation of relief valves and flare networks. Technical shortcomings of legacy steady-state tools are noted, then shown to be overcome by new, fully rigorous steady-state methods. Additionally, the benefits of rigorous dynamic performance validation are reviewed. Now, a reimagined tool, inspired by demands of industry experts, delivers a consistent, rigorous, fully integrated, and much more cost-efficient steady-state to dynamic flare system modeling workflow. Therefore, these most rigorous and respected methods for hydraulic design of safe and operable relief systems can now be extended for the first time to projects with more limited engineering budgets.

The Effect of Relative Volatility, Enthalpy, Density and Surface Tension on Distillation Column Design

Southwest Process Technology Conference

Conference: Southwest Process Technology Conference

Year: 2017

Proceeding: 9th AIChE Southwest Process Technology Conference

Group: Southwest Process Technology Conference

Session: Thermodynamics & Process Simulation

Time: Thursday, October 12, 2017 - 2:00pm-2:30pm

Authors:

Haddad, H.
Wenta, R., URS Washington Division

The effect of errors in relative volatility on distillation column design has been discussed in the past. In this paper we offer a review on this subject from a different perspective. The effect of errors in enthalpy, an area not emphasized in the past, is discussed in detail. We will, graphically and numerically show the effect of errors in enthalpy have a direct effect on the number of stages, reflux ratio and column traffic. The effect of errors in densities and surface tension are also discussed.

Trends in Industry 4.0 and IIOT and their Impact on Process Simulation

Southwest Process Technology Conference

Conference: Southwest Process Technology Conference

Year: 2017

Proceeding: 9th AIChE Southwest Process Technology Conference

Group: Southwest Process Technology Conference

Session: Thermodynamics & Process Simulation

Time: Thursday, October 12, 2017 - 2:30pm-3:00pm

Authors:

Bonner, P.
Ohaco, R.

Industry 4.0 and IIOT have the potential to be the biggest change and truly a paradigm shift in the chemical and hydrocarbon processing industries since the invention of the computer. This presentation will provide an overview of some terms and trends related to Industry 4.0 and provide more detail on the role of simulation models in the new paradigm. As the chemical and hydrocarbon processing industry adopts new technology and moves towards Industry 4.0 and IIOT, some people question the need for simulation models at all. In the future, simulation models are going to be symbiotic with âbig dataâ. The myth that âbig dataâ combined with data analytics will replace or supersede detailed simulation models for operational analysis will be debunked.

Axens
IFP Group Technologies

Hydroconversion of Renewable Lipids into Biojet and Biodiesel

AIChE Southwest Process Technology Conference, 10/13/17, Galveston, TX

Cristal Suarez

RFS Impact on Renewables

- **Renewable Fuels & demand for RINS continue to grow**
- **Ethanol has peaked at the 10% "blend wall"**
- **Biodiesel (BD) is near a B5 'soft' limit** (engines & infrastructure)
- **Renewable Diesel (RD) is about 20-25% of total BD pool.**
- **About 70% of RD is imported**
- **RD can be used for bio-jet**

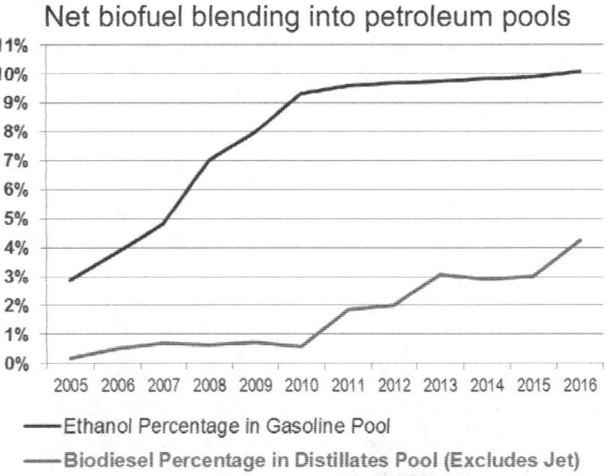

argus consulting services

Source: EIA, EPA and Argus
Note: EPA RD data combined with EIA diesel data.

The Case for Renewable Diesel

Under the RFS, the best way for Obligated Parties to minimize RINs supply risk and maximize profits is to vertically integrate through biofuels production.

From a Refiner's perspective, HVO Renewable Diesel is:
- **The most "expandable" source of RINs**
- **Shares RIN economics with the dominant RIN price-setter**
- **The most easily integrated**
- **The most familiar production technology**

Axens Technology: a Focus on Clean Fuels

Technology, Catalysts, Services
Proven scale up of processes and catalysts: Lab – Pilot – Industrial

Over 2500 process licenses
- **Refining**
- **Petrochemicals**
- **Bio-Chemicals**
- **Bio-Fuels**

Axens' Renewable Porfolio

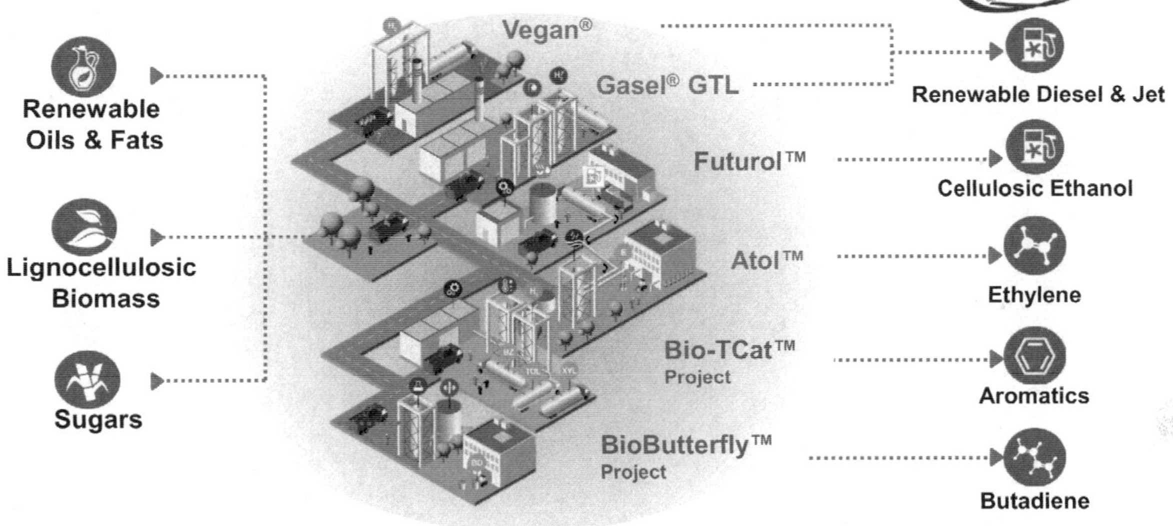

Renewable Fats & Oils

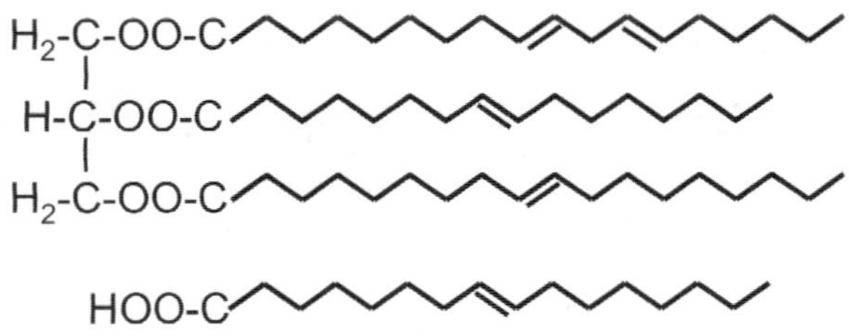

Monoglycerides
Diglycerides
Triglycerides

& Free Fatty Acids

- **Source of C_{16}-C_{18} hydrocarbons**
- **Low sulfur & nitrogen content**

- **High oxygen content**
- **Specific impurities: Metals, gums, free fatty acids, …**

Axens
IFP Group Technologies

Two Renewable Lipids Upgrading Routes

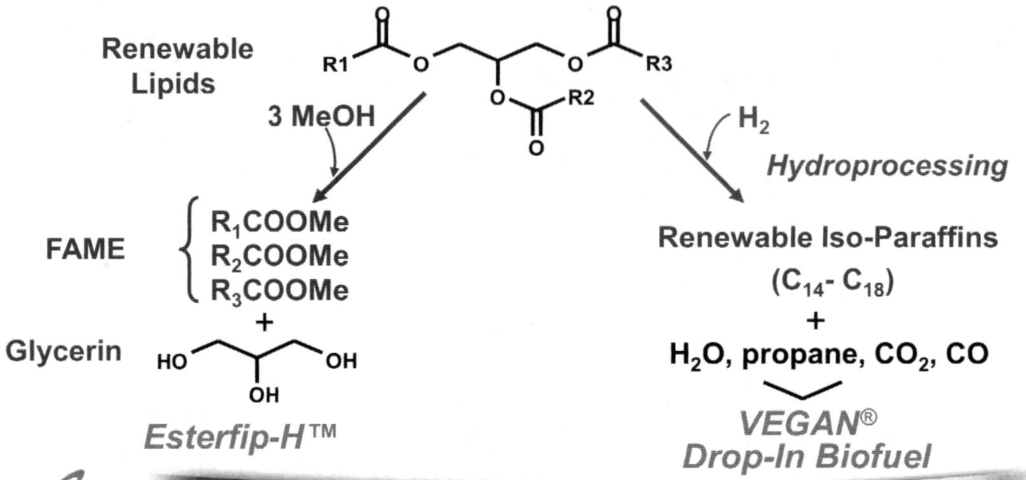

Axens
IFP Group Technologies

Axens Biodiesel Milestones

1992 - first FAME biodiesel plant start-up
2004 - world's 1st solid catalyst biodiesel process *Esterfip-H*

- Consistently high quality products
- Solid catalyst replaces caustic & acid
- Simple separations, No soap formation
- Higher quality glycerin byproduct
- Long catalyst life

Remaining challenges:

- Inherent FAME blend limits with diesel
- Product storage, cold flow properties

Axens
IFP Group Technologies

Vegan™ Renewable Iso-paraffins Route

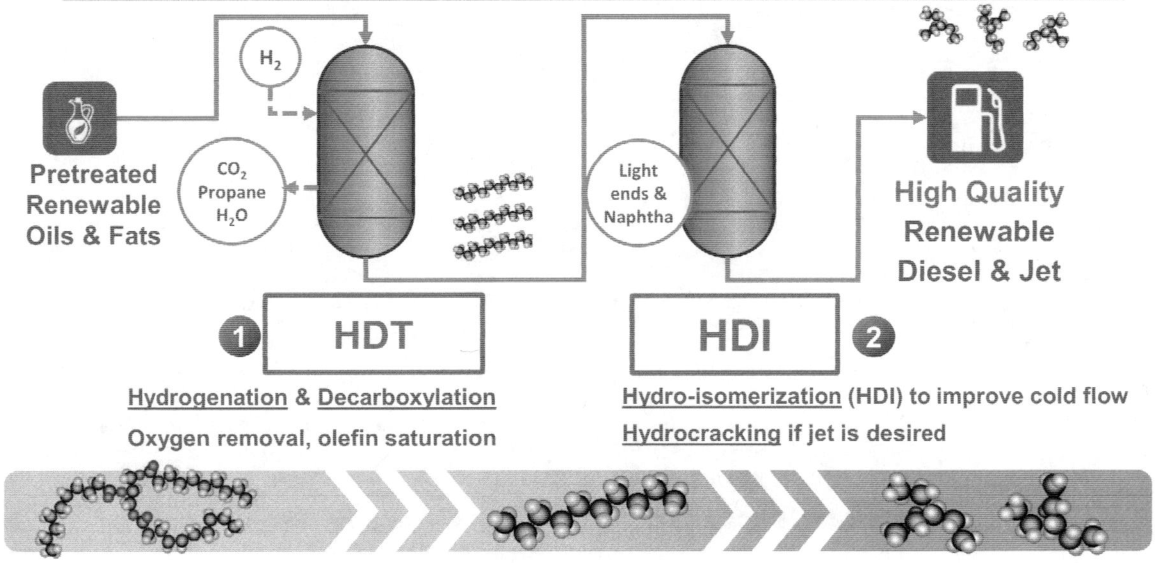

Step 1: Hydrogenation

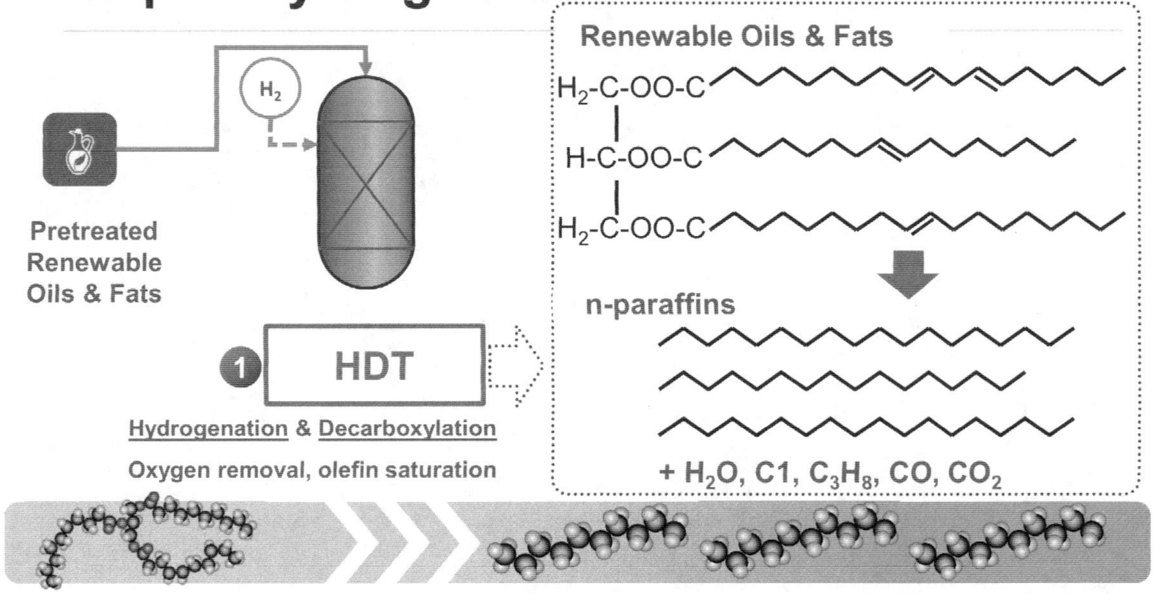

HDT Product n-Paraffins Properties

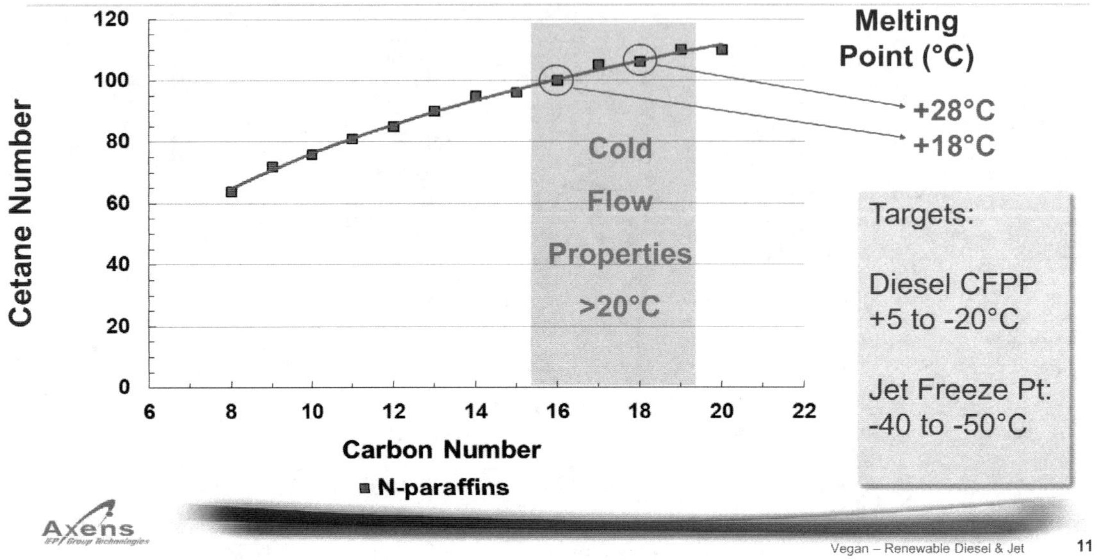

Step 2: Hydro Isomeration

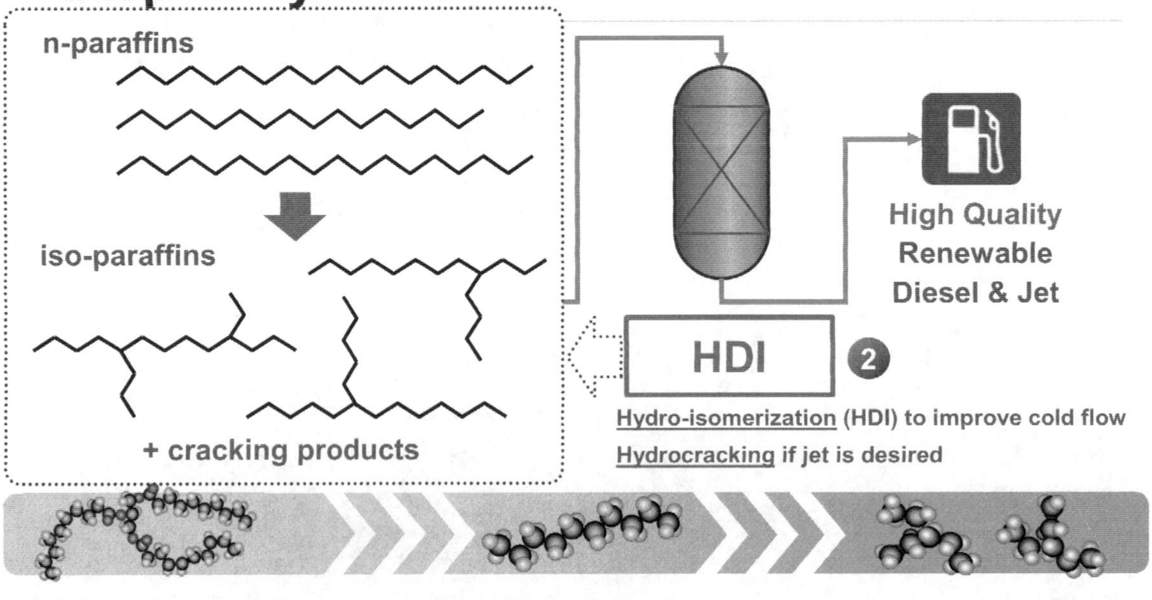

Hydro-Isom Tuned to Diesel & Jet Specs

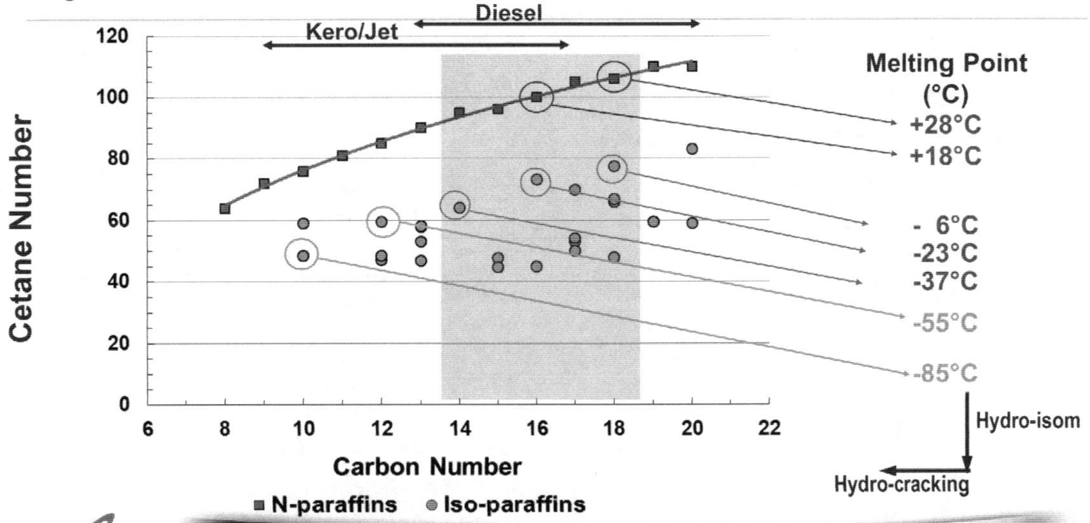

Axens
IFP Group Technologies

Vegan®: Selective Iso-paraffins Route to On-Spec Renewable Diesel & Jet

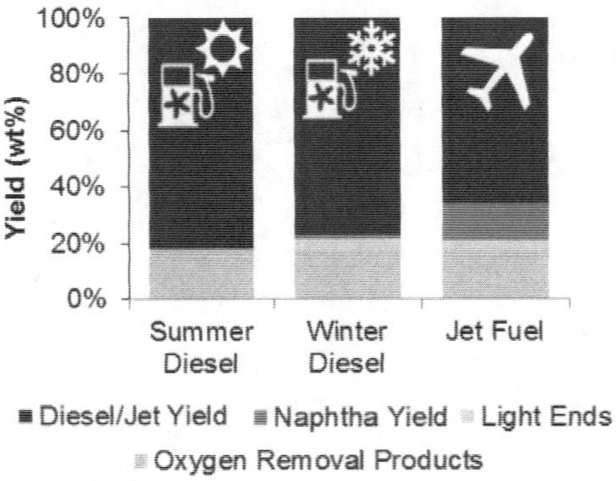

Property	Unit Feed	Summer Diesel	Winter Diesel	Jet Fuel
Freezing Pt/ CFPP (°C)	30	0	-20	< -40
Density(kg/m³)	915	779	773	757
Cetane Index	50	> 80	> 75	

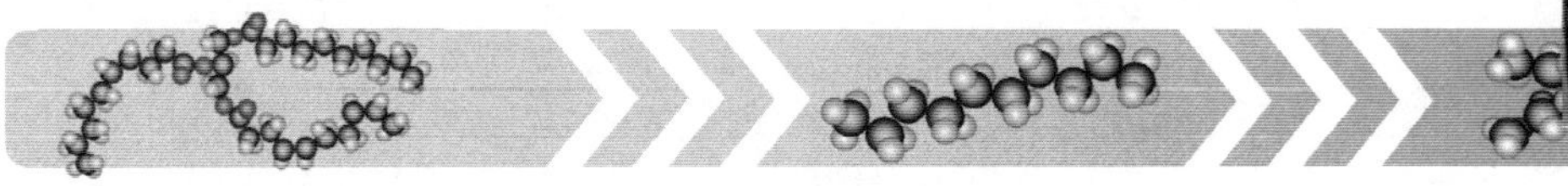

Economics: Feedstock Price Driven

TIC ~ $200 - 250 / ton/y feed

- **Feed pretreatment**
- **Hydroprocessing units ISBL**
- **SMR Hydrogen unit**
- **Off sites & Utilities**
- **Excluding storage & transport facilities**

Feedstock, product & RIN pricing are the determining factors for profitability

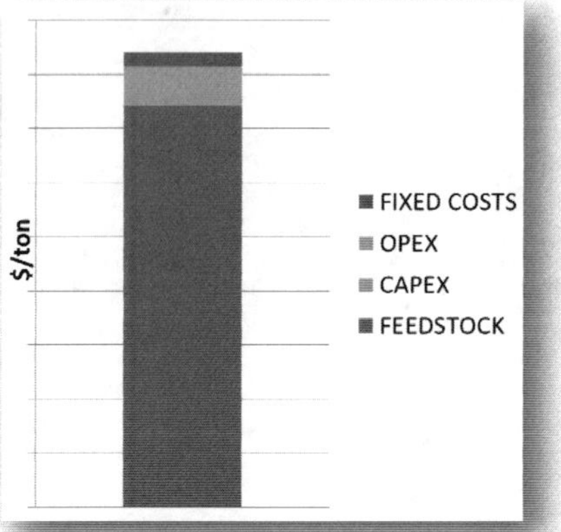

Managing Risk & Volatility:
Stop buying costly RINS, Generate your own !

Axens
IFP Group Technologies

Leveraging Assets: Co-processing in Existing HDT Unit

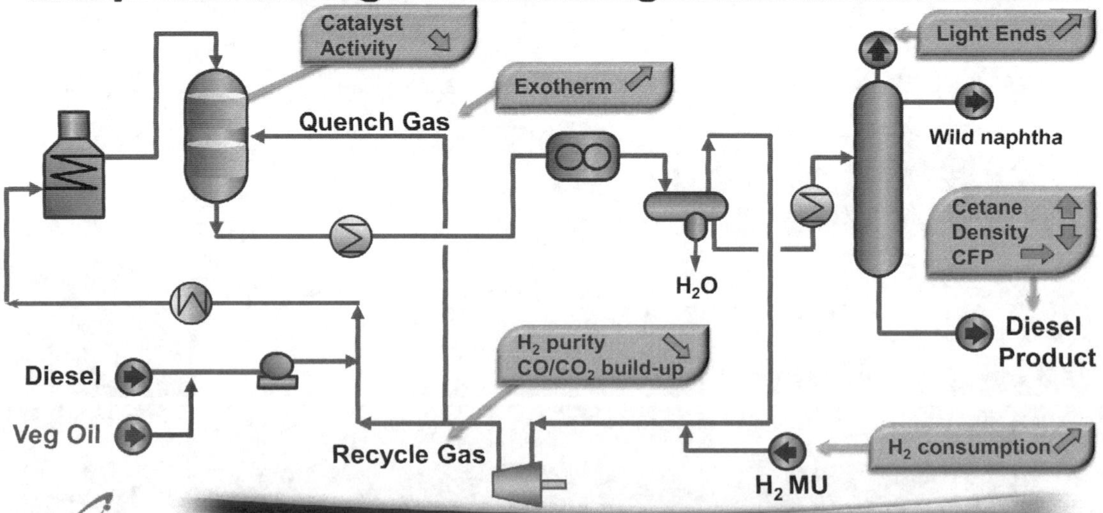

Axens
IFP Group Technologies

La Mède Bio-Refinery Project

TOTAL
COMMITTED TO BETTER ENERGY

- **Conversion of a Conventional Refinery to a Bio-Refinery**
- **Production of 500,000 T/y of Renewable Diesel**
  - **Can also produce Bio-jet**
- **Retrofit of 2 HDT Units into a Vegan™ Unit**
- **Fast-track project: Start-up in 2017**

Two Routes – One Drop-In Solution

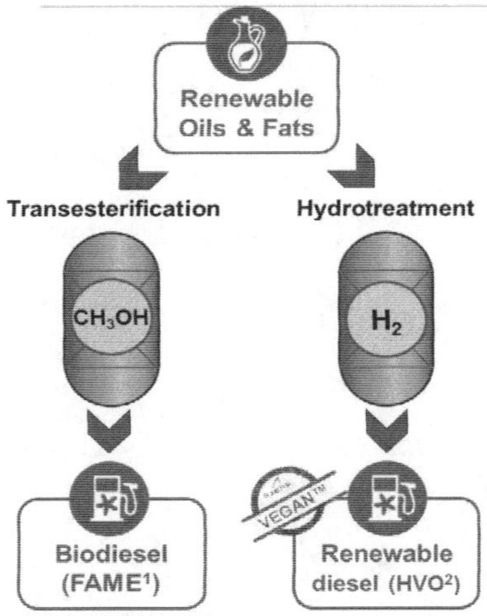

¹: Fatty Acid Methyl Ester
²: Hydrotreated Vegetable Oil

Advantages of Renewable Diesel (HVO)

- ☑ Fully compatible fuel, **oxygen-free**
- ☑ **Higher** energy density
- ☑ Adjustable cold flow properties
- ☑ **ASTM D7566** Biojet blend stock
- ☑ **Higher** cetane **number paraffinic diesel**
- ☑ **Better blended product** stability
- ☑ **Reduced waste & by-products production**
- ☑ **Higher** feedstock flexibility **(UCO, FFA, …)**
- ☑ **True** drop-in biofuel

Conclusions

- **Renewable Fuels and demand for RINS continue to grow**
- **The E10 "blend wall" has arrived, so BD & RD have to grow**
- **Biodiesel is approaching a B5 'soft' limit in many places**
- **Incremental demand is best met with Renewable Diesel & Jet, a 'drop in' fungible fuel with no technical blend limit**
- **Vegan® Technology is available for new process units or reuse of existing hydrotreating units and infrastructure**
- **Production of RINS mitigates risk, maximizes returns**

Axens
IFP Group Technologies

Axens Succeed Together Blog
axens.net/blog

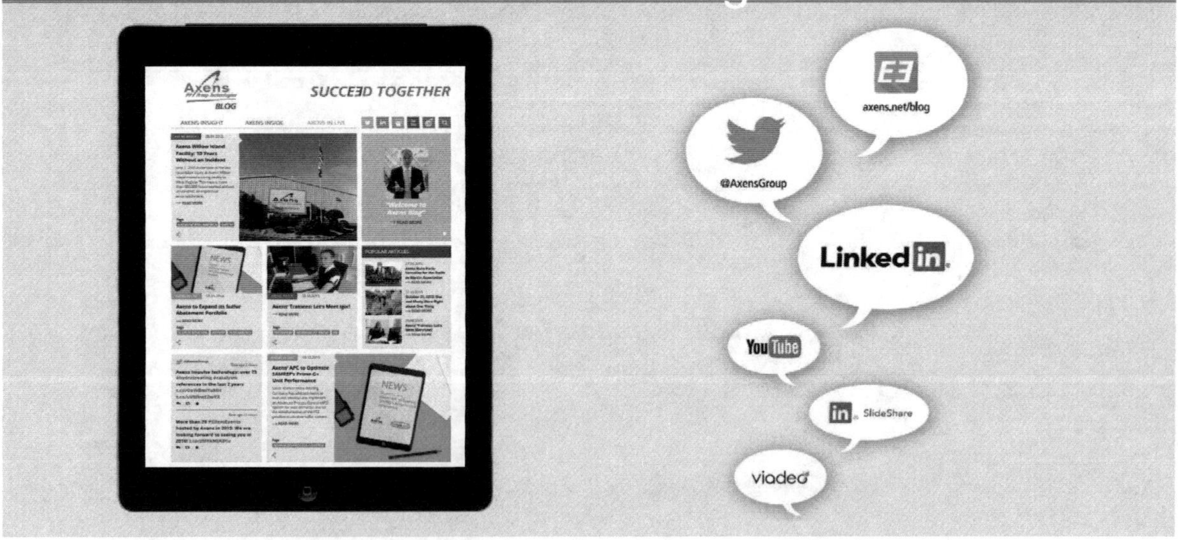

BURNS McDONNELL

Processes for Refinery Propylene

presented by

Dominic Varraveto

October 2017

Agenda

- Overview of Refinery Propylene Processing Options
- Separation and Purification
- Polymerization
- Alkylation of Isobutane
- Alkylation of Benzene

BURNS McDONNELL

Overview of Processes for Refinery Propylene

- Separating Propane and Propylene
  - Ethane (BP = -128.2°F)
  - Propylene (BP = -53.7°F)
  - Propane (BP = -43.6°F)
  - Butylenes (BP= xx to xx°F)
  - Butane (BP = 30.2°F)

Propylene Purity Grades, %

- Refinery Grade, 70
- Chemical Grade 92-96
- Polymer Grade 99.5

- Properties of Transport Fuel produced from Refinery Propylene

BURNS & McDONNELL

Refinery Propylene

Thermally/Catalytically "Cracked" Gas recovered in an "UnSats" Gas Plant is the primary source of Refinery Propylenee

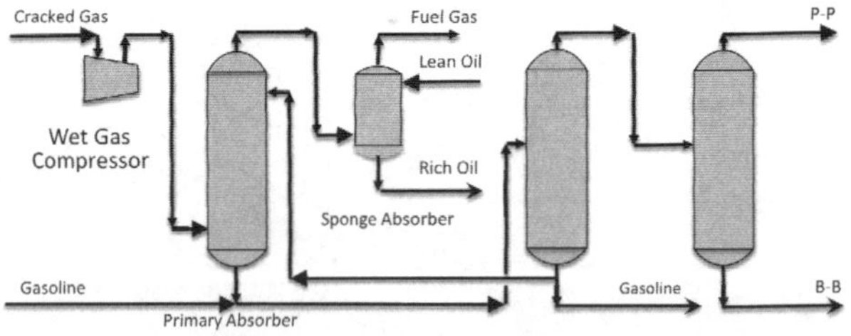

BURNS & McDONNELL

Propylene – Propane Separation

"P-P Splitter"

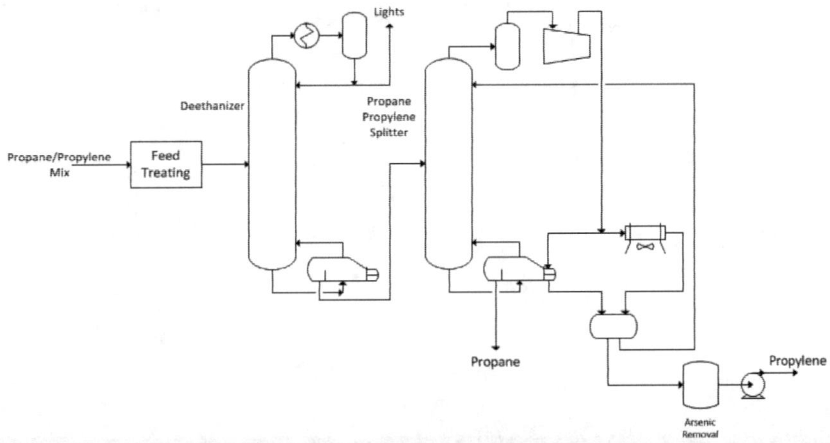

Polymerization Process Refinery Propylene

"Cat Poly"

- Feedstock
- Reactions
- Products
- Catalyst
- Reactor Designs

Polymerization Process for Refinery Propylene

"Cat Poly"

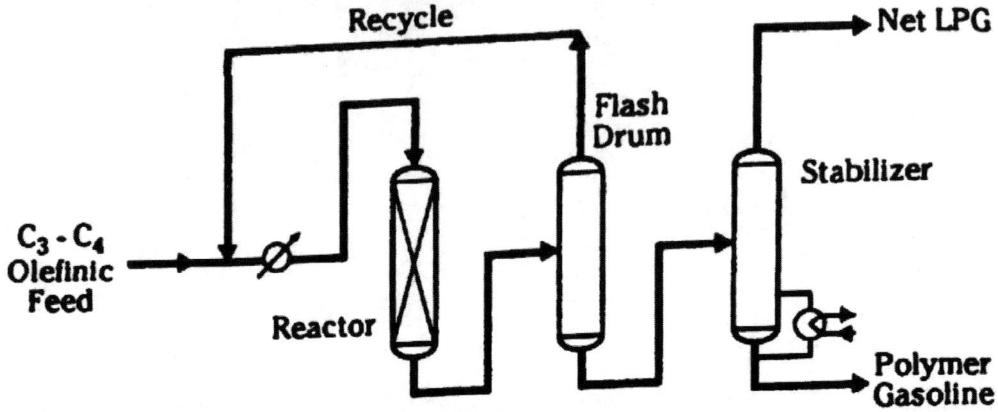

Specialized Polymerization Processes for Refinery Propylene

- Polymerization combined with Hydrotreating
- Selective Feedstock Polymerization
- Oligomerization for Distallates

Process Description of Refinery Alkylation

- Turn LPG into Alkylate, a relatively high octane gasoline blendstock
- Utilizes a strong liquid acid catalyst, HF or H_2SO_4, to react olefins including propylene, with isobutane.
- Reactions occur at moderate pressure and less than ambient temperature

Isobutane Feed

- Contaminants
  - Sulfur, caustic, water
- Source
  - Refinery straight run
    - iC4 ~25%
  - Refinery DIB
    - iC4 ~80-95%
  - Pipeline (NGL Fractionator)
    - iC4 ~95-99%

BURNS & McDONNELL

9

Discussion of Refinery Alkylation Catalyst Options

- Sulfuric Acid, H_2SO_4
  -
- Hydrofluoric Acid, HF
  -
- Solid Catalyst
  -
- Ionic Fluids
  -

BURNS & McDONNELL

10

Sulfuric Acid Alkylation Block Flow & Feedstocks

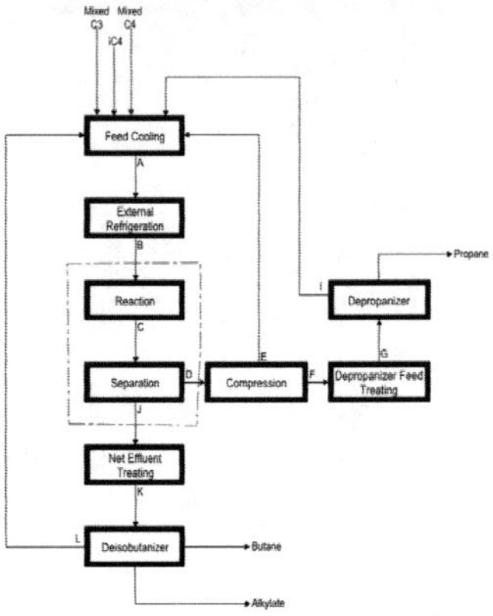

Feed Contaminants: Dienes, sulfur, aromatics, naphthenes, water

- **Propylene, $C_3^=$**
  - ~75% C3 / 25% C_3=
  - **Typically mixed with $C_4^=$**
  - **Requires highest acid spend strength**
  - **Relatively low product octane**
  - **Net propane from Hydrogen transfer reaction**
- **Butylene, $C_4^=$,**
  - ~35 iC4 / ~10% nC4
  - **Most desirable olefin feedstock**
  - **Results in highest product octane**
- **Amylene $C_5^=$,**
  - ~35 iC5/~5% nC5
  - **Least severe operating conditions**
  - **Can be left in unsats gas plant debut bottoms aka cat gasoline**
  - **Least common alkylation feed**
  - **Net isopentane from Hydrogen transfer reaction**

Alkylation Equipment Piping and Metallurgy

- Operating conditions
  - Low pressure (<300 psig), low temperature (0-300 °F), but not cryogenic, almost all hydrocarbons lines are LPG
- Equipment
  - Traditional heat exchangers, reboilers, and pumps
  - Large compressor (HP = alkylate * 0.60)
    - Centrifugal with low suction pressure (~15 psia and low discharge ~150 psia)
    - Single machines with no interstage cooling
- Piping
  - Large lines sizes and equipment (internal rate is ~6x alkylate rate) (6 – 10"), some large lines around the compressor (30 – 60")
- Controls
  - Some mass flow meters, Some very large control valves around compressor,
  - Compressor protection and antisurge system
- Materials
  - Mostly carbon steel, some alloy 20 and stainless

Alkylation Reactions

C5=														
	Selectivity													
Rxn 1	0.80	1	C5=	+	1	iC4	=	1	iC9					
Rxn 2	0.20	1	C5=	+	2	iC4	=			1	iC5	+	1	iC8
BBL (norm)		1.00	C5=	+	1.16	iC4	=	1.32	iC9 +	0.22	iC5	+	0.30	iC8
C4=														
Rxn 1	1.00	1	C4=	+	1	iC4	=	1	iC8					
BBL (norm)		1.00	C4=	+	1.10	iC4	=	1.71	iC8					
C3=														
	Selectivity													
Rxn 1	1.00	1	C3=	+	1	iC4	=	1	iC7					
Rxn 2	0.00	1	C3=	+	2	iC4	=			1	C3	+	1	iC8
BBL (norm)		1.00	C3=	+	1.28	iC4	=	1.77	iC7 +	0.00	C3	+	0.00	iC8

- Reaction #2 are side reactions are dependent on:
  - Catalyst
  - Temperature
  - iC5 content
  - iC4 content
- Reaction isobutane is kept at ~10:1 excess

Alkylation Products

- True Alkylate
  - Mixture of C_7 , $C_{8,}$ & C9 molecules
  - ~ 40 to 50 individual molecules
  - 90 - 94 Octane ~ 2 psia RVP
    - Gasoline Octane, (R+M)/2 = 84-91
    - Gasoline RPV, 7 – 15 psia (seasonal)
  - Contains no toxics (olefins or aromatics)
- Whole Alkylate
  - Alkylate including $C6^-$

Methyl heptane, (29)

Di-methyl hexane, (72)

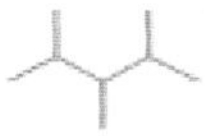

Tri-methyl pentane, (100)

MSAT II STRATEGY: POST FRACTIONATION & EXTRACTION

Background for Benzene Alkylation

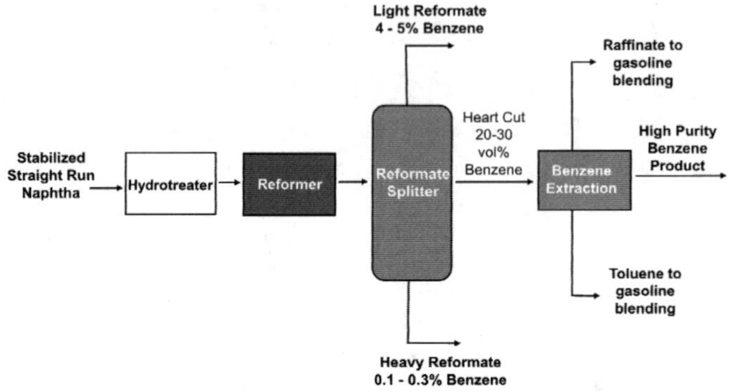

BURNS & McDONNELL

MSAT II STRATEGY: POST FRACTIONATION & SATURATION

Background for Benzene Alkylation

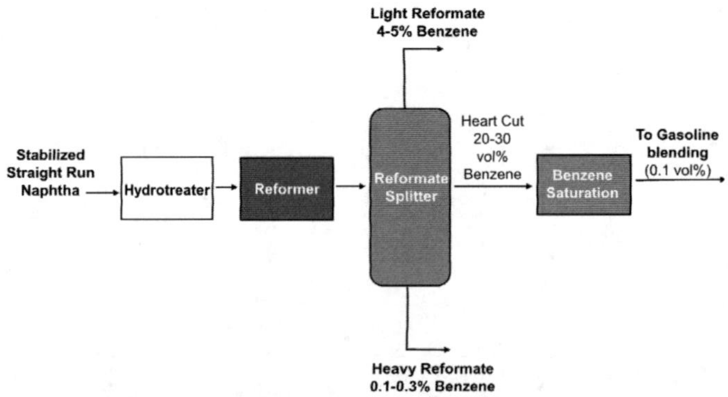

BURNS & McDONNELL

ALTERNATIVE: DILUTE BENZENE ALKYLATION

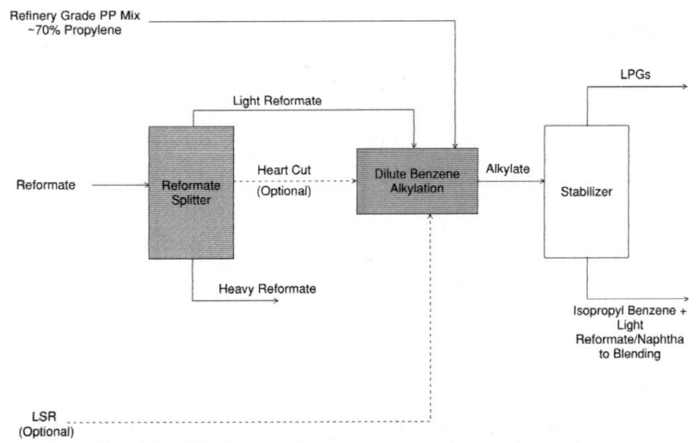

BURNS & McDONNELL

FEEDS

LIGHT REFORMATE

- C4-C6 Range
- 10-15 vol% Benzene

AND/OR

REFORMATE HEART CUT

- C6 Range
- 20-30 vol% Benzene

AND/OR

LSR

- C4-C6 Range
- 1-5 vol% Benzene

REFINERY GRADE PP

- 70% Propylene, 30% Propane

	BENZENE	ISOPROPYL BENZENE
RON	103	112
MON	105	102

BURNS & McDONNELL

REACTION

CATALYST

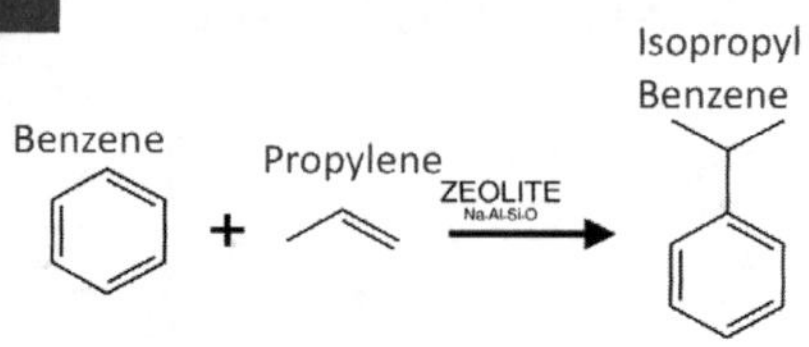

- Zeolite Catalyst (solid fixed bed)
- Example:
  ZSM-5 (Zeolite Socony Mobil-5) $Na_nAl_nSi_{96-n}O_{192}\cdot 16H_2O$ (0<n<27).
- Lewis Acid – Friedel-Crafts Alkylation Reaction.
- Regenerable.

CONDITIONS

- Liquid Phase Reaction.
- Reaction conditions: 350-450 °F @ 800-850 psig.
- Exothermic.

BURNS & McDONNELL

REACTION (CONT'D.)

SECONDARY/SIDE REACTIONS

- Main undesired product: Di-isopropyl Benzene.
- Propylene polymerization, tri-isopropyl benzene, other reactions much less common.
- Heavily alkylated single-ring aromatics have relatively high octane numbers (compared to other gasoline components) but octane numbers generally get lower as MW increases.

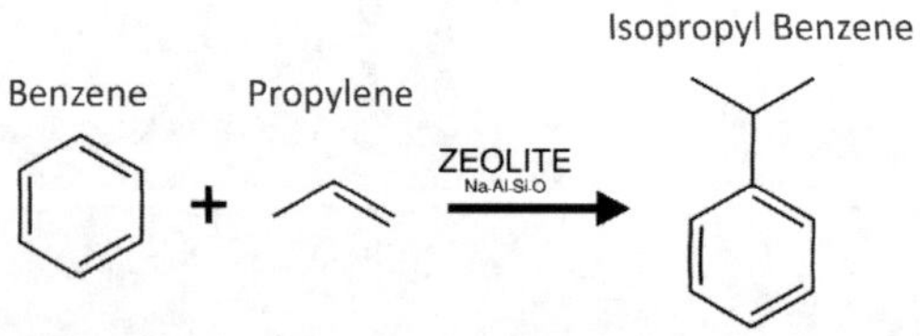

BURNS & McDONNELL

CAPITAL STRATEGIES

POSSIBLE TO CONVERT BENZENE SATURATION UNIT TO BENZENE ALKYLATION?

- ✗ Reaction section and heat conservation exchangers likely occur in the vapor phase at a much lower pressure than the high-pressure liquid phase benzene alkylation (as currently developed technology).
- ✓ Stabilizer easily re-purposed for benzene alkylation.
- ✓ Mothball/repurpose hydrogen recycle compressor, heater, flash pots, etc. for another service.

BURNS & McDONNELL

Alkylation of Benzene

Primary Advantages

- Isolation from chemicals markets makes benzene/propylene export costly and impractical.
- Refinery throughput not big enough to achieve economies of scale – PP splitter/Benzene extraction large capital investments.
- No hydrogen demand. No access to pipeline, and on-purpose hydrogen production is an additional investment.

BURNS & McDONNELL

Alkylation of Benzene

ADDITIONAL ADVANTAGES

- Reduction of Benzene in Gasoline Pool.
- Octane upgrade of 2-3 numbers.
- Upgrade propylene to high-octane gasoline blendstock (from fuel gas).
- HD-5 quality propane stream is a product of stabilizer.
- Reformer/FCC feed and severity flexibility.
- Alkylation capital cost comparison:

BENZENE ALKYLATION	ISOBUTANE ALKYLATION
BASE, PER BARREL PRODUCT (ISOPROPYL BENZENE + LIGHT NAPHTHA)	~4 x BASE, PER BARREL ALKYLATE PRODUCT

BURNS & McDONNELL

WORKS CITED

- Ghosh, Prasenjeet, Karlton J. Hickey, and Stephen B. Jaffe. "Development of a detailed gasoline composition-based octane model." Industrial & engineering chemistry research 45.1 (2006): 337-345.
- Do, Phuong TM, et al. "Catalytic strategies for improving specific fuel properties." Catalysis 20 (2007): 33-64.
- BUSINESS WIRE. (2012, March 15). Badger Congratulates Calumet on Start-Up of World's First BenzOUT™ Process Unit. Retrieved March 10, 2017, from http://www.businesswire.com/news/home/20120315005052/en/Badger-Congratulates-Calumet-Start-Up-World%E2%80%99s-BenzOUT%E2%84%A2-Process
- BUSINESS WIRE. (2014, September 24). Badger Licensing to Provide Proprietary BenzOUT™ Technology for Grassroots Plant in Newcastle, Wyoming. Retrieved March 10, 2017, from http://www.businesswire.com/news/home/20140924005114/en/Badger-Licensing-Provide-Proprietary-BenzOUT%E2%84%A2-Technology-Grassroots
- Varraveto, D., & Phillips, B. (2012, February 7). Gasoline: from MSAT2 to Tier 3 [Tech Tuesday Presentation]. Burns and McDonnell, Kansas City, MO.

BURNS & McDONNELL

BURNS McDONNELL

Digital industrial maturity model

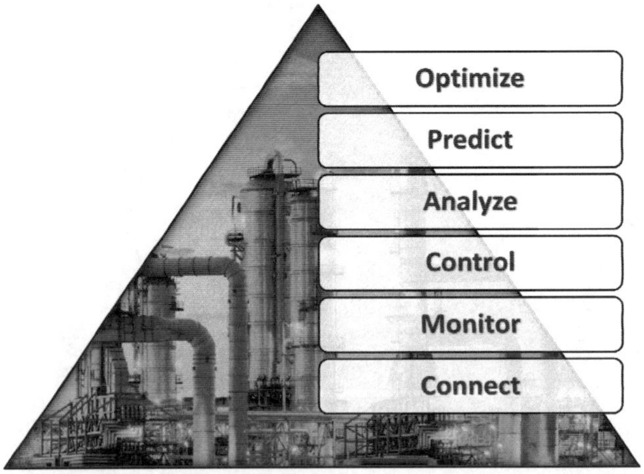

suez

2

O&G investment in digital technology

Digital technology investment over the next 3-5 years

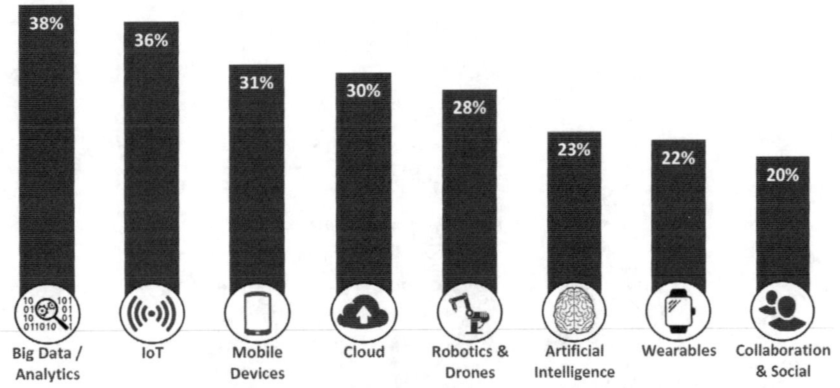

Source: Accenture, The 2016 Upstream Oil and Gas Digital Trends Survey

Are big data analytics a priority?

How important is Big Data analytics relative to other priorities in your company?

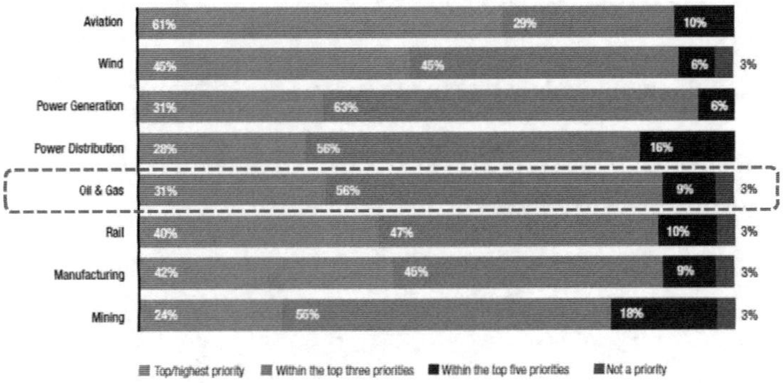

Source: GE & Accenture, Industrial Internet Insights for 2015

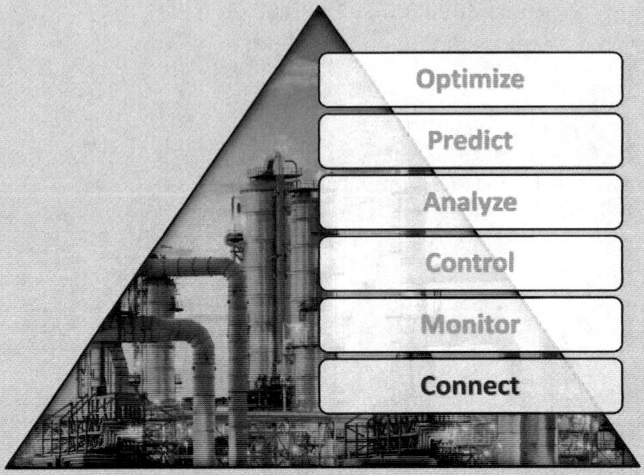

Connect – types of connected data

- Real-time (streaming)
- Near real-time
- Edge computing
- Manual, or on-demand, data upload

Connect – oversight & communication

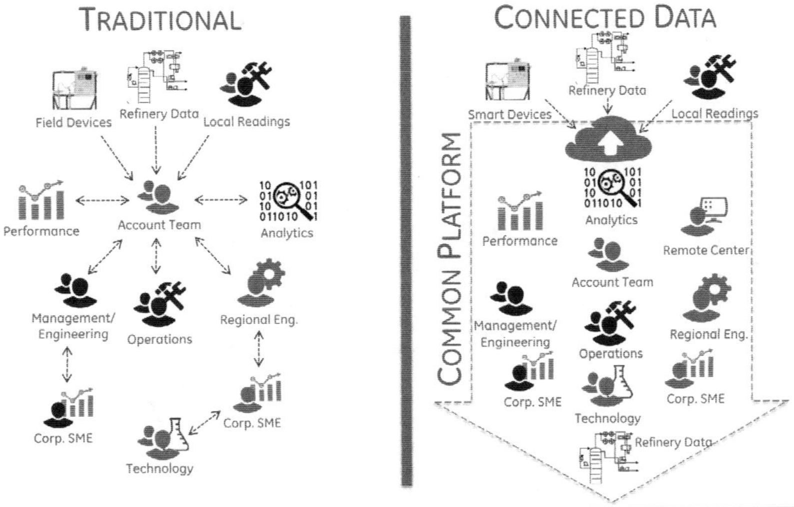

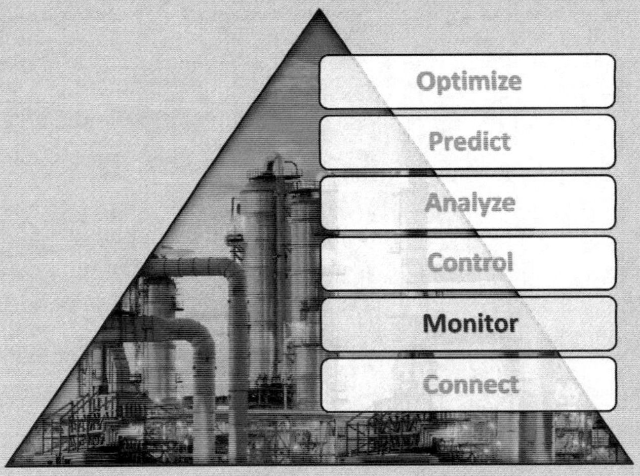

Monitor – faster system response

Refinery cooling tower example

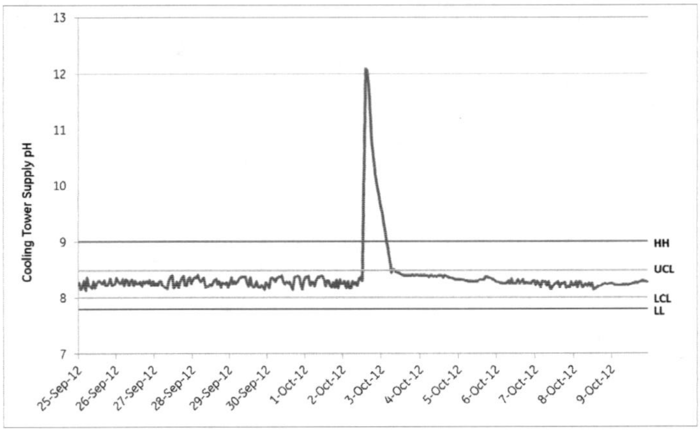

 9

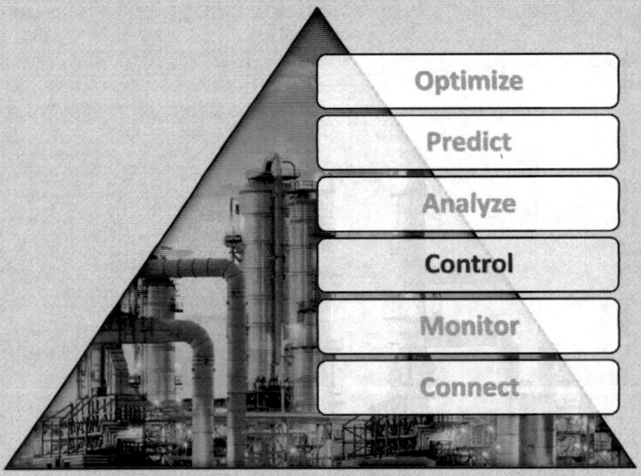

 10

Control – changing performance expectations

Crude tower overhead chloride control automation example

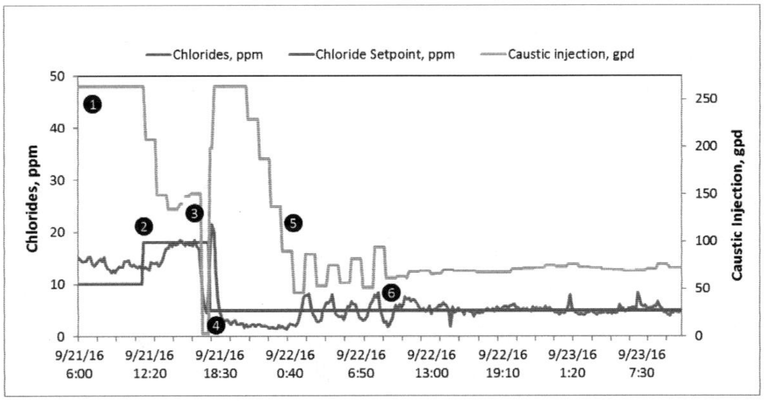

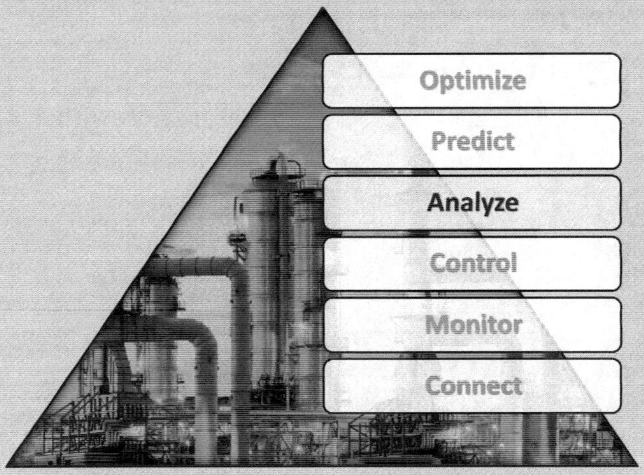

Analyze – system insights and root causes

Caustic injection example - baseline

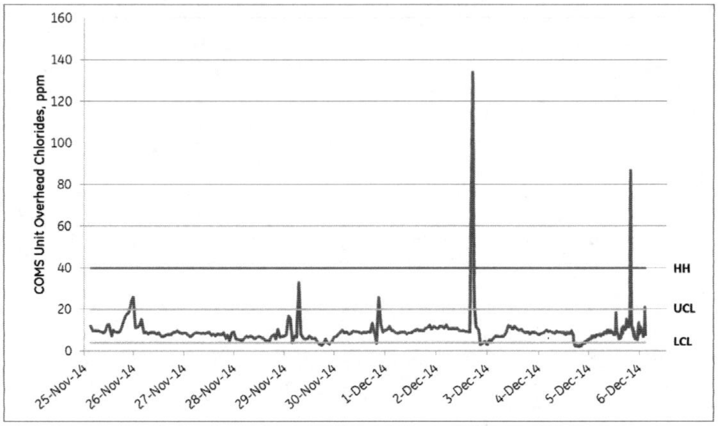

 13

Analyze – system insights and root causes

Caustic injection example - outcome

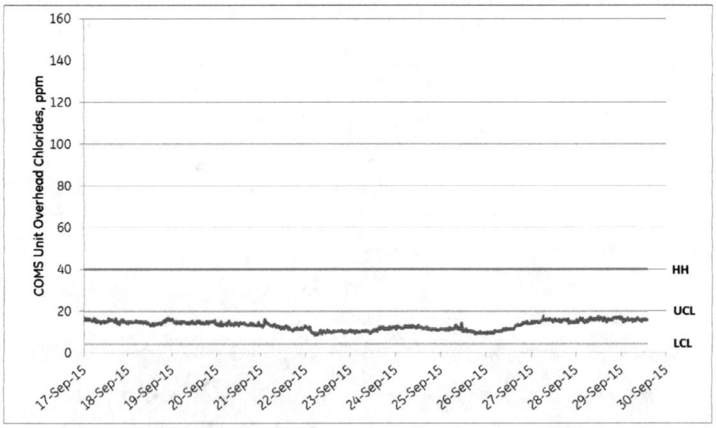

 14

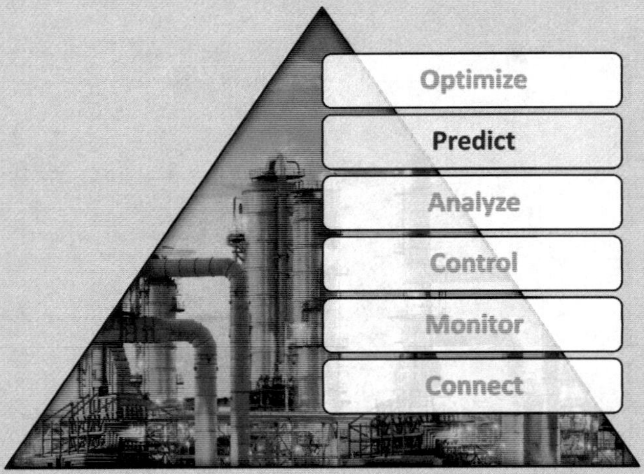

Predict – correlating system performance

Desalter & waste water treatment example

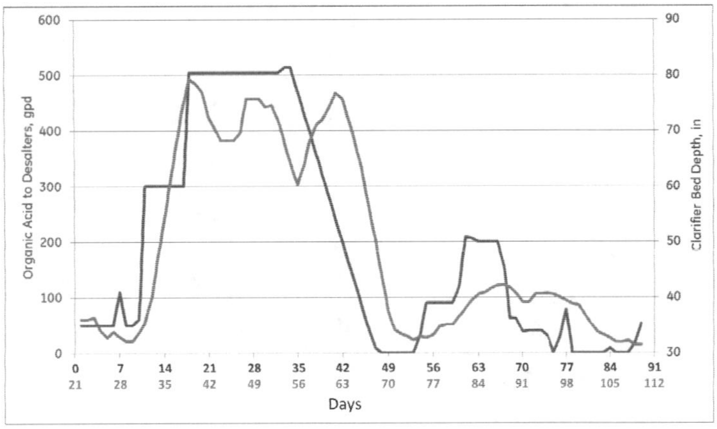

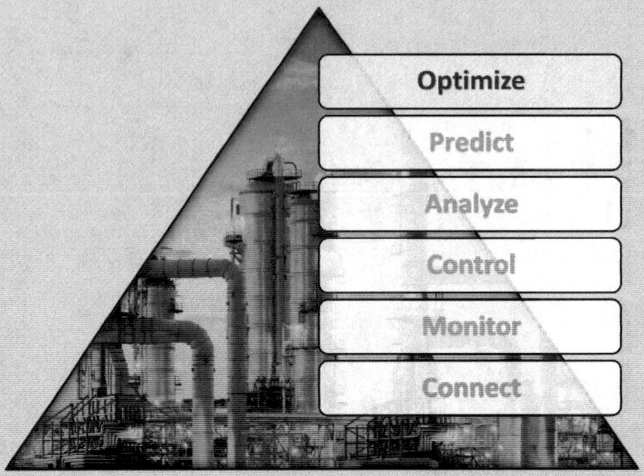

suez

Optimize – improve profit & reliability

Crude tower overhead salting example

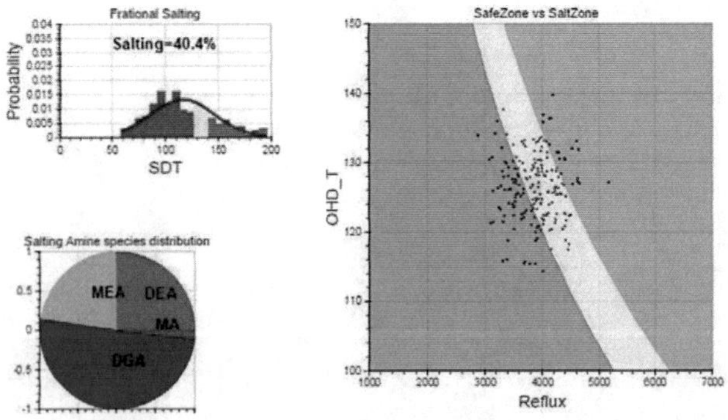

suez

Overhead amine speciation

Crude tower overhead salting example

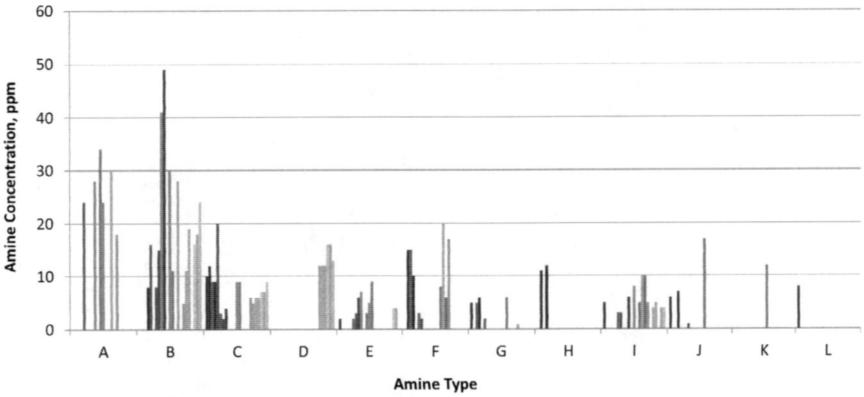

Optimize – modify operating points & reduce variation

Crude tower overhead salting example

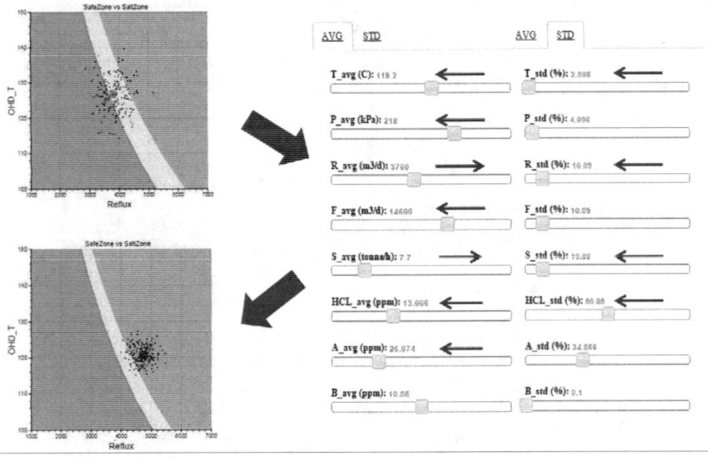

Optimize – improve profit & reliability

Crude tower overhead salting example

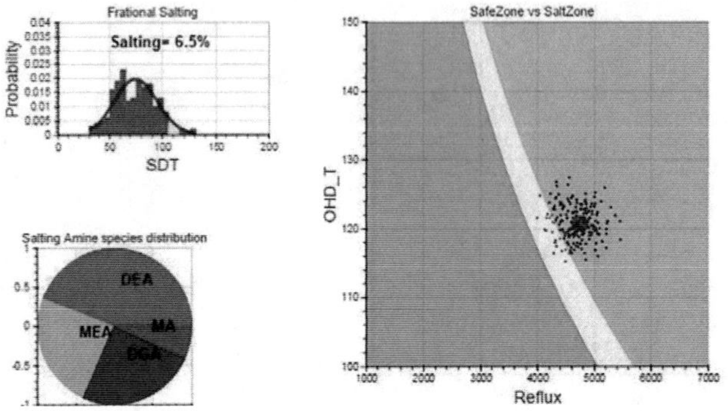

Connecting data - benefits

- Improve visibility
- Accelerate responses
- Quantify optimization opportunities
- Increase reliability, safety, compliance
- Improve profit potential

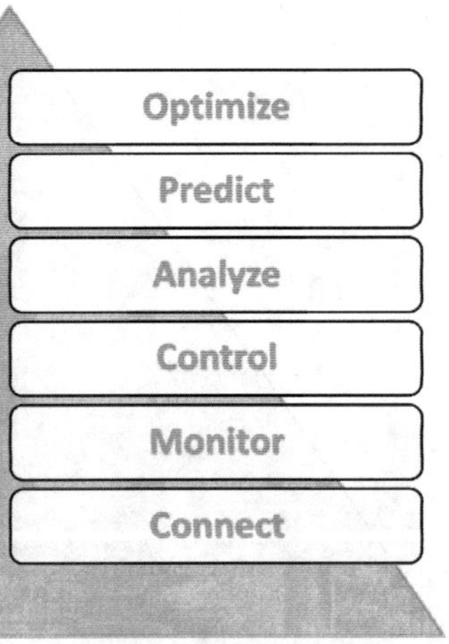

Thank you!
46
62
D MAINTENANCE
suez

Strategies for Managing Contaminant Metals in the FCC Unit

AIChE Southwest Process Technology Conference – October 2017

Rebecca Kuo, Technical Service Engineer

BASF
We create chemistry

1

Agenda

BASF
We create chemistry

- Feedstock variations
- Contaminant metals and their effects
  - Case studies
  - Mitigation strategies
- Advanced passivation technologies

2

Diverse FCC Feeds

- Residue-containing and tight oil feeds have become much more prevalent in FCC
- Resids typically contain high Ni, V, CCR, N
- Tight oils typically contain high Fe, Ca, Na

Oil Quality Variability from One Field

Feed Property	Global Average	Global Minimum	Global Maximum
API	24.0	10.4	31.7
CCR, wt%	1.5	0.02	8.87
Ni, ppm	2.3	0	13
V, ppm	2.7	0	15
Fe, ppm	4.8	0	35
Na, ppm	1.3	0	6
Basic N, ppm	377	5	1058

3

Gasoil-Resid Unit Split

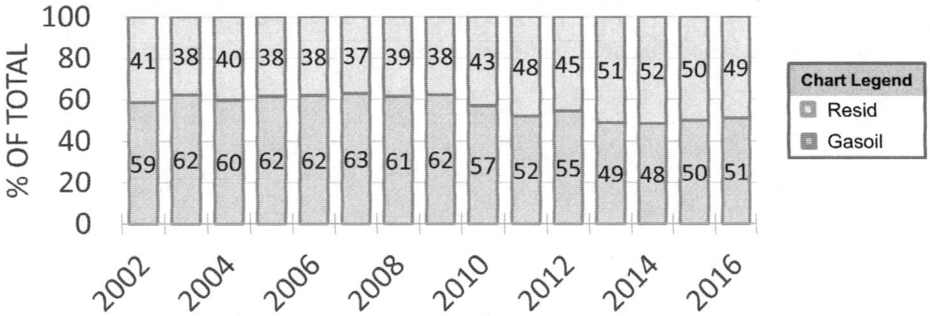

- Resid >3000 ppm ecat Ni + V; Gasoil <3000 ppm ecat Ni + V
- Globally, more units are processing resid with the average moving from ~40% resid in the early 2000s, to today ~50%

4

Feedstock Contaminant Effects

BASF
We create chemistry

- **Why do we care about metal contaminants?**
- Metal contaminants are harmful to the catalyst and can cause unwanted reactions in the FCCU
  - The metals deposit on the circulating catalyst and cause competing chemistries to the desired reactions
  - Salts (Na, Ca, Mg) act as a base and attack the acid sites on the catalyst, lowering activity and conversion

5

Elements of Concern

BASF
We create chemistry

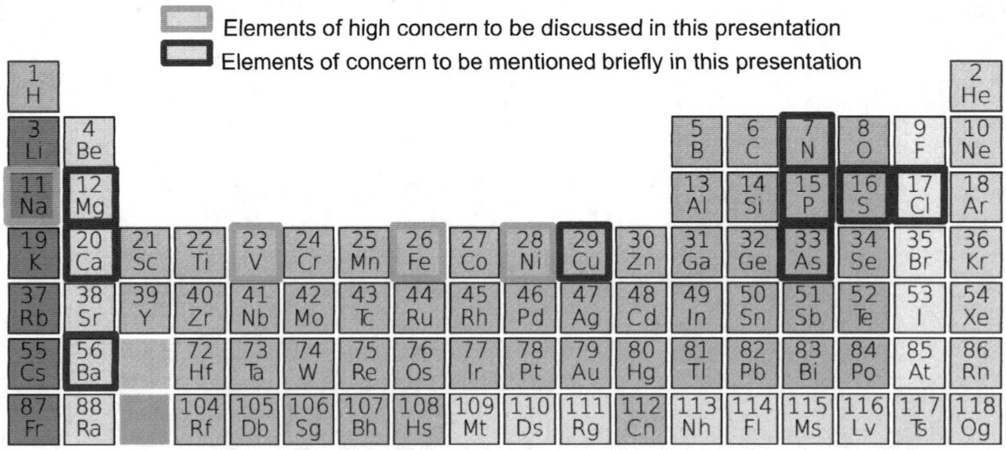

6

Elements of Concern: Nickel

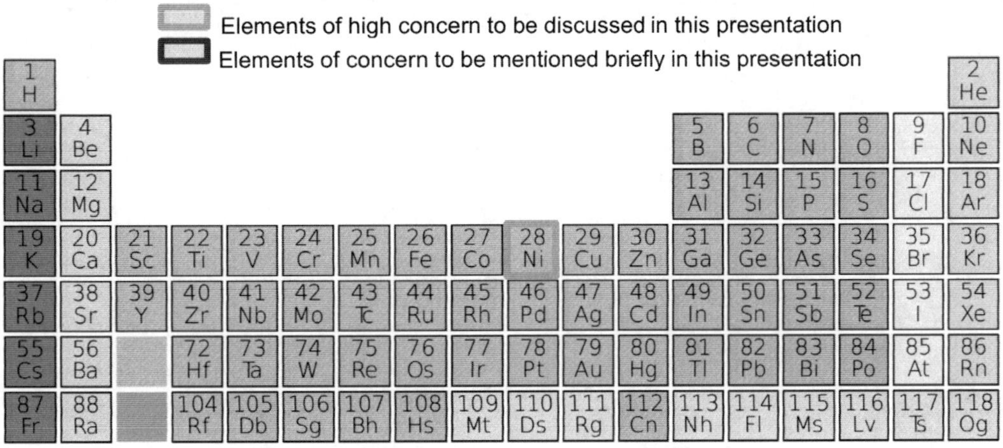

7

Nickel Sources & Effects

- Ni comes in with the feed and deposits on catalyst particles
  - Undergoes redox cycles
  - Nickel oxides do not migrate
- Catalyst activity is not significantly affected
- Ni acts as a dehydrogenation catalyst

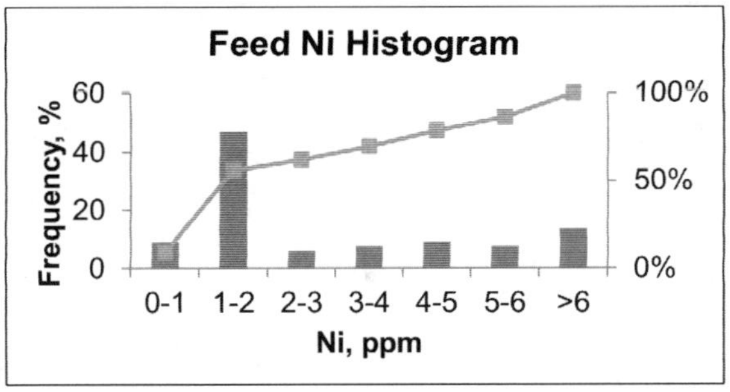

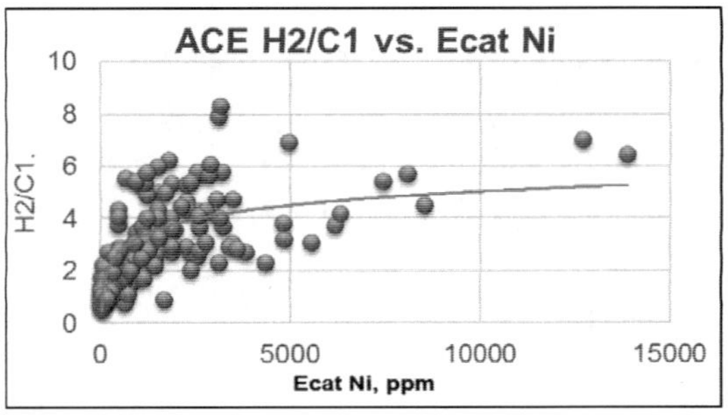

8

Nickel Effects
Case Study

BASF
We create chemistry

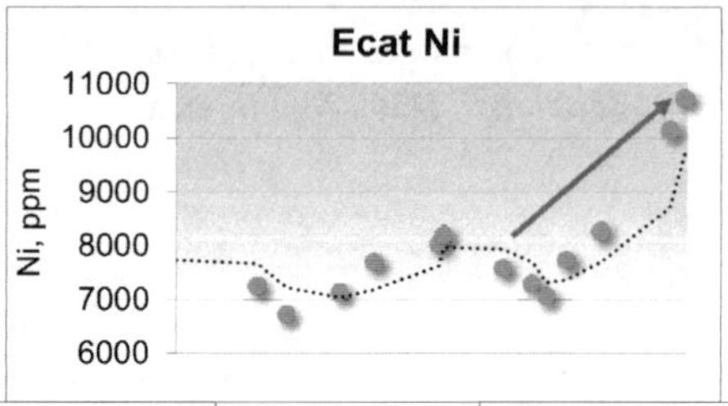

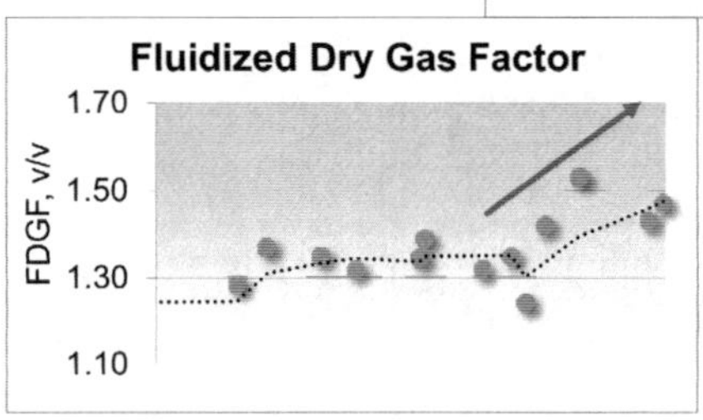

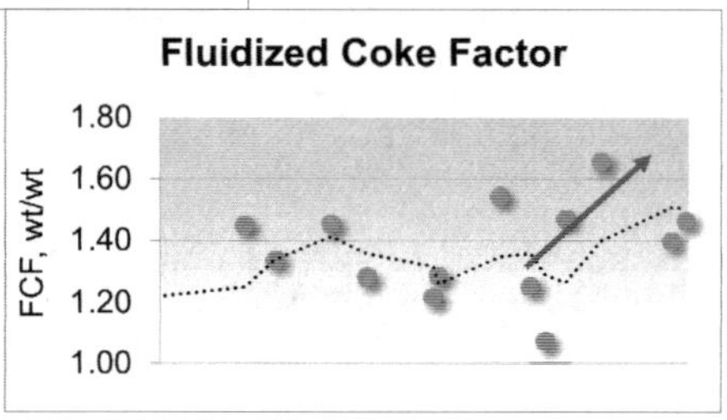

9

Nickel Mitigation Strategies

BASF
We create chemistry

- Typically, Ni becomes a concern >800 ppm on the ecat
- Some refiners use antimony (Sb) injection to passivate Ni
  - Can increase NOx emissions
  - Can increase bottoms fouling
- A more permanent solution is to use catalyst with specialty alumina (such as Flex-Tec)
  - Ni deposits on alumina and becomes passivated

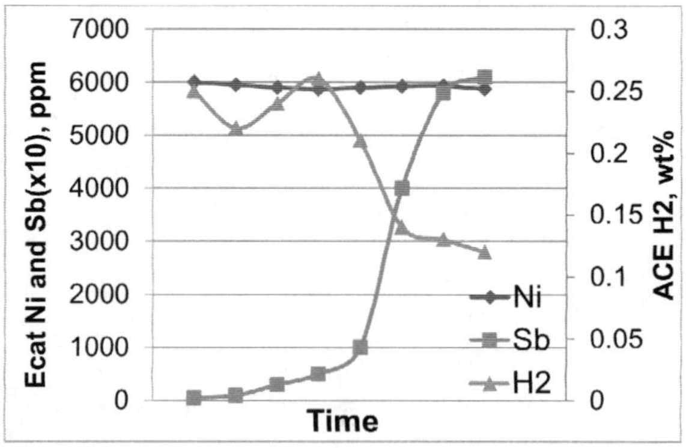

10

Elements of Concern

BASF We create chemistry

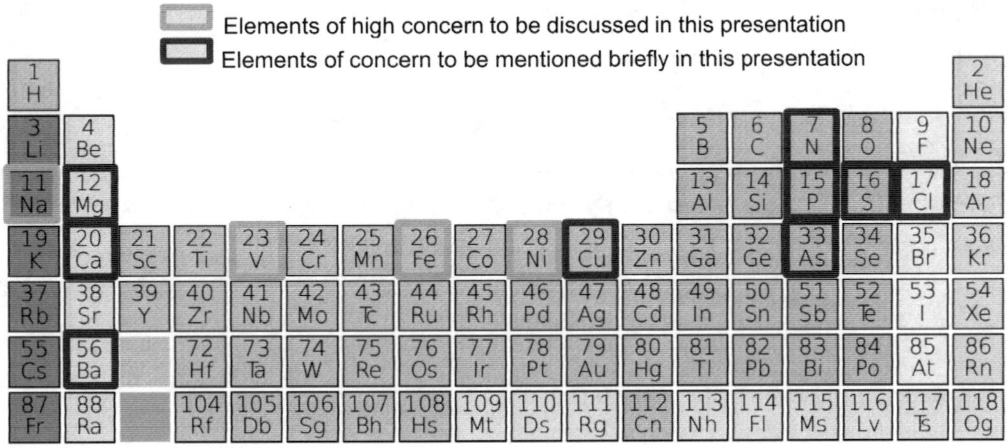

11

Elements of Concern: Vanadium

BASF We create chemistry

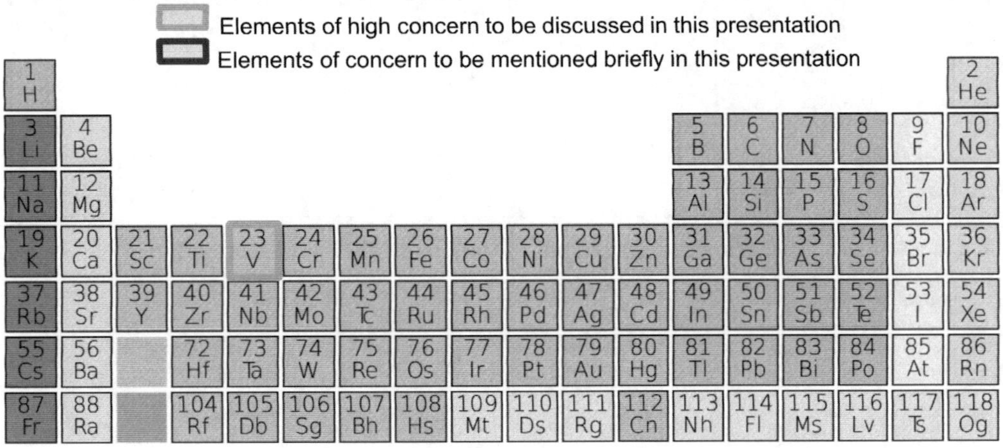

12

Vanadium Sources & Effects

BASF
We create chemistry

- Vanadium comes in with the feed and deposits on the catalyst particle
- V is converted into an oxide
  - V is very mobile – can migrate from **particle to particle** & **within the particle**
- Destroys zeolite, especially in the presence of high Na
- V causes significant reduction in activity and has some dehydrogenation activity

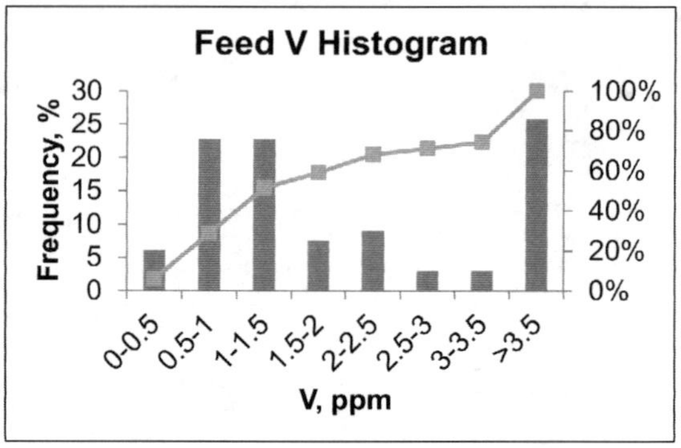

13

Vanadium Effects Case Study

BASF
We create chemistry

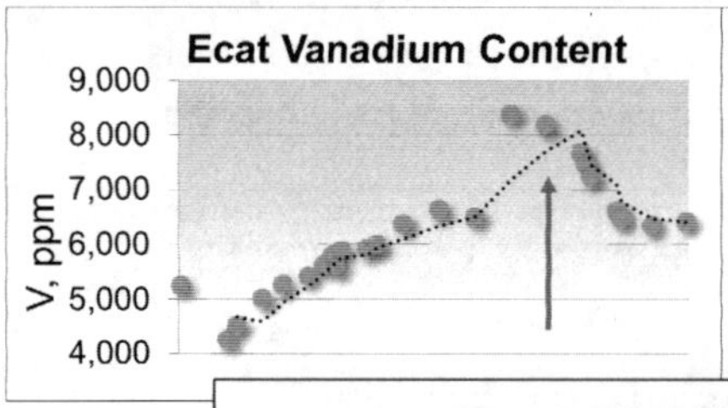

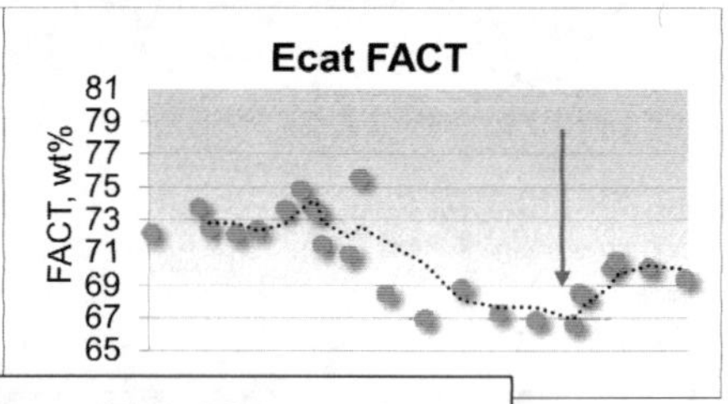

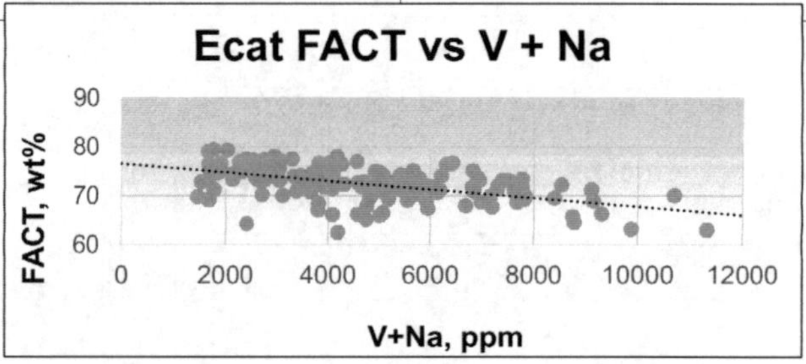

14

Vanadium Mitigation Strategies

BASF
We create chemistry

- Typically, V becomes a concern >1500 ppm on ecat
- Refiners can use a V-trap additive to passivate
  - Can be REO, Mg, or Ca/Mg based
  - Can be added separately or blended into the catalyst formulation
- Because the effect of V is exaggerated with high Na, use a low-Na fresh catalyst
- Increase catalyst additions or use purchased catalyst to flush out the V in the circulating inventory

23
V
Vanadium
50.9415

15

Elements of Concern

BASF
We create chemistry

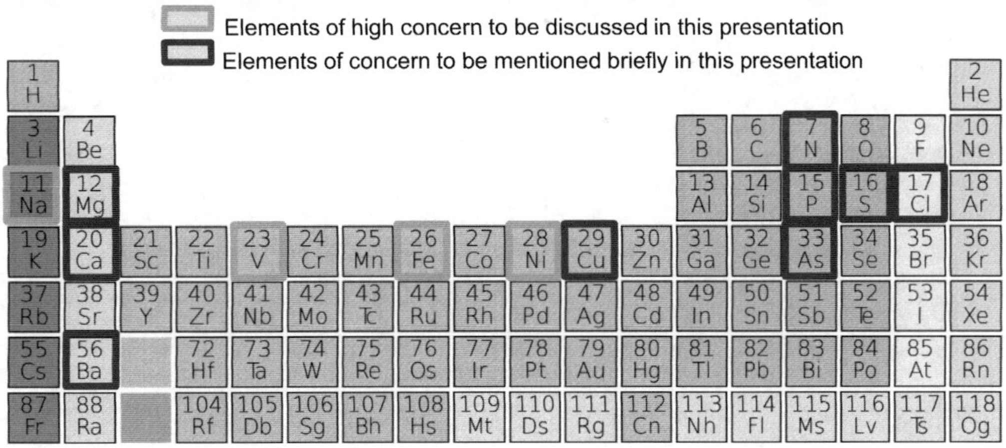

16

Elements of Concern: Sodium

BASF
We create chemistry

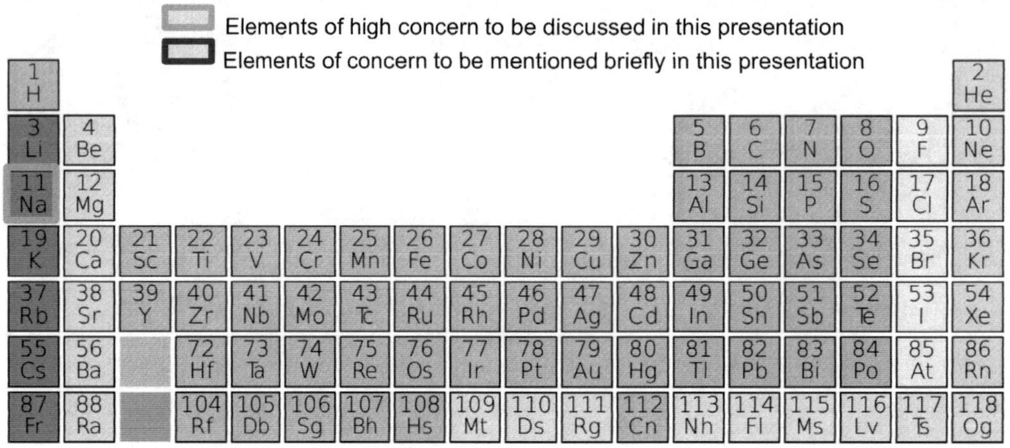

17

Sodium Sources

BASF
We create chemistry

- Sodium comes from various sources:
  - Fresh catalyst – typically between 0.15-0.3 wt%
  - Feed – deposits on the surface of catalyst particle
    - Does not migrate within the particle or to other particles

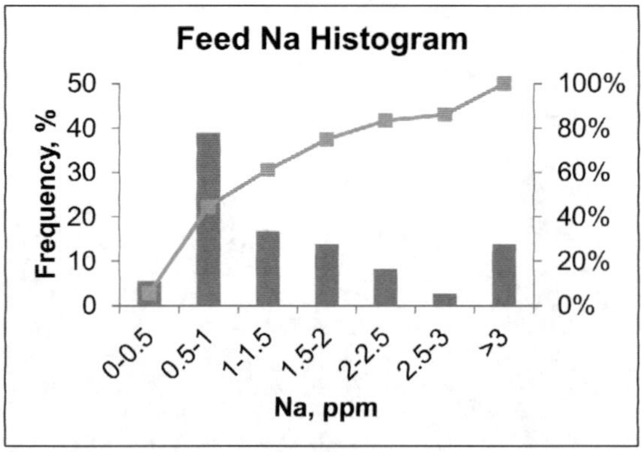

18

Sodium Effects

- Na acts as a permanent catalyst poison
  - Neutralizes acid sites
  - Also forms a **low melting point eutectic** with vanadium to lower ecat activity and conversion
  - Activity loss exaggerated when ecat V and regen temps are high
  - Can impact gasoline octane
- Ca and K have similar effects

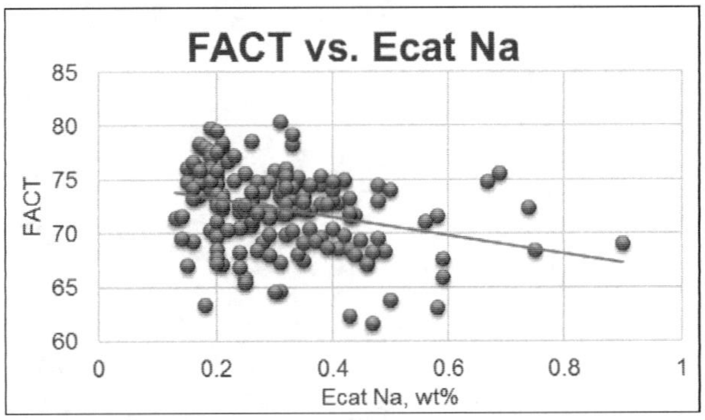

19

Sodium Mitigation Strategies

- Employ low-Na fresh catalyst
- Improve and optimize desalter operation – remove Na from the feed
- Increase catalyst additions or use purchased catalyst to flush Na out of circulating inventory
- Increase fresh catalyst activity to combat activity loss from Na

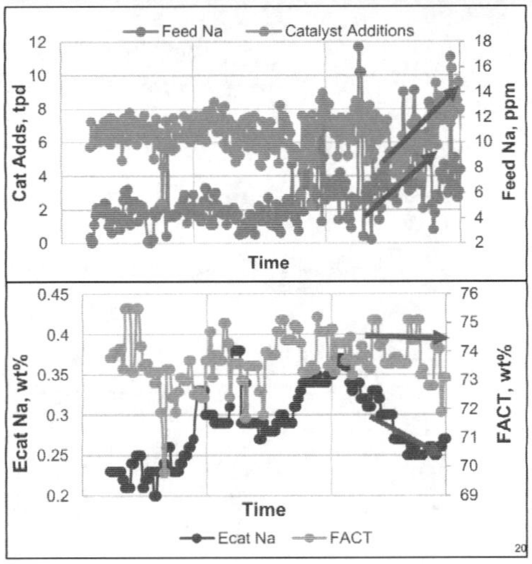

20

BASF
We create chemistry

Elements of Concern

Elements of high concern to be discussed in this presentation

Elements of concern to be mentioned briefly in this presentation

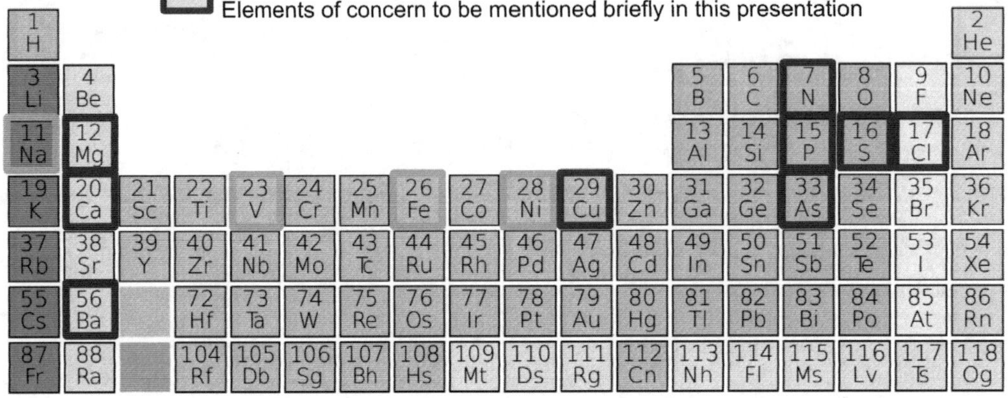

21

BASF
We create chemistry

Elements of Concern: Iron

Elements of high concern to be discussed in this presentation

Elements of concern to be mentioned briefly in this presentation

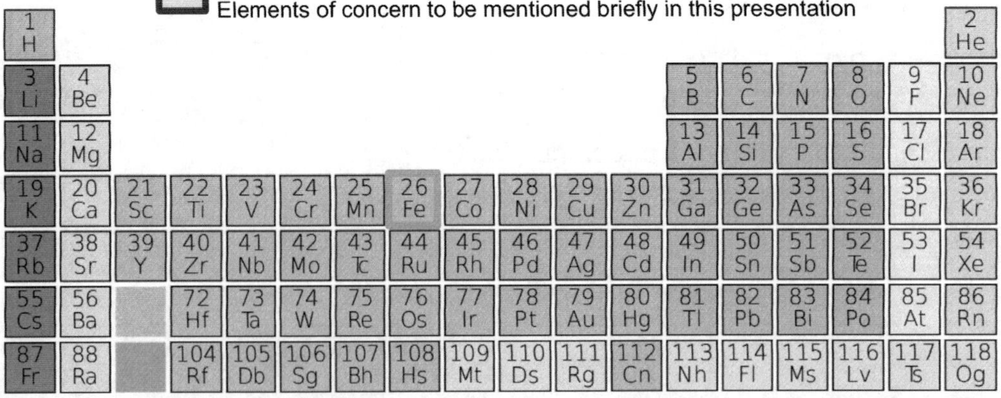

22

Iron Sources

- Fe comes from various sources:
  - Fresh catalyst – typically between 0.25-0.75 wt%
    - Incorporated within the silica/alumina framework → does NOT impact surface accessibility or side chemical reactions
  - Organic Fe from the feed
  - Inorganic Fe from equipment corrosion
- Refiners should focus on “Added Fe” which deposits on the catalyst surface
- $Fe_{(Added)} = Fe_{(Ecat)} - Fe_{(Fresh)}$

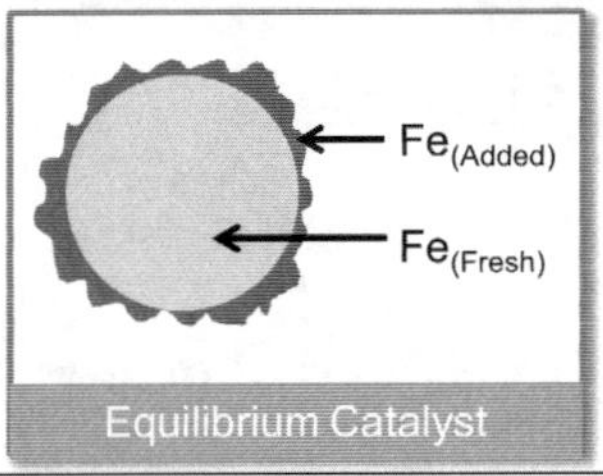

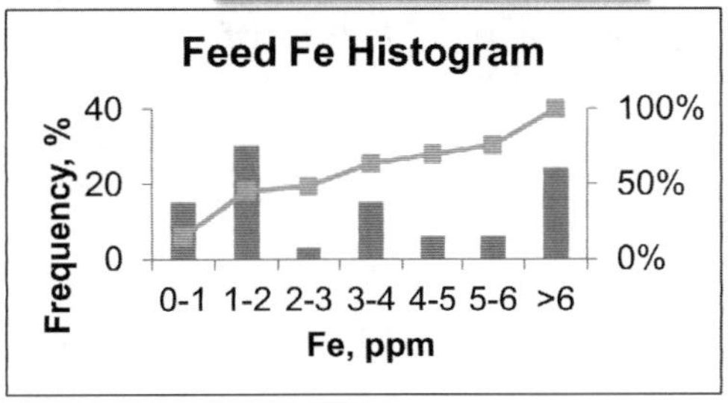

23

Iron Effects

Physical Effects

- Surface **nodule** formation, which has been reported to cause catalyst **circulation** issues
- **Vitrification** on catalyst surface, loss in surface area
- Severe poisoning leads to surface blockage and reduced **conversion** and high slurry yield with non-BASF catalyst
  - Slurry density becomes unusually light due to uncracked feed

24

Iron Effects

Chemical Effects

- **Dehydrogenation**: some refiners have reported increased H_2 make
- Transfers S from reactor to regenerator for increased **SOx**
- Acts as a **CO promoter**: can be an issue in partial-burn units

25

Iron Imaging: Iron deposits on the surface of the catalyst

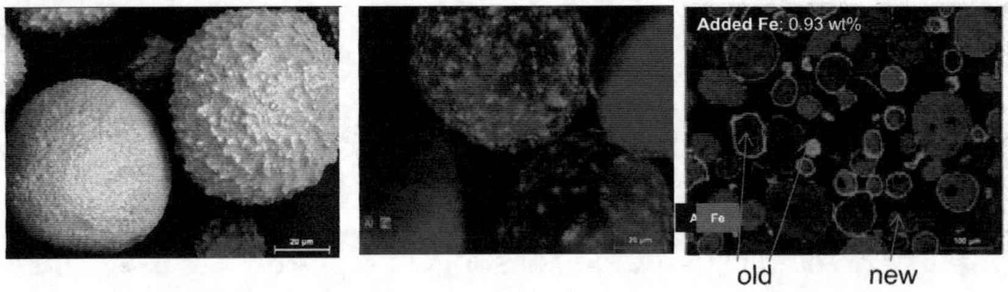

- Fe deposits on the catalyst surface, with formation of very clear surface nodules.
- Old and new catalyst particles can be easily distinguished
- Dissimilar iron coating on each catalyst → indicative of limited mobility from one catalyst particle to another.

26

Fe Effects
Case Study #1

BASF
We create chemistry

- Resid unit in Middle East
  - Feed API: 16 – 19
  - Ecat Ni: 6200 ppm; Ecat V: 5100 ppm
  - Using non-BASF catalyst
- Added Fe increased from 0.23 to 0.43 wt%
- Resulted in a conversion loss of **4 vol%**

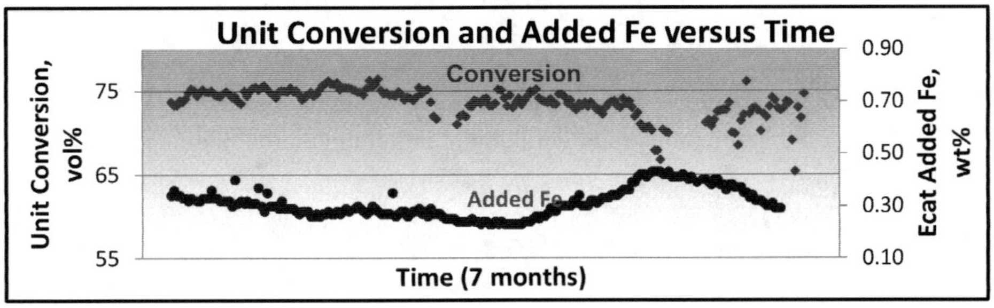

27

Fe Effects
Case Study #2

BASF
We create chemistry

- R2R unit in Asia Pacific
  - Feed API: 21 – 26
  - Ecat Ni: 2200 ppm
  - Ecat V: 2000 ppm
- Long history of using BASF catalyst since 2002
- Refiner started processing high-Fe feed
- Ecat Fe increased up to 1.3 wt% (or Added Fe of 0.7 wt%)
- Noticed change in color of the ecat around ~0.2 wt% added Fe

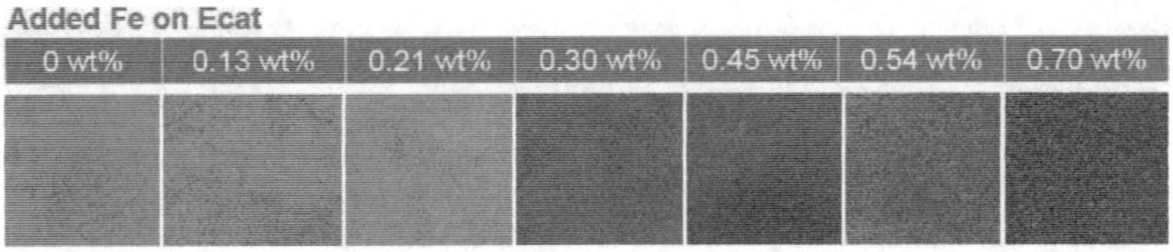

28

Fe Effects
Case Study #2

- Added iron increase from 0.15 to 0.70 wt% with no loss in unit conversion
- During the same period, other contaminants (Ni, V) decreased / stayed same

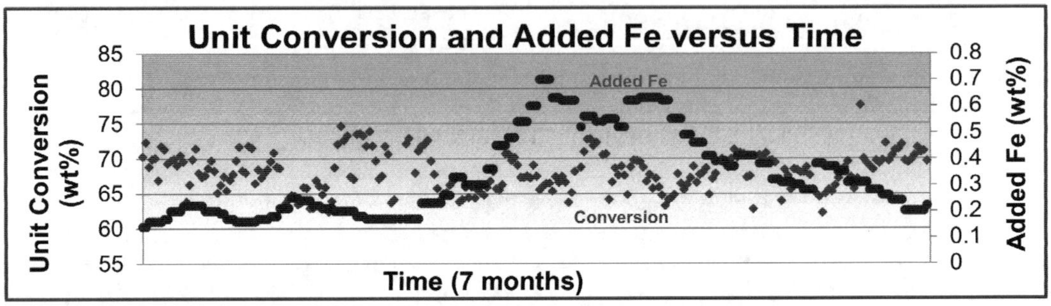

Result: No detrimental impact on unit conversion and yields at significantly high Added Fe

29

Iron Mitigation Strategies

- Employ fresh catalyst with optimized surface porosity
- Increase catalyst additions or use purchased catalyst to flush Fe out of circulating inventory

Surface Pores	Minimal	Optimized
Diffusion of feed	poor	excellent
Threshold to added Fe	Low (e.g. 0.3 wt%)	High (e.g. >1.5 wt%)
Resulting liquid yield	low	high

30

Mitigation strategies should make sense!

BASF We create chemistry

- Design of the catalyst passivation technology: passivator mobility should complement metal's mobility

Metal	Metal mobility	Passivator mobility	Passivator technology
	Mobile	Immobile	REO or separate particle
	Immobile	Immobile	Specialty Alumina
		Mobile	Boron Based Technology

31

Mitigation Strategies – Boron-Based Technology (BBT)

BASF We create chemistry

- New BASF catalyst technology designed to passivate Ni
- Boron is mobile under FCC conditions and migrates to Ni on the catalyst

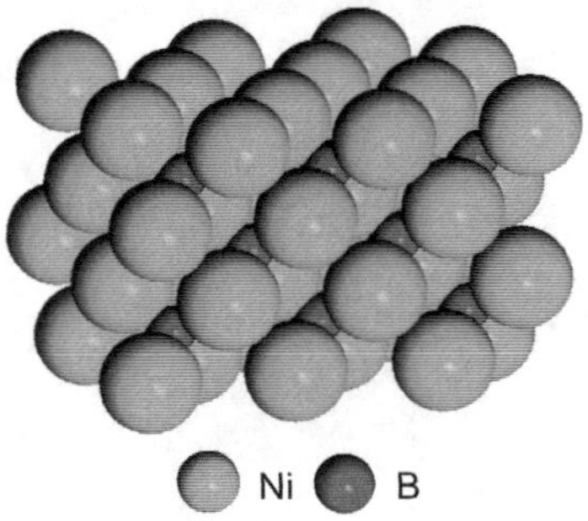

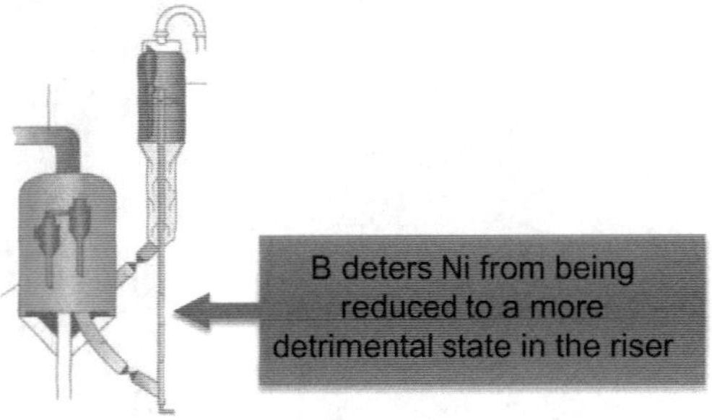

32

BOROCAT™ Reduces Hydrogen and Coke in the FCCU

BASF
We create chemistry

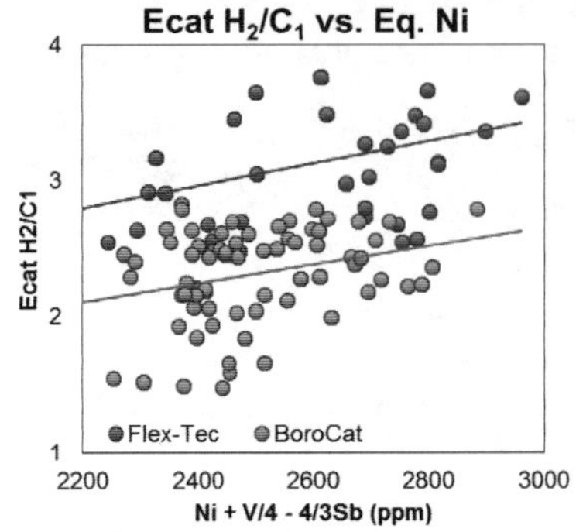

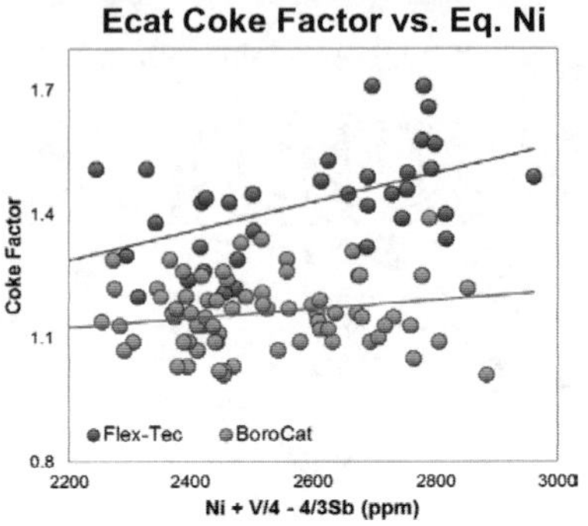

- BoroCat™: lower hydrogen, lower coke, higher liquid products
- Catalyst architecture modifications allow for dramatic alleviation of constraints

33

Mitigation Strategies Summary

BASF
We create chemistry

Nickel (Ni)
- Employ catalyst incorporated with specialty alumina
- Boron-based technology
- Antimony (Sb) injection

Vanadium (V)
- Employ catalyst designed for high V – high REO, stable zeolite
- Employ low-Na fresh catalyst
- Incorporate V-trap additive

Sodium
- Employ low-Na fresh catalyst
- Efficient and optimized desalter operation

Iron (Fe)
- Employ catalyst with optimized surface porosity
- Use of purchased catalyst to lower ecat Fe

34

Mitigation Strategies Summary

BASF We create chemistry

Feed Metal	Deposition Method	Primary Effect	Secondary Effect	What to Watch for in the Unit	Solutions
Vanadium	Deposits evenly and migrates in the regenerator	Permanent catalyst poison	Dehydrogenation agent	• Increased H2 and coke production • Decreased catalyst activity	• V-trap additive • Flush catalyst
Sodium	Deposits evenly and does not migrate	Permanent catalyst poison	Forms low melting point eutectic with V	• Decreased catalyst activity	• Flush catalyst • Catalyst change • Optimize desalter operation
Nickel	Deposits and binds on the outside of the particle. Does not migrate	Dehydrogenation agent (creates coke and H2)	--	• Increased H2 and coke production	• Antimony injection • Catalyst change • Flush catalyst
Iron	Deposits and binds on the outside of the particle. Low mobility between particles	Creates nodules on the surface of the particle at very high levels	Dehydrogenation agent CO promotion Transfers S from reactor to regenerator	• Circulation issues and lower ABD • Over-promotion in partial burn units • Slightly increased SOx emissions	• Flush catalyst • Catalyst change

Additional Feed Contaminants

BASF We create chemistry

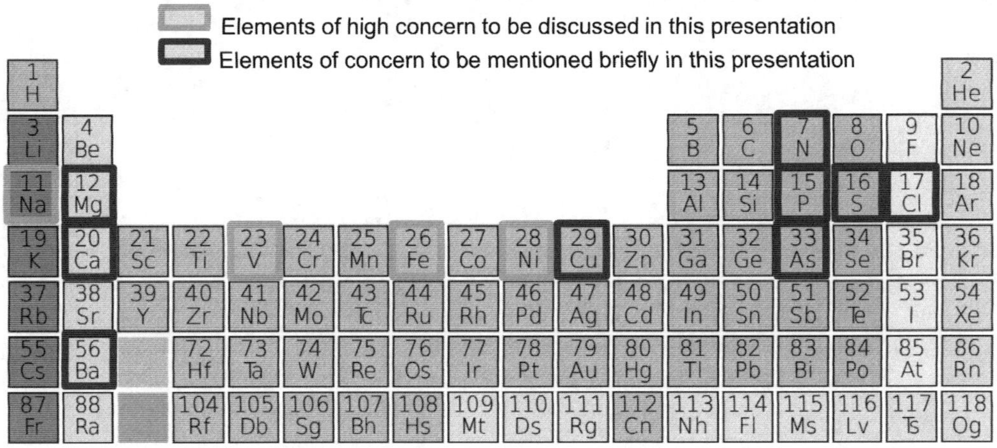

36

Additional Feed Contaminants

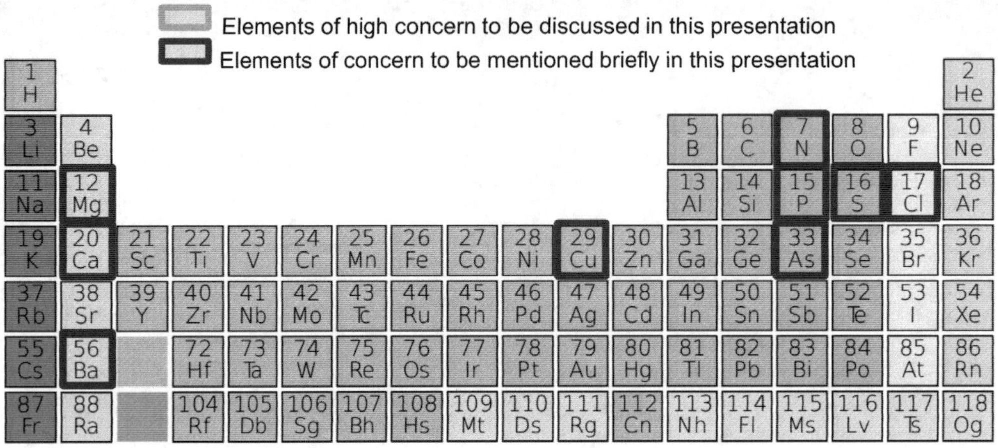

37

Additional Feed Contaminants

Calcium

- Comes in with feed or some V-trap additives
- Neutralizes acid sites (less active than Na); can also plug pores in combination with high added Fe
- Use flush catalyst or employ catalyst with optimized surface porosity

Sulfur

- Comes in with feed
- No negative effects on conversion or yields
- Use hydrotreating or gasoline sulfur additives to remove for environmental purposes

Barium

- Used in some V-traps or chemicals for tight oil extraction
- Has no effect on catalyst activity or yields when REO is present

Arsenic

- Comes in with feed
- Volatile at FCC conditions and does not stay on the FCC catalyst
- Arsine (AsH_3) does go out with the C3= stream which must be treated

38

Additional Feed Contaminants

BASF We create chemistry

Chlorine
- Comes with feed and poor desalter operation
- Re-disperses and reactivates "old" metals on the catalyst, especially Ni
- Use low-Cl catalyst or flush catalyst to remove from inventory

Phosphorus
- Used to stabilize zeolite – typically an indicator of ZSM-5 additive
- Can also be used in chemicals for tight oil extraction – no effect on activity

Copper
- Comes in with feed or NOx reduction additives
- Has 5x more dehydrogenation activity than Ni
- Use low-Cu low NOx promoter such as CONQUERNOX

(Basic) Nitrogen
- Comes in with feed and neutralizes acid sites, lowering conversion
- Temporary poison – burns off in the regenerator as coke

39

Summary: Metals Contamination in FCC

BASF We create chemistry

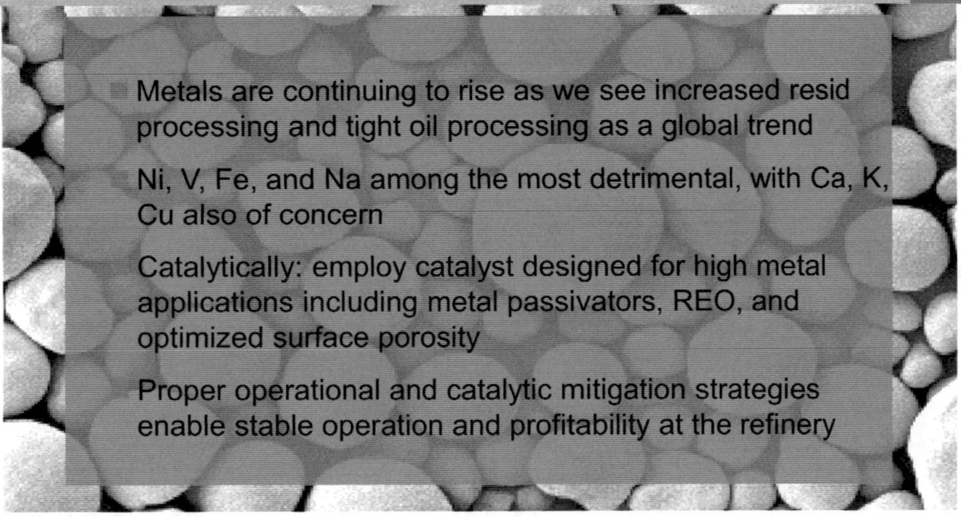

- Metals are continuing to rise as we see increased resid processing and tight oil processing as a global trend
- Ni, V, Fe, and Na among the most detrimental, with Ca, K, Cu also of concern
- Catalytically: employ catalyst designed for high metal applications including metal passivators, REO, and optimized surface porosity
- Proper operational and catalytic mitigation strategies enable stable operation and profitability at the refinery

40

BASF

We create chemistry

Effective Interviewing

Greg Yeo
ExxonMobil Chemical Company

Agenda

- Before the Interview
  - Preparing as if it's an important class
    - Understand the industry and company: What are they looking for?
    - Be prepared to tell your story in a relevant way
- At the Interview
    - Be authentic, relevant and motivated
- After the Interview
    - Followup, build a network, build a career

Interviewer Interaction

- What's the interviewer looking for? *The "right" balance of:*
  - **Demonstrated Technical ability**
  - **Demonstrated Leadership**
  - **Demonstrated Collaboration**
    - *Engagement*
    - *Energy, Interest*
    - *Flexibility*
    - *Communication*
  - ***"Grit"***
- What does the interviewer have to go on?
  - **Academic record**
  - **Resume**
  - **Face to face interview**
  - *Internship (maybe?)*
  - *Campus interview (maybe?)*
  - *Phone interview (maybe?)*

"Does this person make my team better? Is this an engineer that I want on my next project?"

AIChE
Southwest Process Technology Conference

Prepare

- Research the company in advance: *Most candidates do not research companies!*
  - Understand the company's business
  - What they are looking for?
- Prepare an interesting resume: *Most resumes look alike!*
  - Emphasize *your* experience and skills
  - *Make it a platform to tell your story*
- Analyze yourself and ***prepare***: *Most candidates do not prepare!*
  - Know your strengths and weaknesses
  - What do they need? Do your skills fit? Do you know what you want? Are you flexible?
  - Would ***you*** hire you for this position?
- Prepare examples, stories, questions
  - This is a future colleague. Be interesting, memorable, able to tell an engaging story.
  - ***I remember engaging stories from energized, capable candidates!***

AIChE
Southwest Process Technology Conference

If You Have A Pre-interview Phone / Lunch / Dinner Meeting

- This is part of the interview process – it matters
  - Be yourself! This person is trying to get to know you.
  - Are you the kind of person they would want to work with?
- Ask lots of questions
  - Show genuine interest. Let this meeting be part of your prep
  - This person got a job – learn from them
- Be energetic, interested, professional

Before I interview a student,
I always talk to the person who took them to dinner.

AIChE Southwest Process Technology Conference

At the Interview

- Dress for Success
  - If you don't know, ask or overdress. And be comfortable with it.
- Silence your phone in advance
- Use positive body language
- Be enthusiastic, engaging, and authentic.
  - Make eye contact. Show interest. And mean it.
- Create a two way conversation
  - You are interested in this interviewer and company
  - Ask questions. Take the lead when it's offered.
  - ***Be able to tell an <u>interesting story that relates to the position/company.</u>***
    - Leadership
    - Working with / connecting with people
    - Technical depth
    - Ability to stick with tough challenges

AIChE Southwest Process Technology Conference

At the Interview

- Direct the conversation about your qualifications, previous accomplishments, and interest in the job
  - Highlight how previous experiences prepared you for this position
  - *If you understand the position, you can tell the right story about yourself*
- Don't be afraid to highlight areas that company may not ask about
  - Language skills, since it may be a part of the position
  - Interest in travel – ***if it's a component of the position***
  - Long term career goals that are consistent with the company
  - Your innate skills and energy:
    1. Technical excellence
    2. Leadership
    3. Team success
    4. Grit

Southwest Process Technology Conference

Make it a Conversation!!

- What are the primary sources of revenue?
- Ask/talk about the key products or services
- How many of your senior leaders are engineers?
- What is a typical career path for a ChE like me?
- Do you see my experience and skills as a good fit for this position?
- What opportunities are there for training, travel, continuing education?
- What kinds of things to employees get involved with outside of the plant/office?
- Is the company growing? What drives growth? Are there global operations?

Southwest Process Technology Conference

After the Interview

- Before you leave, ask, "What's the next step?"
  - When can you expect to hear from them?
- Thank everyone that you talk to
- Send a follow-up note or email
- Analyze your performance
  - What should I do differently next time?
- Follow up

Southwest Process Technology Conference

Be Careful, Be Thoughtful, Be Professional

- Reputation and ethics matter
- Be careful with social media – LinkedIn, Facebook, Twitter
- Build bridges – don't burn them
- Learn about industries, companies, cultures interviewing experience
- Find ways to stay current

Southwest Process Technology Conference

Questions?

Or conversation later today

AIChE
Southwest Process Technology Conference

Big Data Analytics Skills – A Revolution Lacking in Revolutionaries

Southwest Process Technology Conference

Conference: Southwest Process Technology Conference

Year: 2017

Proceeding: 9th AIChE Southwest Process Technology Conference

Group: Southwest Process Technology Conference

Session: Friday Keynote Address

Time: Friday, October 13, 2017 - 11:15am-11:40am

Authors:

Colegrove, L., The Dow Chemical Company

Since the advent of the common use of the term âBig Data[i]â in 2010 to describe large, unruly sets of data, and the artful analysis thereof, industry has felt besieged by opportunity. The chemical and refining industry has been faced with âBig Dataâ long before it was cachet. Yet within industry there is still a struggle to get hearts and minds around what the Big Data Journey means to an organization and how to take advantage of the data opportunities that lie within. The added complication to the chemical industry is that it must add âcontextualizationâ to the data analysis because itâs not enough just to know there is a signal, we need to know what the signal means[ii]. Yet during this data revolution there is an underlying challenge that Academia, Industry and Governmental Agencies must tackle together to ensure the US Chemical industry retains its competitive position in the world. Together we need to provide education for the higher level skills and abilities required across roles in manufacturing to assure that no one gets left behind in the new Big Data age. Indeed, the next great technological âstepâ made may not be in engineering design or IoT connectedness (reference Industry 4.0), it may be in designing and creating a better human capability in âdata acuityâ out of our university system; operators, engineers and scientists alike who are better prepared for the data age than their predecessors. This talk will take a look at what Industry requires in skills sets and perspective to be âdata readyâ, and what Academia and Government might do to improve the output of data-ready skills.

Advances in Big Data Analytics at The Dow Chemical Company

Leo Chiang, Associate Technology Director
The Dow Chemical Company

2017 AIChE Southwest Process Technology Conference
October 12, 2017.

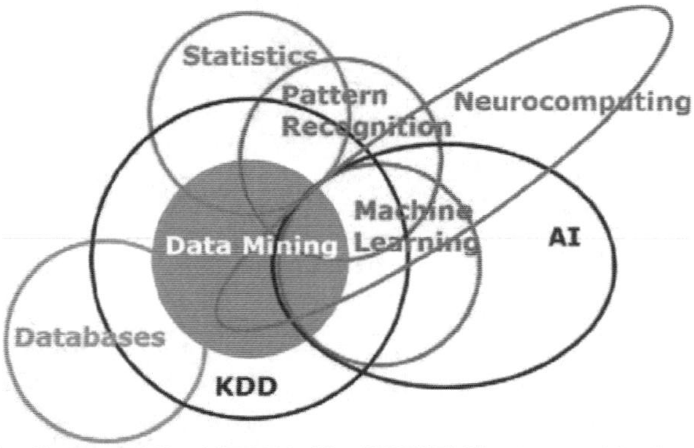

https://www.analyticsvidhya.com/blog/2015/07/difference-machine-learning-statistical-modeling/

How do we turn data into knowledge and value?

2

Driving force of Big Data growth

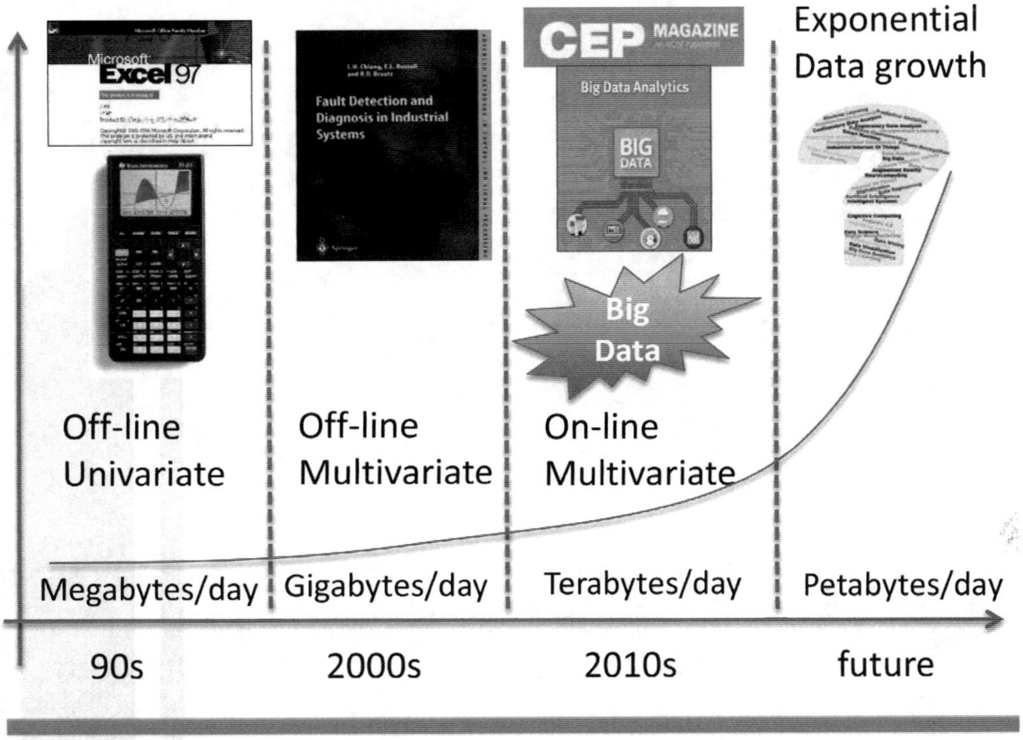

3

From univariate to multivariate

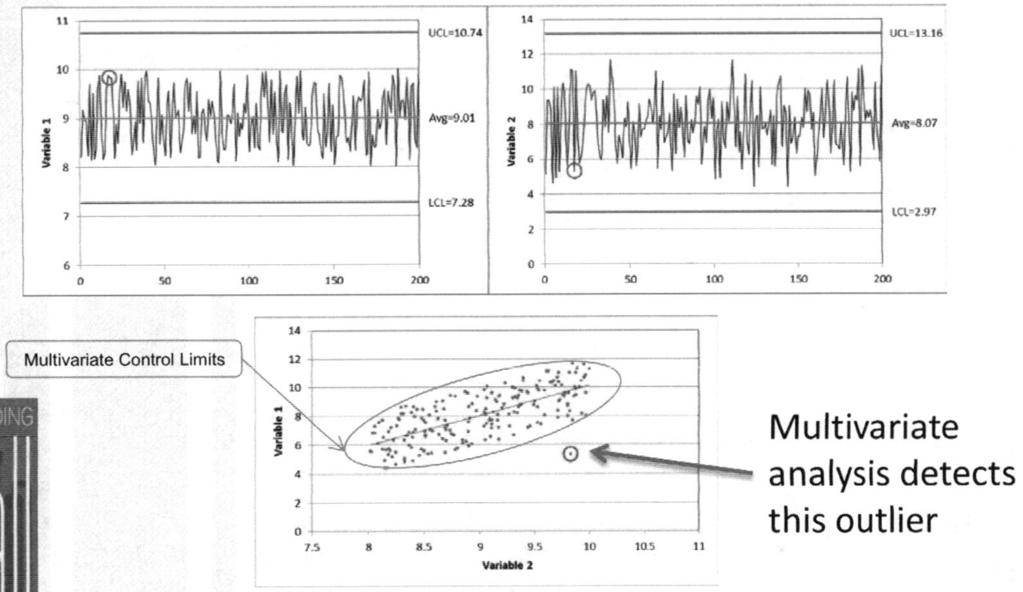

4

Multivariate Analysis for continuous processes

It is overwhelming to analyze 50+ variables in real time

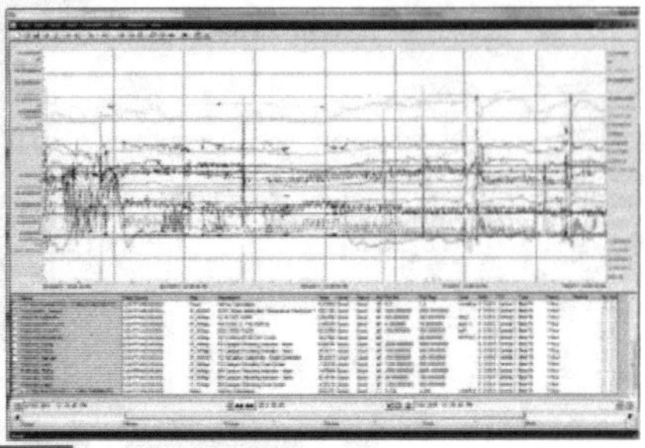

Process conditions lead to fouling

Normal conditions

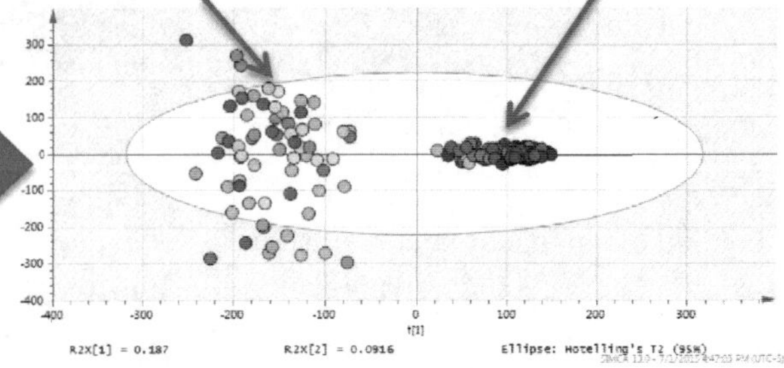

Multivariate analysis increases **speed of problem solving**.
We use existing data to support hypotheses, eliminate hypotheses, and generate new hypotheses

STANDING Tall OPERATIONS

5

Multivariate Analysis for batch processes

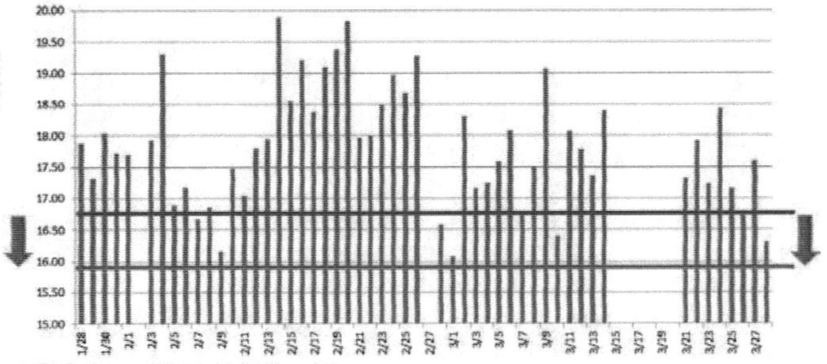

500 batches,
55 millions data points

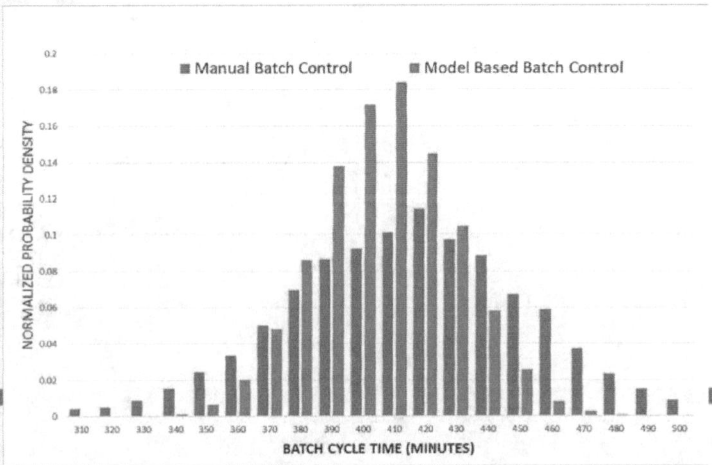

STANDING Tall OPERATIONS

6

On-line fault detection

From Reactive To Proactive

From Reactive	To Proactive
• Troubleshoot and fix problems off-line	• Prevent and predict problems on-line
• Process Optimization & Understanding	• Improve Reliability & Quality
• End user: Process Chemometrics experts	• End user: operators and engineers

- **Anticipate and prevent future problems**
  - Online monitoring of key variables, multivariate fault detection and monitoring

Inferential sensors

400+ inferential sensors applications (process monitoring, sample plan reduction, and AC&O) at Dow

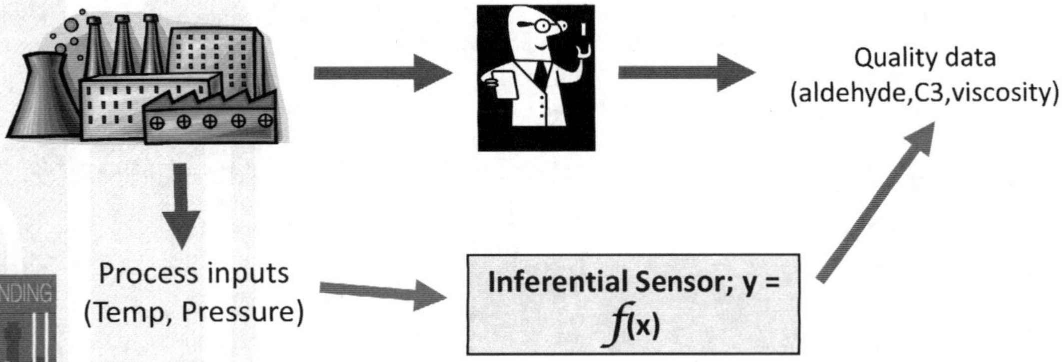

Where is Big Data Analytics research going?

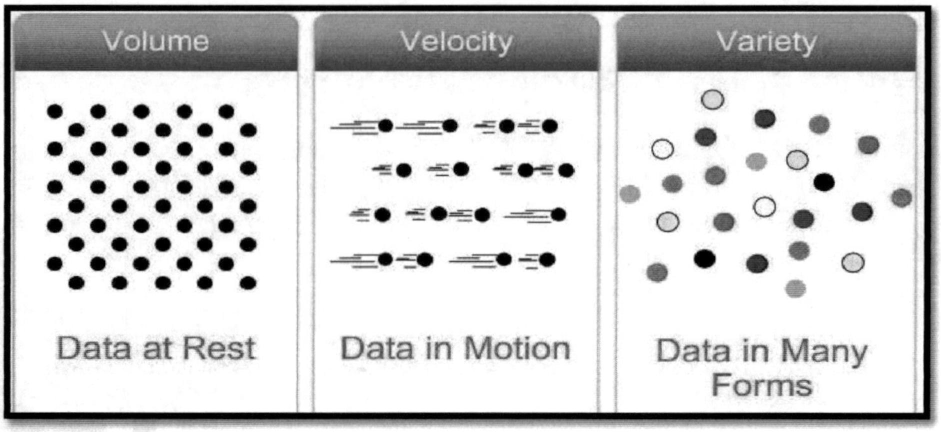

Not all data are created equal, need to filter out noise

Real time model adaptation

Integrate structured and unstructured data; Knowledge integration

10

Workforce development: Opportunities at all levels

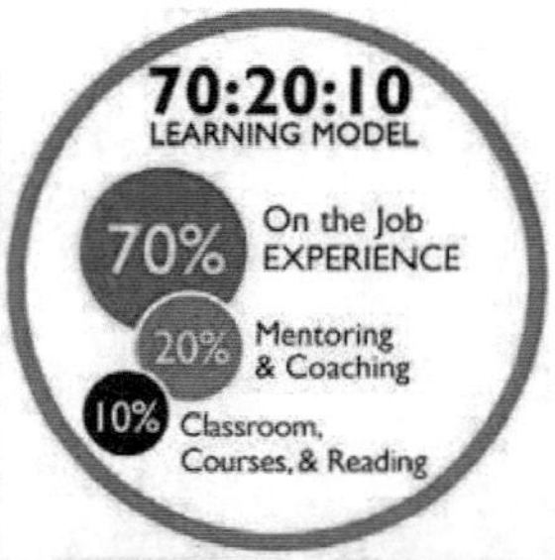

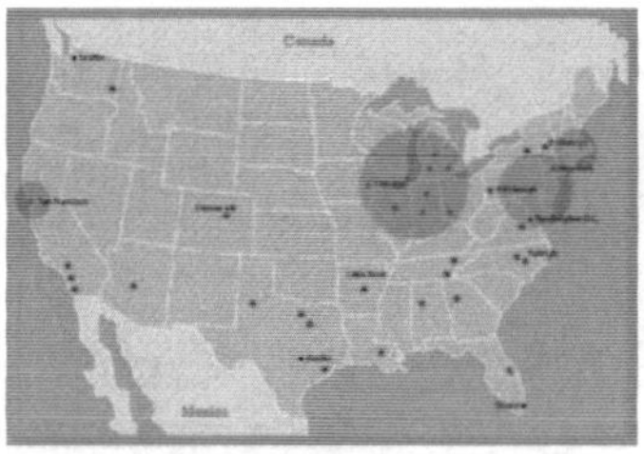

http://data-informed.com/bigdata_university_map/

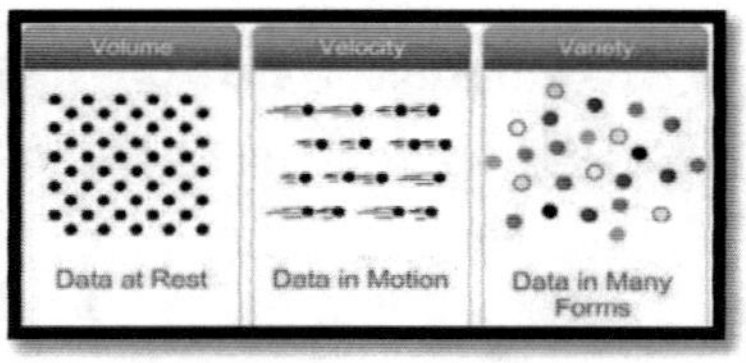

B.S.

- Need more statistics, programming, and data-driven decision making training

M.S.

- Need to balance breath vs. depth on technical topics

Ph.D.

- Need to incorporate Big Data research into Chemical Engineering and Chemistry

11

This is the Big Data world today, each company claims that they have a solution for us

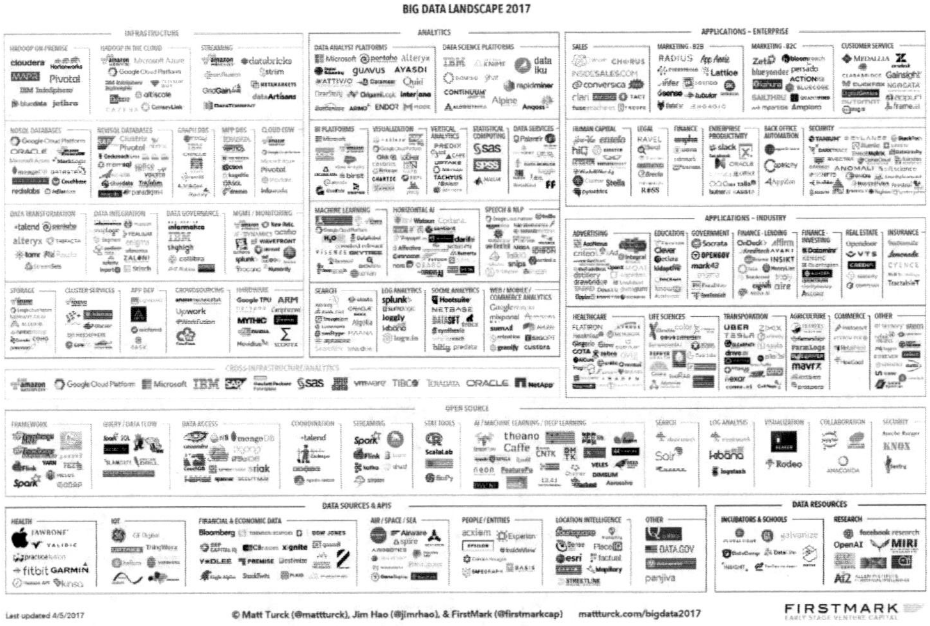

12

Additional resources

CH09CH04-Chiang ARI 18 February 2017 11:22

Big Data Analytics in Chemical Engineering

Leo Chiang, Bo Lu, and Ivan Castillo

The Dow Chemical Company, Freeport, Texas 77541; email: hchiang@dow.com

Annual Review of Chemical and Biomolecular Engineering, June 2017

Annu. Rev. Chem. Biomol. Eng. 2017. 8:4.1–4.23

The *Annual Review of Chemical and Biomolecular Engineering* is online at chembioeng.annualreviews.org

This article's doi:
10.1146/annurev-chembioeng-060816-101555

Keywords

big data analytics, Internet of things, data-driven modeling, Industry 4.0, machine learning

Abstract

Big data analytics is the journey to turn data into insights for more informed business and operational decisions. As the chemical engineering community is collecting more data (volume) from different sources (variety), this journey becomes more challenging in terms of using the right data and the right tools (analytics) to make the right decisions in real time (velocity). This article highlights recent big data advancements in five industries, including chemicals, energy, semiconductors, pharmaceuticals, and food, and then discusses technical, platform, and culture challenges. To reach the next milestone in multiplying successes to the enterprise level, government, academia, and industry need to collaboratively focus on workforce development and innovation.

13

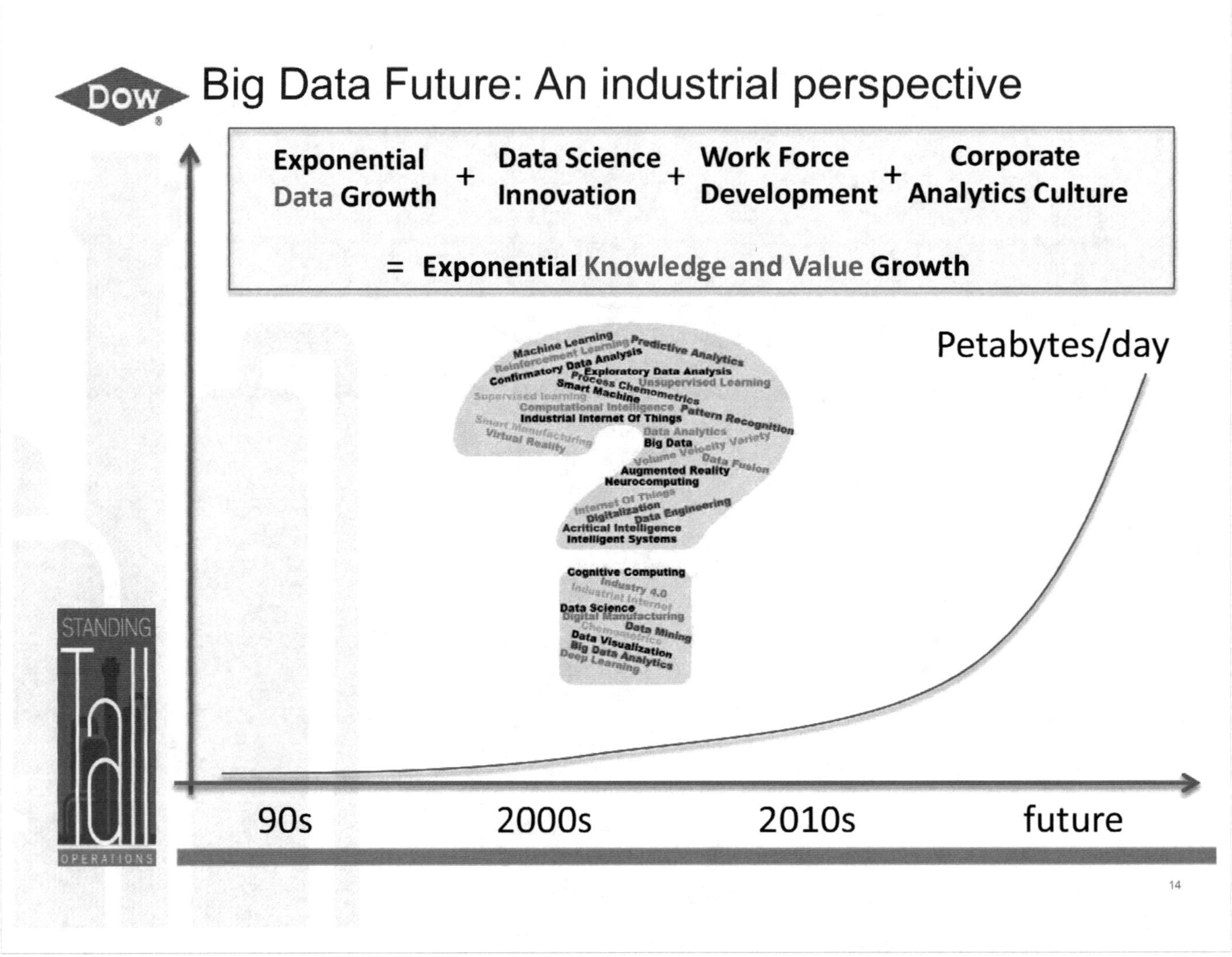
Big Data Future: An industrial perspective
Exponential Data Growth + Data Science Innovation + Work Force Development + Corporate Analytics Culture
= Exponential Knowledge and Value Growth
Petabytes/day
90s
2000s
2010s
future
14

Desalting Fundamentals

A Process Engineer's Perspective

Andrew W. Sloley, Principal Consultant, Advisian

Prepared for Presentation at the
American Institute of Chemical Engineers
2017 Fall Southwest Process Technology Conference
Galveston, TX, October 12-13, 2017

www.**advisian**.com

Advisian
WorleyParsons Group

But First, What We Do

Consulting and Engineering in:

Refinery and Petrochemicals

- Molecule management
- Concept selection
- FEED
- Economics and planning
- Technology selection
- Refinery/petrochemical integration

Operational Excellence

- Benchmarking
- Gap analysis
- Safety and environment

Sulfur and Gas Treating

- Oxygen enrichment
- Gas treating solvents

Infrastructure and Environmental

- Water
- Materials handling

Management, Economics, and Business Operations

EcoNomics™

Advisian / 2

Overview

Advisian / 3

Desalting

- Remove residual water and contaminants from crude oil
- ~90% of crude units include desalters
- Major contributor to extended refinery run-lengths

Advisian / 4

Desalting Benefits

- Reduced corrosion of crude distillation tower and overhead system
- Reduced exchanger fouling downstream of desalter
- Reduced pressure drop in downstream heat exchangers and fired heater
- Reduced energy consumption in fired heater
- Increased capacity in distillation towers
- Improved residue quality
- Decreased fouling and plugging in atmospheric column, vacuum column, and condensing heat exchangers

Desalting Process

Desalting Steps

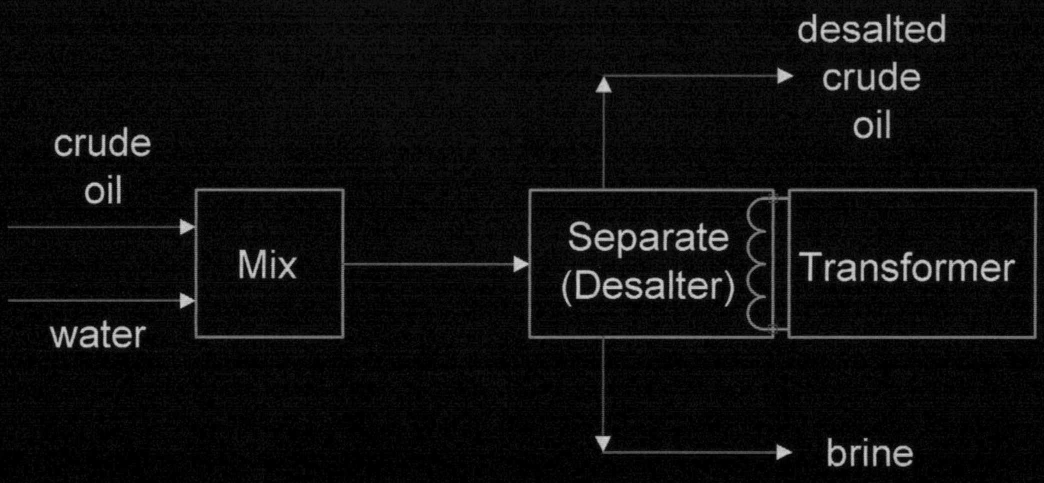

Expected Performance

Removal of	Light Crude °API >30 (SG < 0.875)		Medium Crude °API 20-30 (SG 0.875-0.935)		Heavy Crude °API <20 (SG > 0.935)	
Stages	1	2	1	2	1	2
Water, %	90-95	98-99	85-90	97-98	85-90	95-98
Salts, %	>95	>95	>90	>90	>85	>85
Solids, %	60-90	60-90	50-80	50-80	50-80	50-80

One-Stage Desalting

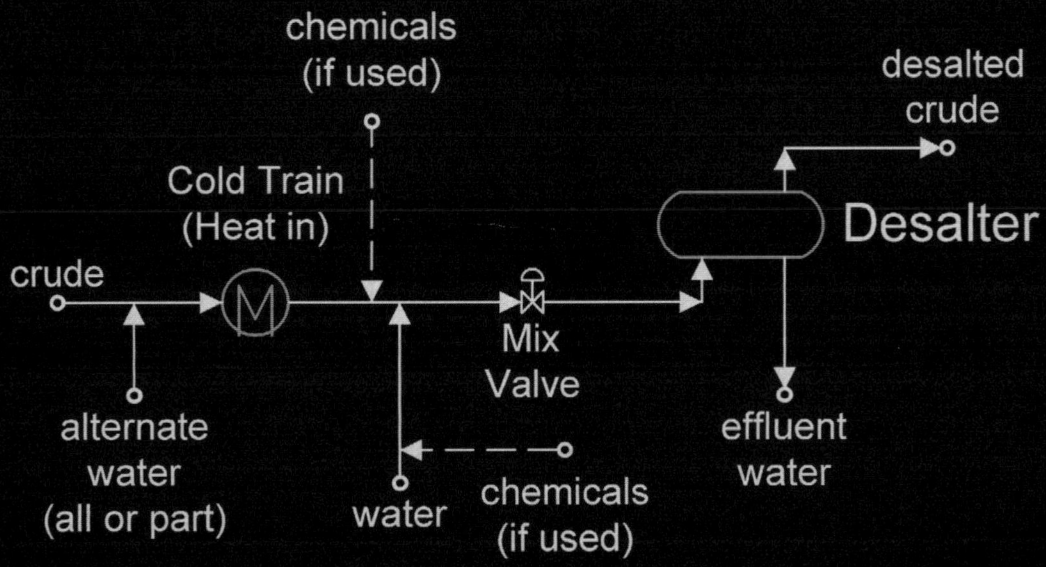

Advisian / 9

Two-Stage Desalting

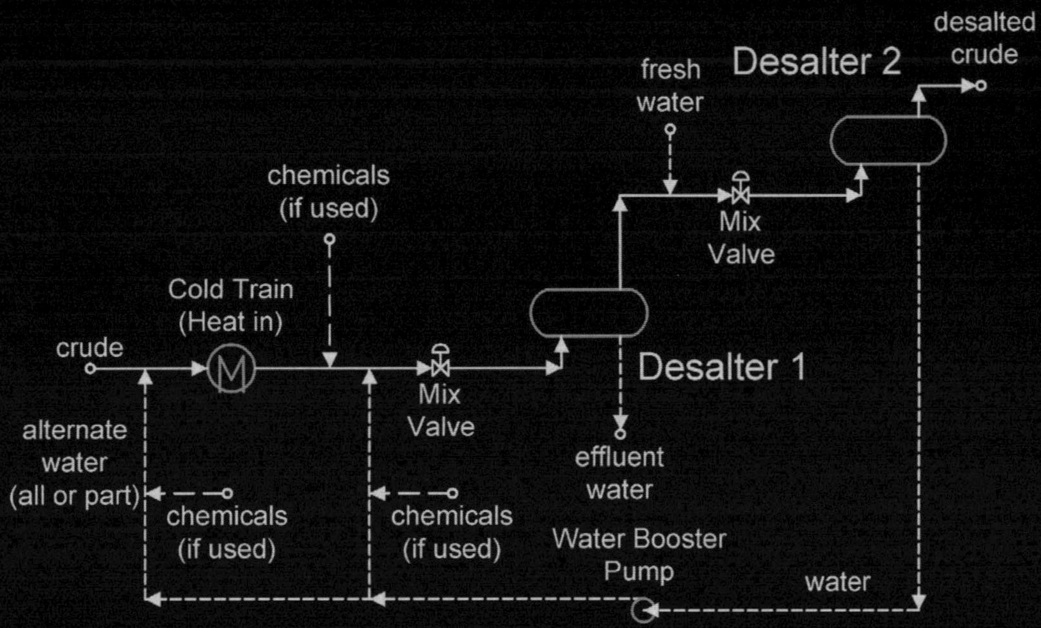

Advisian / 10

Design and Operating Factors

- Crude salt content
- Crude type and properties
- Dehydration efficiency
- Salt removal efficiency
- Solids removal efficiency
- Oil-water separation effectiveness
- Capacity
- Desalter capital and operating costs vs. downstream capital and operating cost

Advisian / 11

Contaminants

Advisian / 12

Contaminant Types

- Oil soluble
- Water soluble
- Insoluble (solids)
  - Oleophilic
  - Oleophobic
  - Complex structures
- Water
- Native vs. added

Water

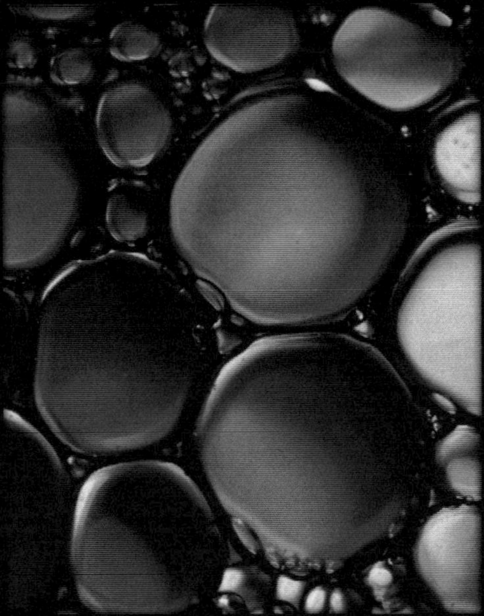

- Increased downstream pressure drop
- Increased energy consumption
- Decreased capacity
- Rule of thumb: 1 barrel of water costs 7 barrels of oil capacity

Salt Stability

Advisian / 15

Calcium Chloride and Magnesium Chloride

- Hydrolyze at low temperatures
- Problems result in:
  - Atmospheric preflash tower
  - Atmospheric tower
  - Atmospheric tower overhead

Advisian / 16

Salt Conversion

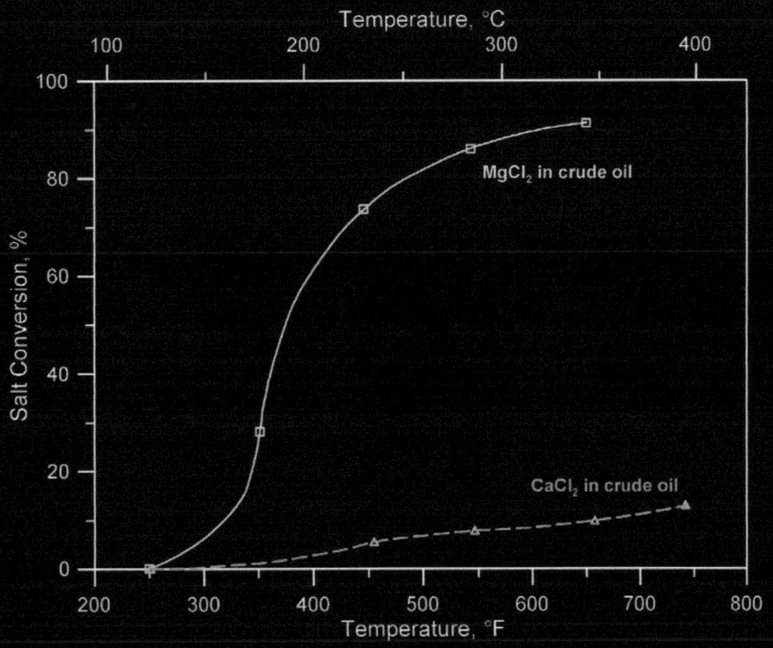

Advisian / 17

Hydrolysis Examples

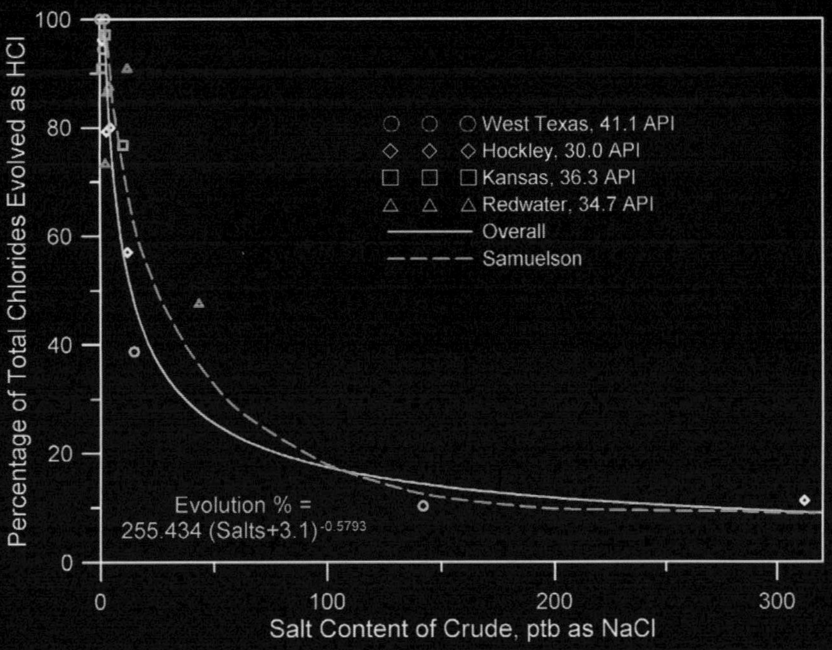

Advisian / 18

Water Addition

Advisian / 19

Water Addition

- Dilutes salt content of un-removed water
- Dilutes surfactants and emulsifiers
- Helps wash solids out
- Helps remove water soluble chemicals
- Controls water content to make settling and coalescence more effective
- Ranges 2% to 10% (most operations in 4% to 8% range)

Advisian / 20

Addition Locations

- Upstream of desalter
- Upstream of exchangers:
  - Can wash exchangers
  - Problems for carbonate precipitations
- Upstream of crude charge pump:
  - Very effective mixing
  - Can create very small water droplets – difficult to remove
- Multiple locations

Advisian / 21

Water Addition Locations

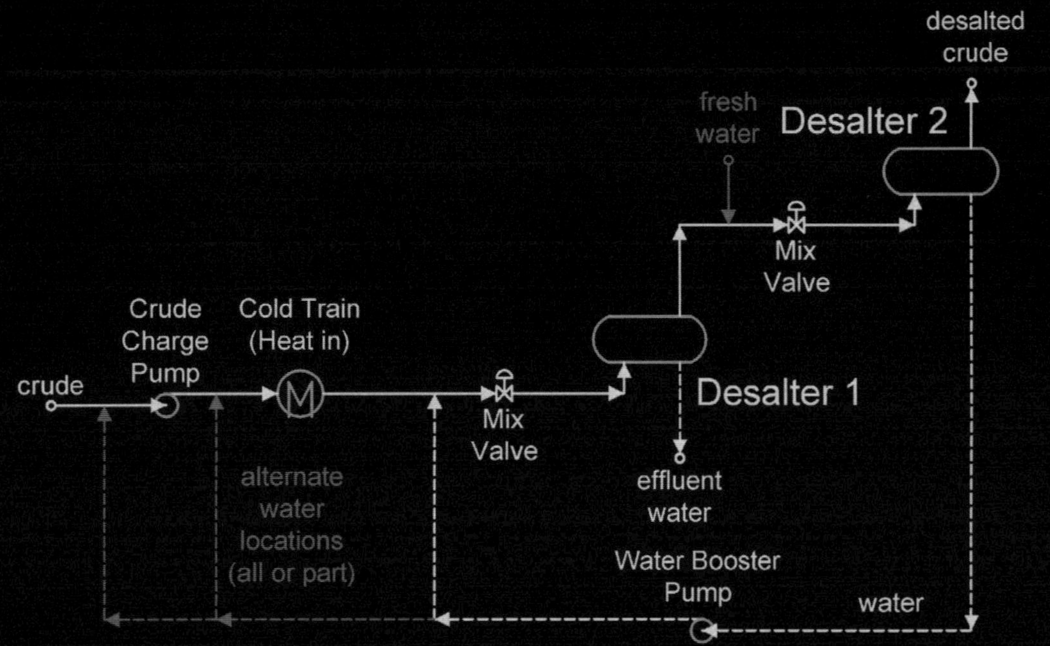

Advisian / 22

Masking the Most of Limited Water

- Cross-flow water
- Water recycle:
  - Not a water-saving measure

Advisian / 23

Water Cross-Flow

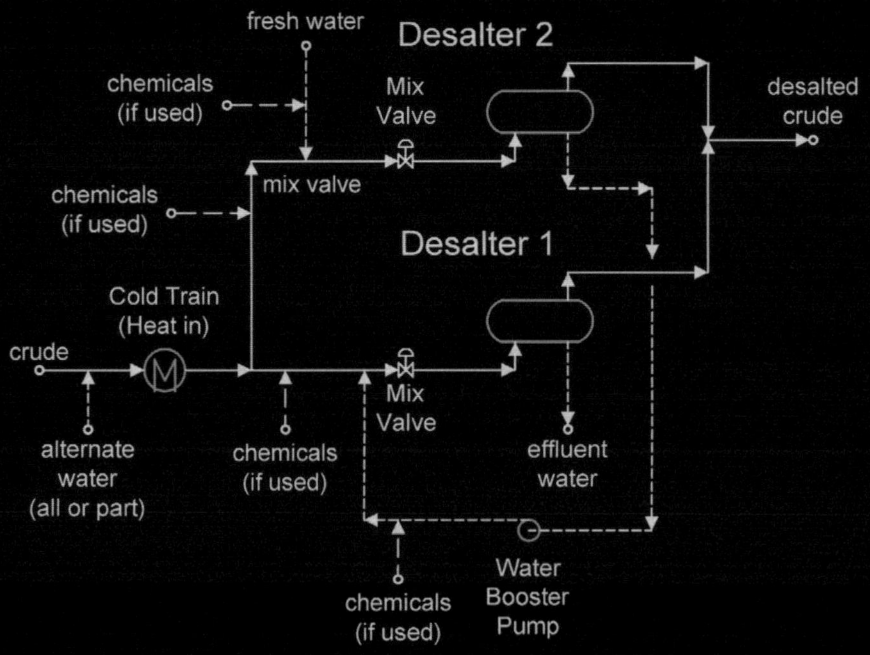

Advisian / 24

Water Recycle

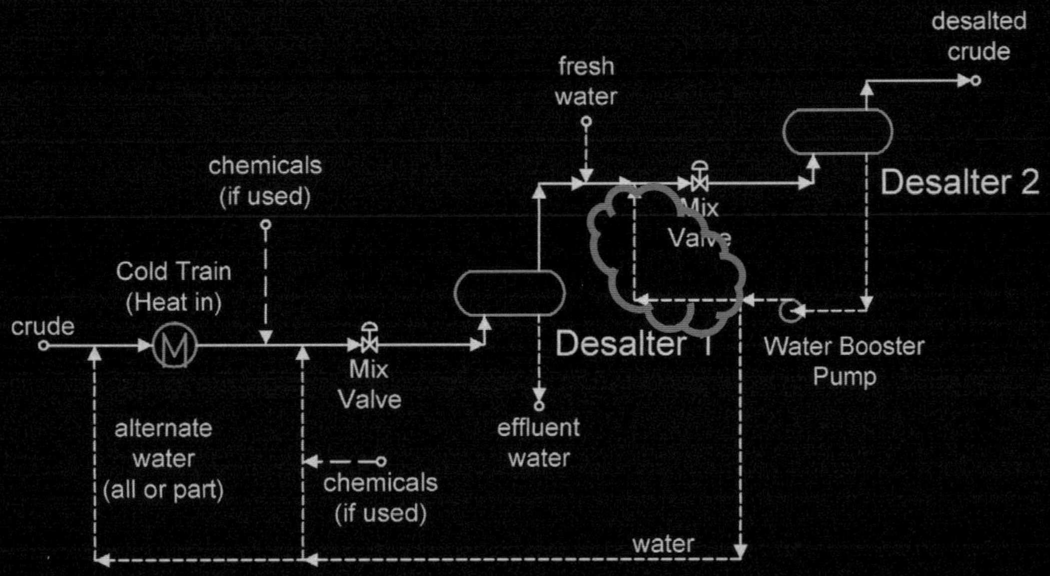

Water Mixing

Mixers

- Mix valves: conventional approach:
  - Pressure drop 5-35 psi, typically 10-20 psi (69-138 kPa)
  - Set experimentally
- Static mixers:
  - Lower pressure drop, not as effective mixing
- Shear devices:
  - Low pressure drop
  - Limited experience

Advisian / 27

Mix Valve Pressure Drop and Performance

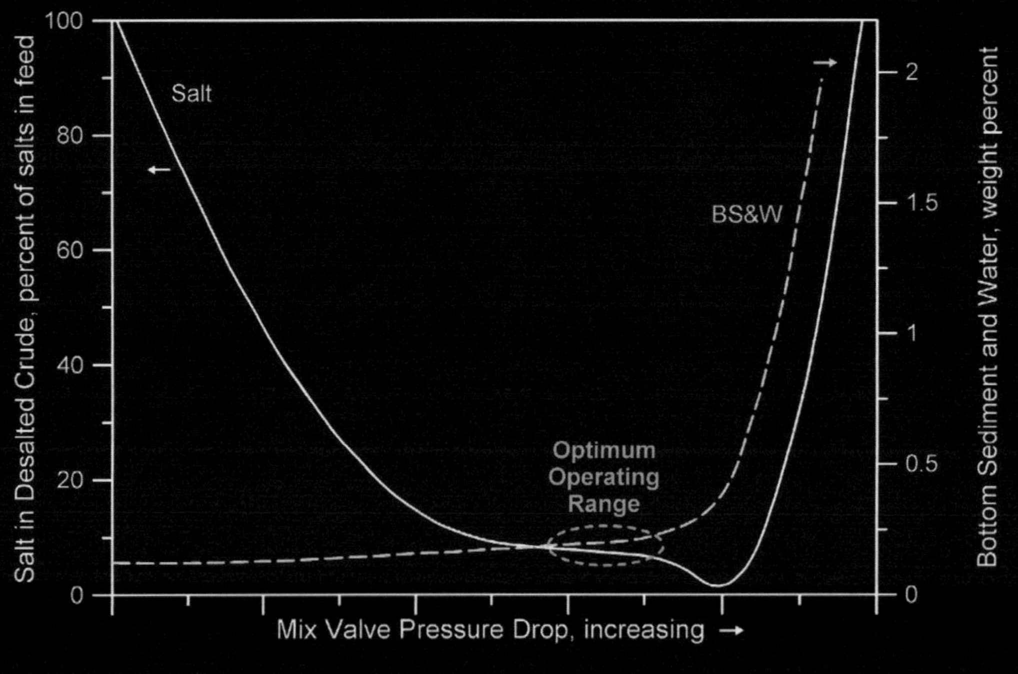

Advisian / 28

Settling

Advisian / 29

Settling

- Heavy droplets (water) settle by gravity
- Bulk phase (oil) rises to outlet
- Velocity of water dropping must be greater than of oil rising
- Stokes' law calculates settling velocity:
  - Low velocity (laminar flow)
  - Spherical droplets

Advisian / 30

What Makes Settling Better?

Fundamentals

- High-density difference
- Larger droplets
- Low viscosity

Applications

- Low interfering chemicals
- Low solids
- Low waxes
- pH range
- Additives

Coalescence

Coalescing

- Desalters use electric field to coalesce water droplets
- Liquid-liquid electrostatic precipitator
- Dipole moment applied on droplets
- Droplets attracted to each other

Advisian / 33

Induced Dipoles

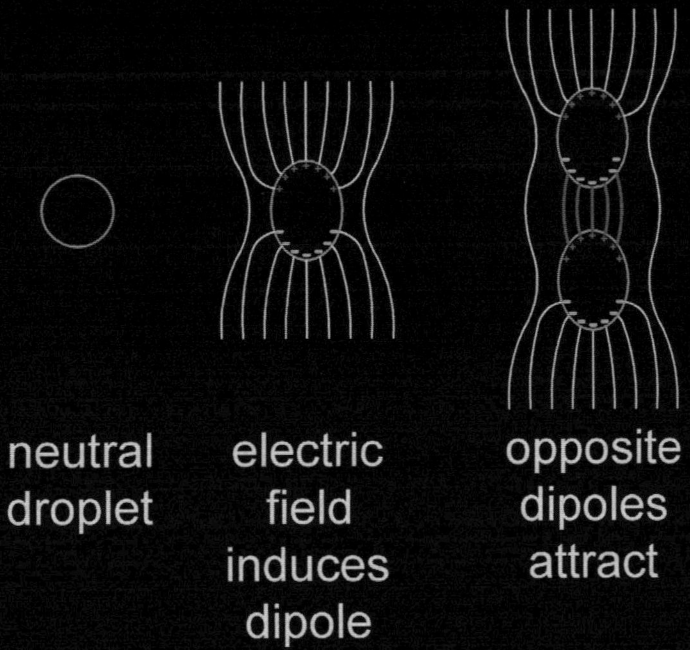

Dynamics

- As oil rises towards top of coalescing zone, water content drops
- Water droplets further apart
- Coalescing force drops
- Sets limit on desalter performance, ~0.1 volume percent water

Advisian / 35

Coalescing Forces and Water Content

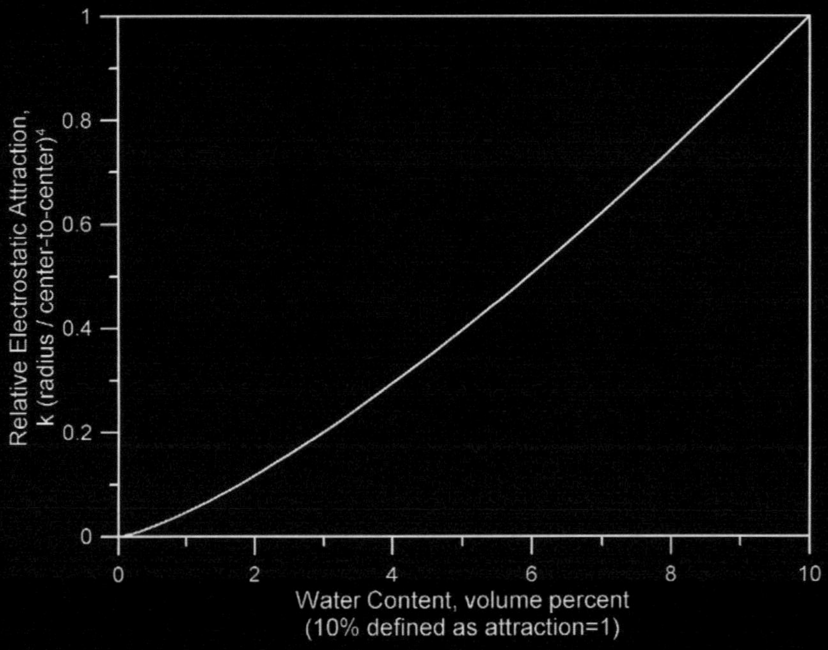

Advisian / 36

Temperature

Temperature

- Can be changed by controlling unit correctly
- Mechanical limit: insulators on bushings for grids, set maximum temperature:
  - Most plants limit to 300°F - 310°F maximum (149°C - 154°C)
  - Some plants run to 325°F (163°C)
  - At higher temperatures, bushings "blow out," grids short circuit

Higher Temperature

Benefits	
Lower oil viscosity	Settling improves (less water in oil)
Lower water viscosity	Separation improves (less oil in water)
Weakens surfactant layers	Coalescence improves
Melts waxes in oil	Coalescence improves

Costs	
Decreased density difference	Settling degrades
Increased soluble water in crude	More water downstream
Increased soluble hydrocarbons in water	More hydrocarbon to water treatment
Increased crude conductivity	Higher electrical consumption

Advisian / 39

Oil-Water Density Difference

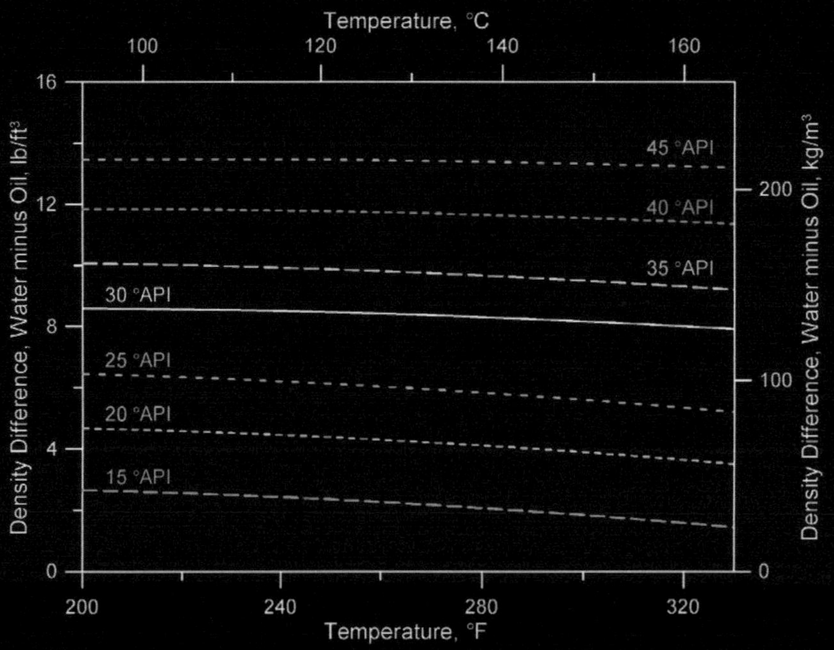

Advisian / 40

Summary

Advisian / 41

Operating Optimization

- Water wash ratio
- Water wash addition location
- Desalter operating temperature
- Mix valve pressure drop
- Upstream improvements (in tank farm)

Advisian / 42

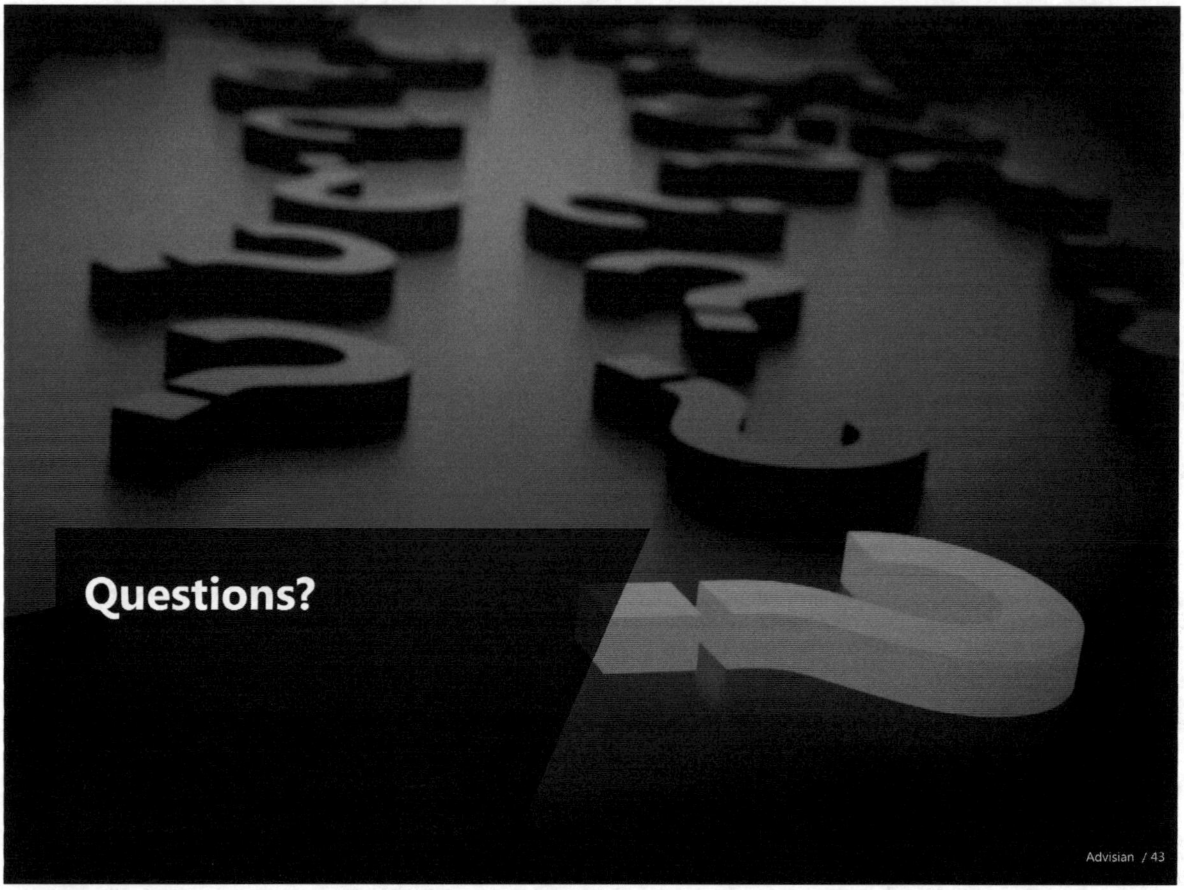

Disclaimers

DISCLAIMER

This presentation has been prepared by a representative of Advisian.

The presentation contains the professional and personal opinions of the presenter, which are given in good faith. As such, opinions presented herein may not always necessarily reflect the position of Advisian as a whole, its officers or executive.

Any forward-looking statements included in this presentation will involve subjective judgment and analysis and are subject to uncertainties, risks and contingencies—many of which are outside the control of, and may be unknown to, Advisian.

Advisian and all associated entities and representatives make no representation or warranty as to the accuracy, reliability or completeness of information in this document and do not take responsibility for updating any information or correcting any error or omission that may become apparent after this document has been issued.

To the extent permitted by law, Advisian and its officers, employees, related bodies and agents disclaim all liability—direct, indirect or consequential (and whether or not arising out of the negligence, default or lack of care of Advisian and/or any of its agents)—for any loss or damage suffered by a recipient or other persons arising out of, or in connection with, any use or reliance on this presentation or information.

Advisian / 44

Advisian

WorleyParsons Group

About Us

Consulting and Engineering in

Refinery and Petrochemicals

- Molecule Management
- Economics and Planning
- Technology Selection
- Refinery-Petrochemical Integration

Logistics

- Hydrocarbon Logistics
- Bulk liquid terminals
- Railway corridors
- Market analysis
- Risk Analysis
- Safety and Environment

Advisian

WorleyParsons Group

Sulfur and Gas Treating

- Oxygen enrichment
- Gas treating solvents

Infrastructure and Environmental

- Water
- Materials handling

EcoNomics™

Advisian / 46

Tracerco
Providing Insight Onsite

On-line Leak Detection Methods for Heat Exchangers and Valves

Presented by: William Mixon
For the 2017 AIChE Southwest Process Technology Conference
October 13, 2017

JM Johnson Matthey
Inspiring science, enhancing life

Tracer Technologies

Tracerco
Providing Insight Onsite

1. Radioisotope Tracers
   - Can be in the form of solids, liquids, or gases.
   - Tracers emit high-energy gammas, meaning that they can be detected using external radiation detectors mounted to piping and vessels (no sampling required).
   - Typically isotopes with short half-lives are used.
   - Very small injection sizes used. Typically 5 grams of tracer or less.
   - Typical applications:
     - On-line leak tests of heat exchangers and valves
     - Flow rate measurement and flow meter verifications
     - Distribution studies through reactors, packed towers, & L/L contactors
     - Residence time & carryover studies of separators

JM Johnson Matthey
Inspiring science, enhancing life

Tracerco Radiation Awareness

Tracerco
Providing Insight Onsite

- Tracerco employ's a fulltime RSO to ensure regulatory and license compliance
- Tracerco is Licensed by State and Federal Agencies and inspected once a year by appropriate agencies
- Tracerco is responsible for the *Transport*, *Use, and Disposal* of any radioactive material we bring onto your site

- Plant does *NOT* inform *EPA, State, or Federal Agencies* about testing
- There is no danger to plant personnel working around process equipment, as long as they stay out of the barricaded area
- Our radiation sources are about 1000 times smaller than radiography sources.

JM Johnson Matthey
Inspiring science, enhancing life

Tracerco
Providing Insight Onsite

Heat Exchanger Leak Testing

JM
Johnson Matthey

Tracerco Diagnostics™ Leak study

Tracerco

Testing for leaks

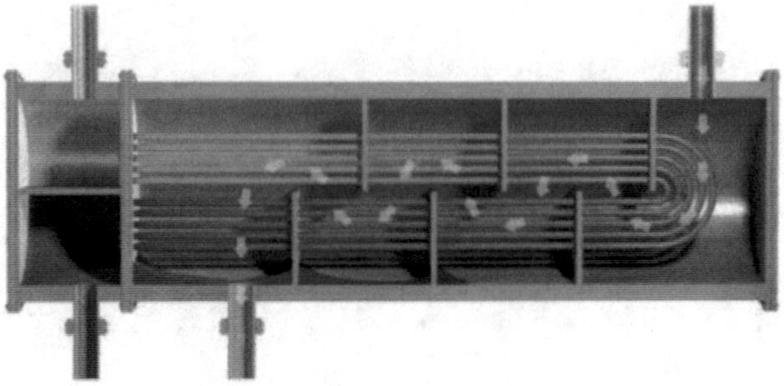

Tracerco can test heat exchangers for leaks using techniques tailored to the severity of leak:

- Using specialized tracers to suit process conditions
- Identify severity of the leak
- Minimize downtime and maintenance costs

JM Johnson Matthey

Inspiring science, enhancing life

Tracerco Diagnostics™ Leak study

Tracerco

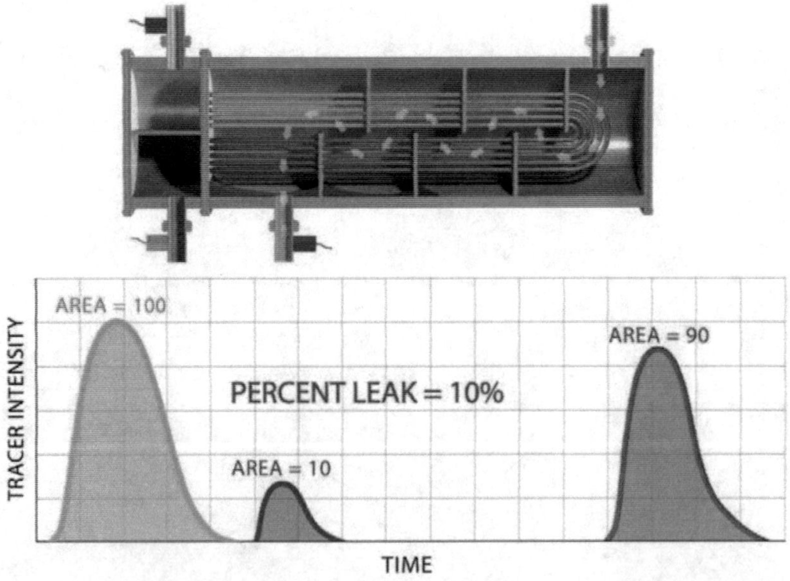

JM Johnson Matthey

Inspiring science, enhancing life

Heat Exchanger Leak Testing

Radioisotope Tracer Technique

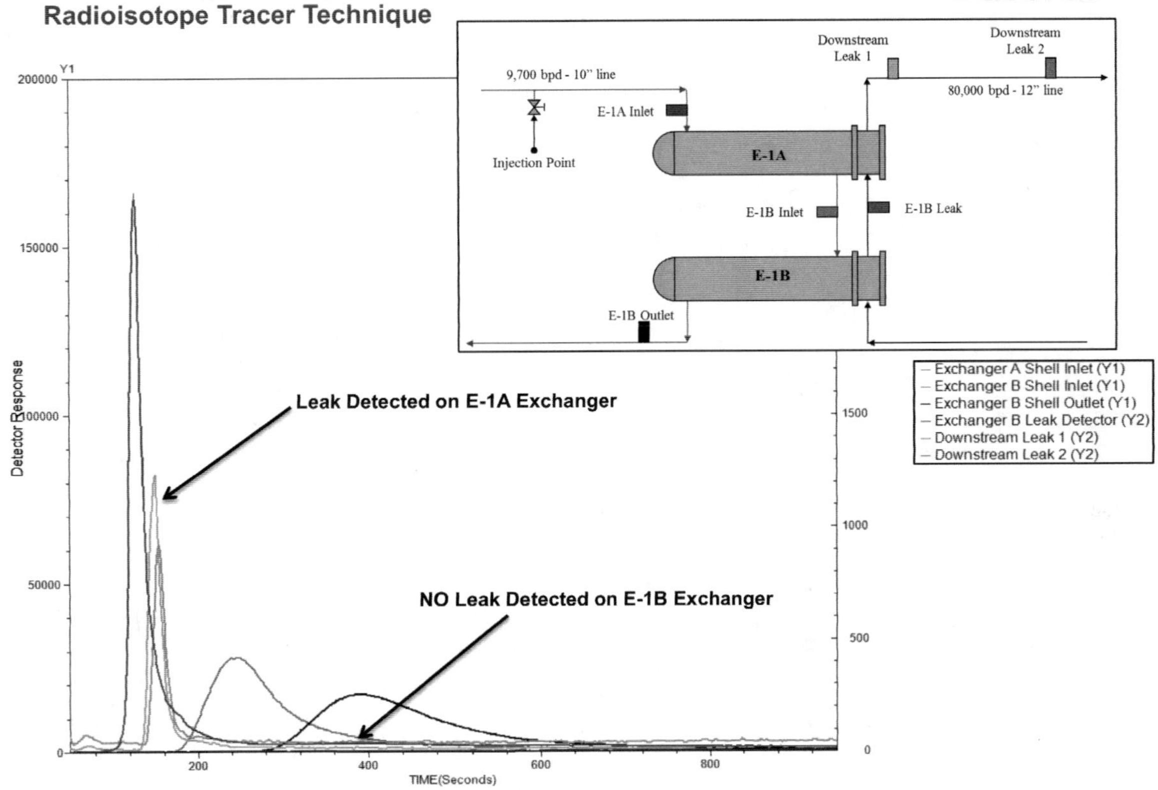

Leak Test of GOHT Feed/Effluent Exchangers

Providing Insight Onsite

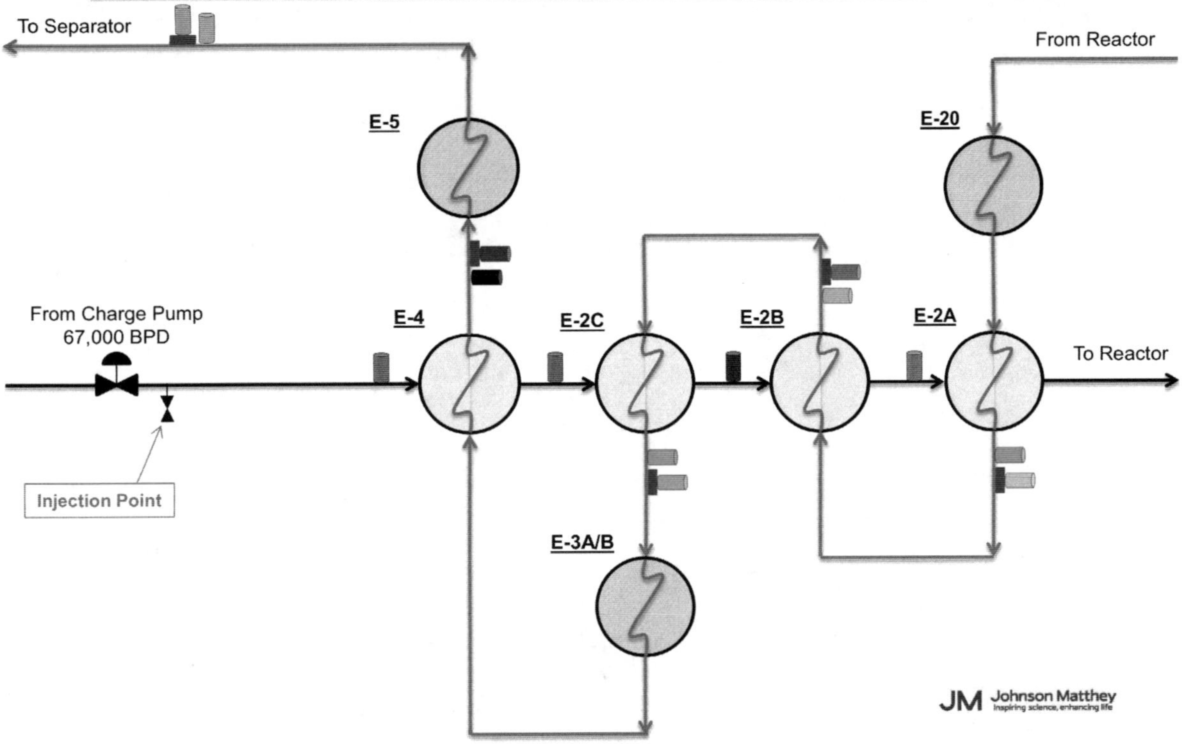

Figure 2
GOHT Feed/Effluent Exchanger Leak Test
June 2, 2016 - Injection #1

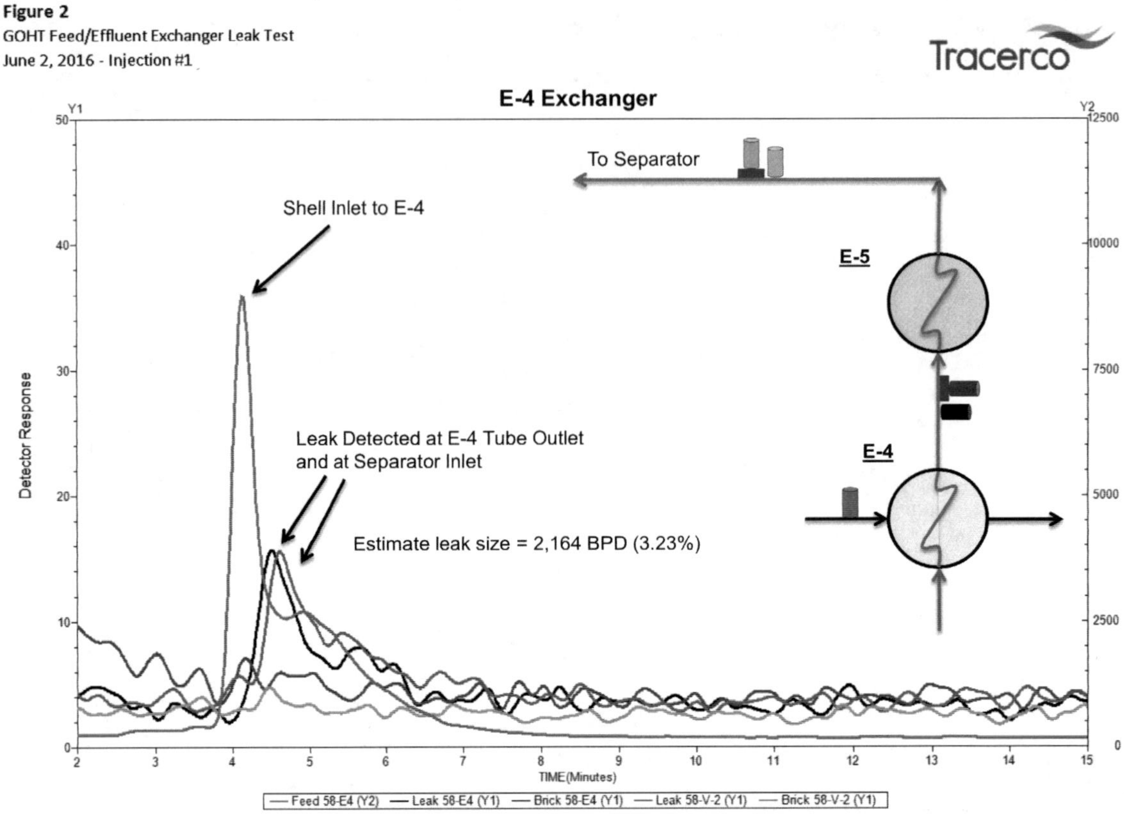

Figure 3
GOHT Feed/Effluent Exchanger Leak Test
June 2, 2016 - Injection #1

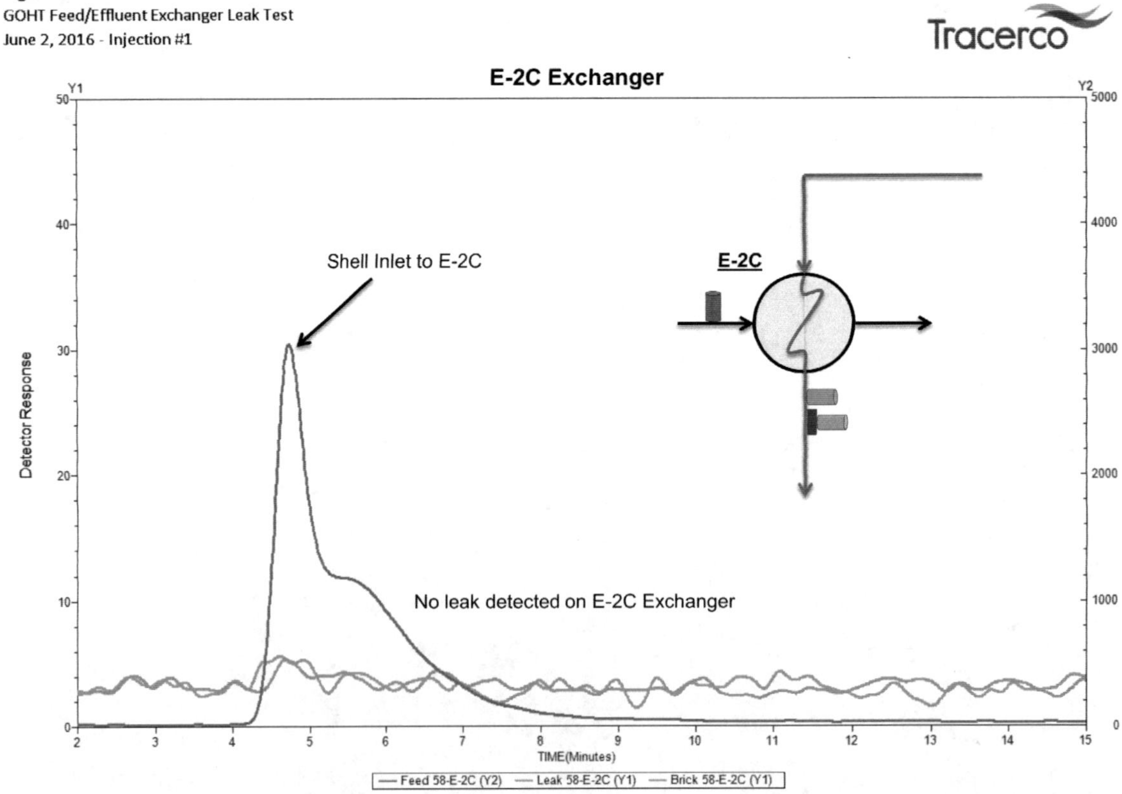

Figure 4
GOHT Feed/Effluent Exchanger Leak Test
June 2, 2016 - Injection #1

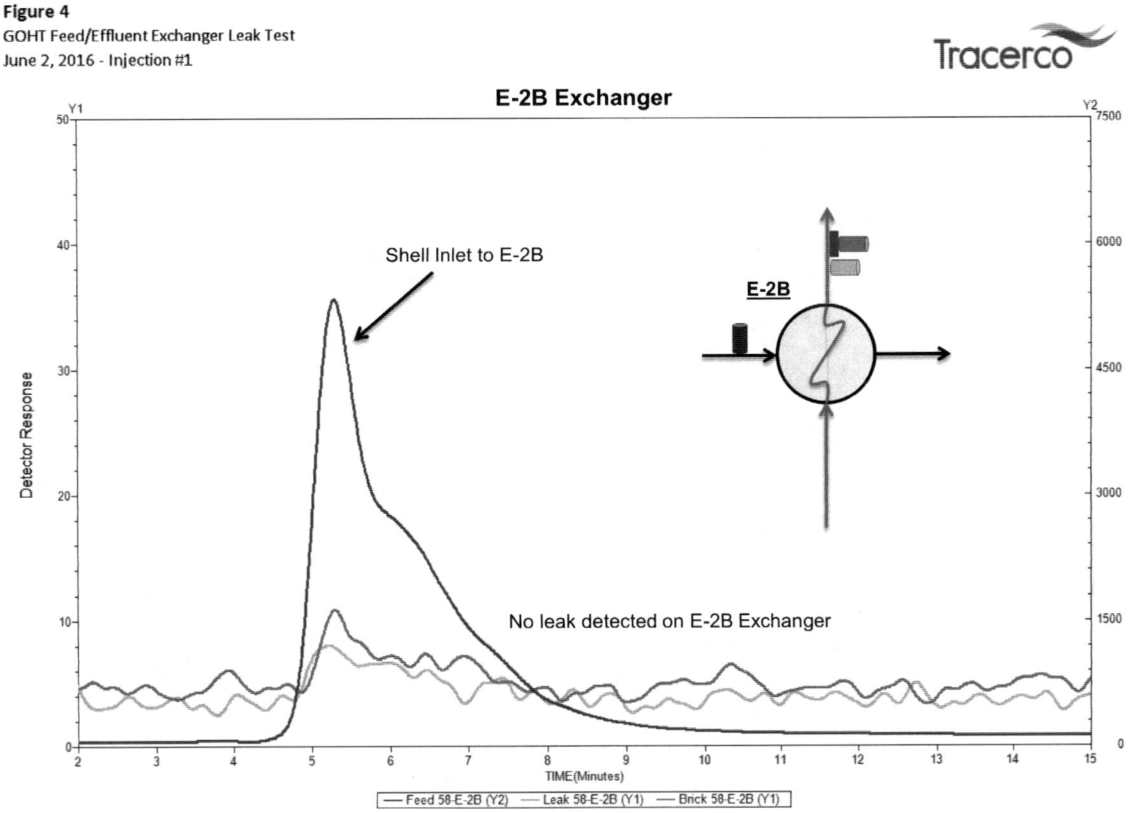

Figure 5
GOHT Feed/Effluent Exchanger Leak Test
June 2, 2016 - Injection #1

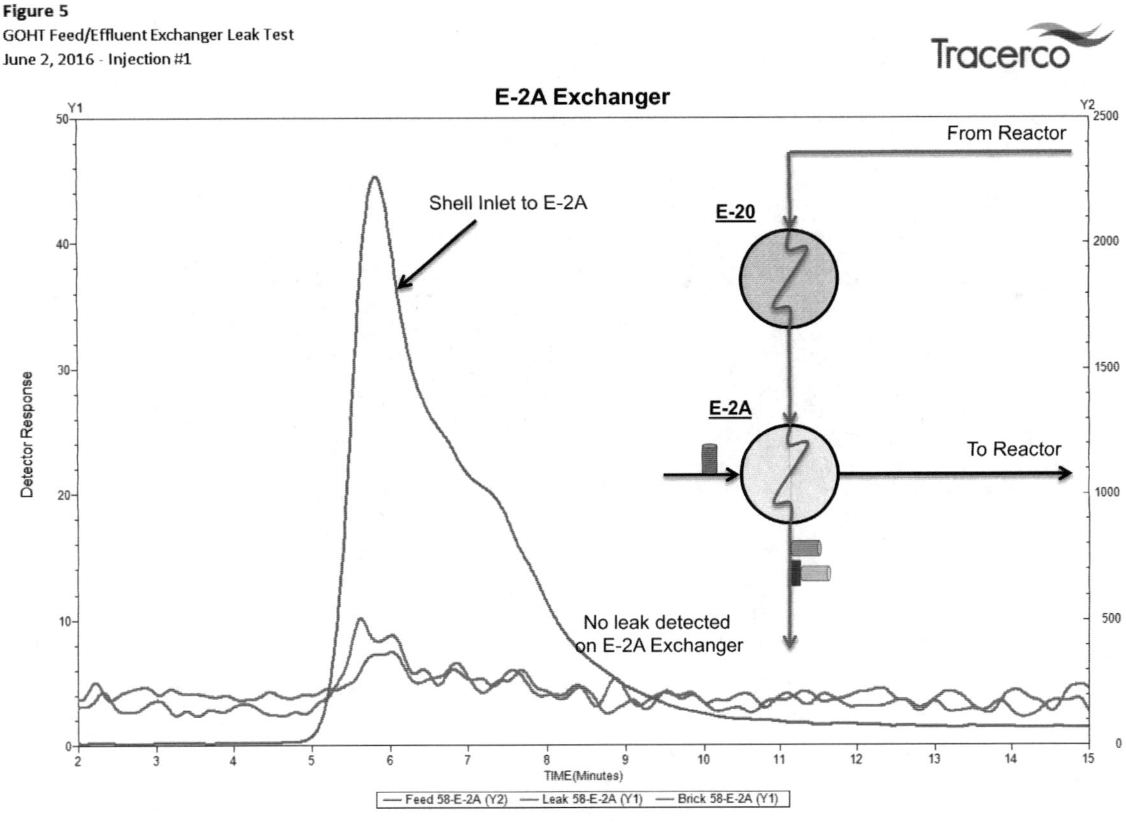

Tracer Technologies

Tracerco
Providing Insight Onsite

2. Non-Radioisotope Tracers
   - Can be in the form of liquids or gases
   - Small injection sizes (usually less than 1 kg)
   - A large library of tracers are available with a wide range of boiling points/physical characteristics
   - Cannot be "seen" with external detection. Sampling is required.
   - Detection sensitivities commonly in the ppb range.
   - Sample analysis can be performed on-site (for helium) or at a Tracerco laboratory.
   - Typical applications:
     - On-line leak tests of heat exchangers
     - Off-line leak tests of heat exchangers (helium)
     - On-line leak detection of vacuum systems (helium)
     - Residence time studies

JM Johnson Matthey
Inspiring science, enhancing life

Tracerco Diagnostics™ Leak study

Tracerco
Providing Insight Onsite

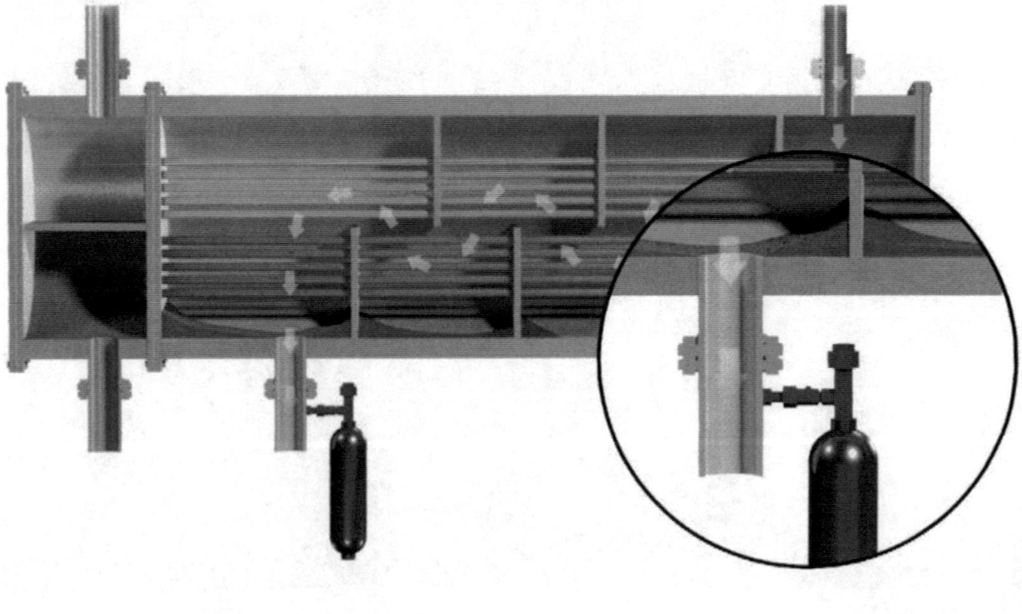

JM Johnson Matthey
Inspiring science, enhancing life

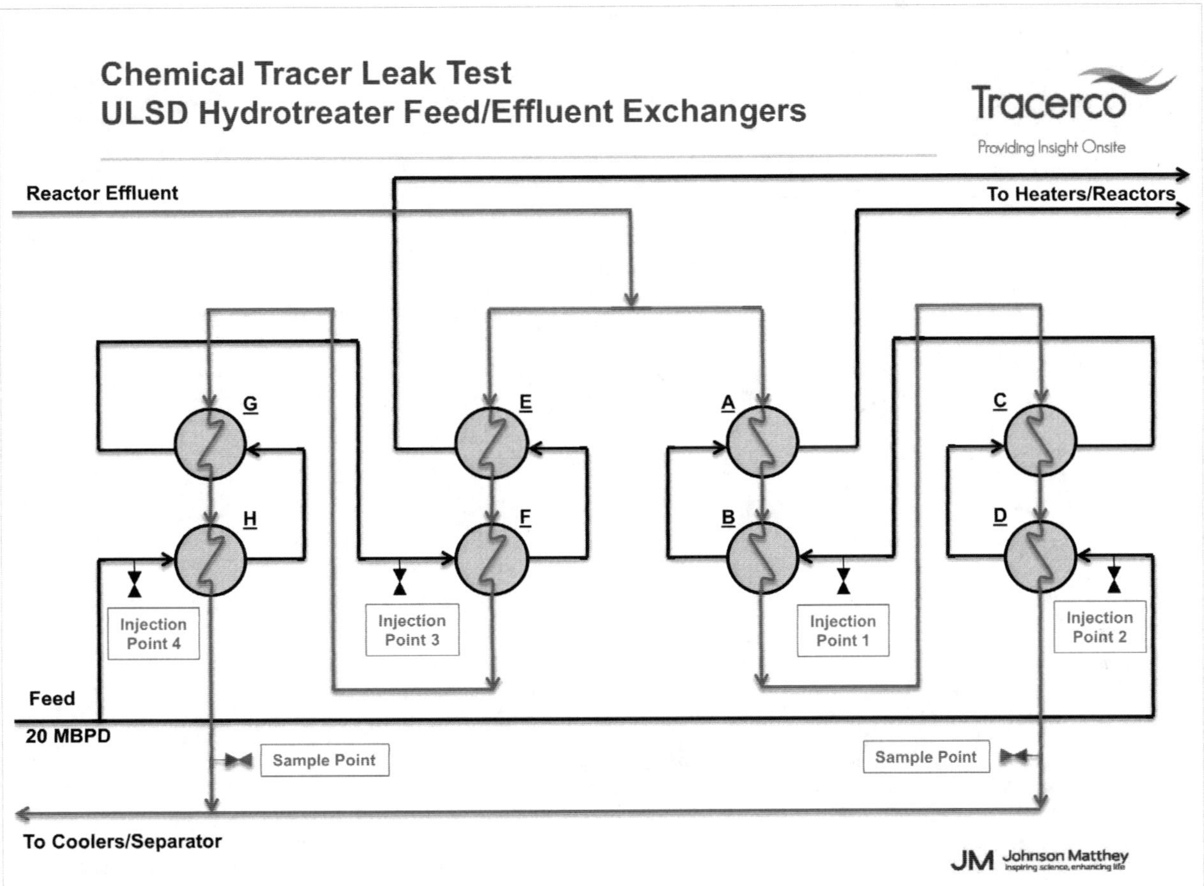
Chemical Tracer Leak Test
ULSD Hydrotreater Feed/Effluent Exchangers
Tracerco
Providing Insight Onsite
Reactor Effluent
To Heaters/Reactors
G
E
A
C
H
F
B
D
Injection Point 4
Injection Point 3
Injection Point 1
Injection Point 2
Feed
20 MBPD
Sample Point
Sample Point
To Coolers/Separator
JM Johnson Matthey

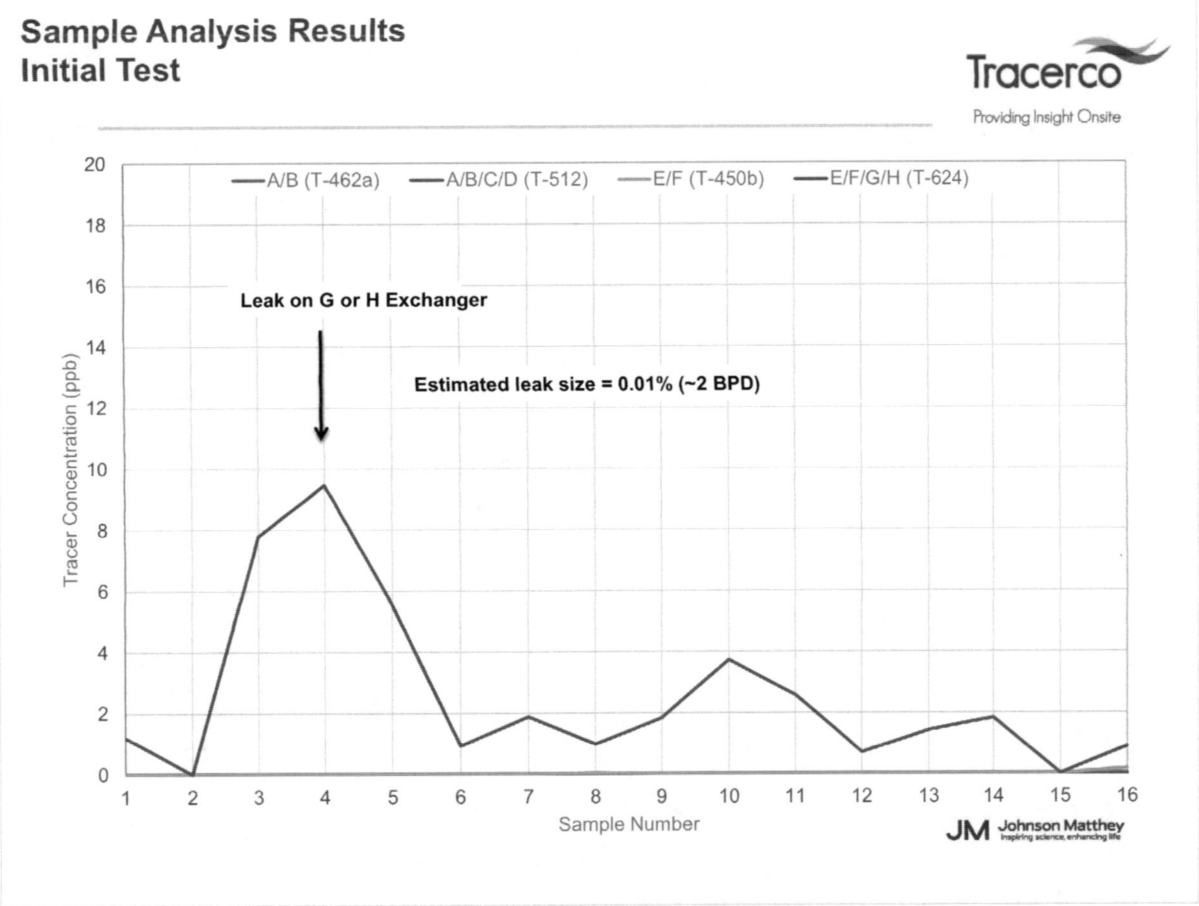
Sample Analysis Results
Initial Test
Tracerco
Providing Insight Onsite
A/B (T-462a)
A/B/C/D (T-512)
E/F (T-450b)
E/F/G/H (T-624)
Leak on G or H Exchanger
Estimated leak size = 0.01% (~2 BPD)
Tracer Concentration (ppb)
0
2
4
6
8
10
12
14
16
18
20
1
2
3
4
5
6
7
8
9
10
11
12
13
14
15
16
Sample Number
JM Johnson Matthey

Sample Analysis Results
Follow-Up Test (1 month later)

Tracerco
Providing Insight Onsite

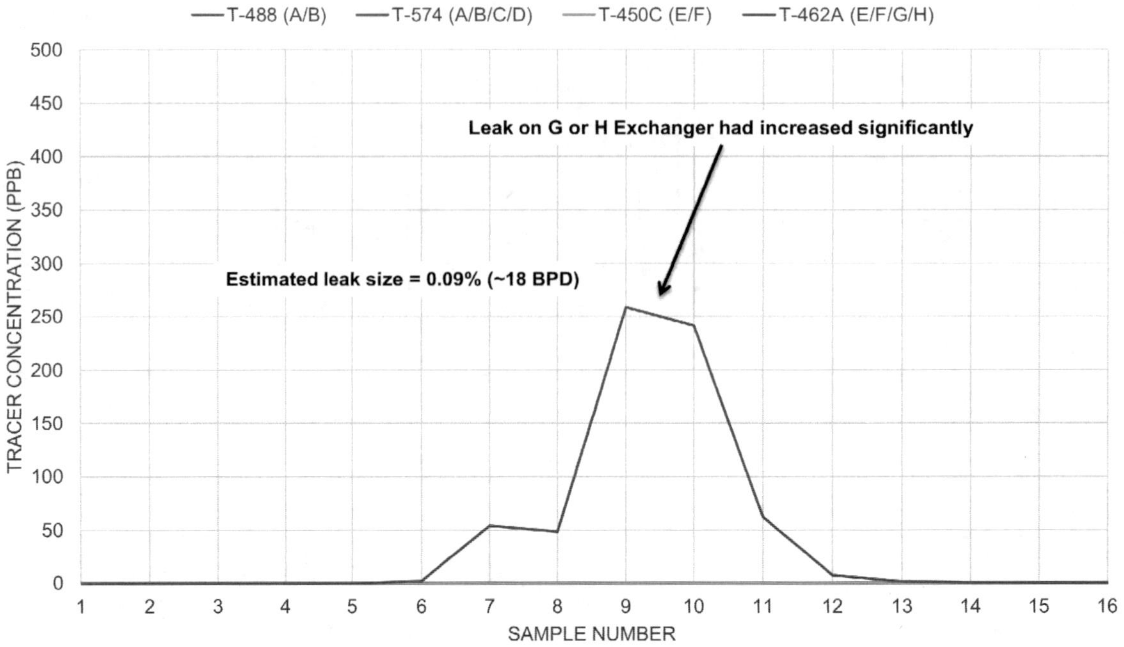

JM Johnson Matthey
Inspiring science, enhancing life

C2 Splitter Reboiler and Interboiler

Tracerco
Providing Insight Onsite

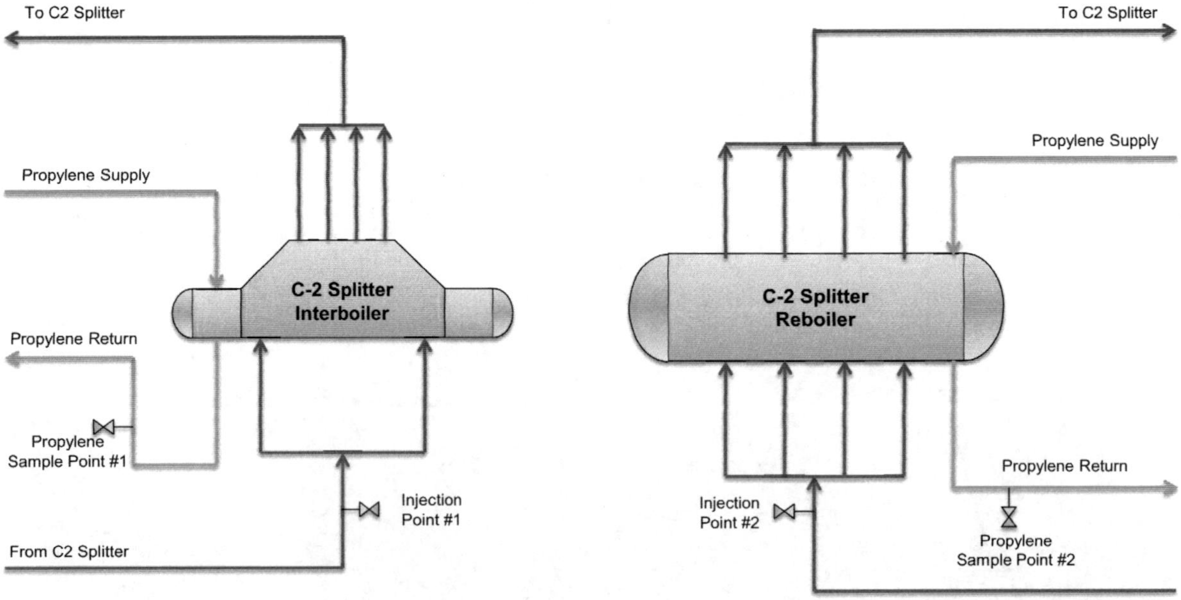

JM Johnson Matthey
Inspiring science, enhancing life

Sample Analysis Results

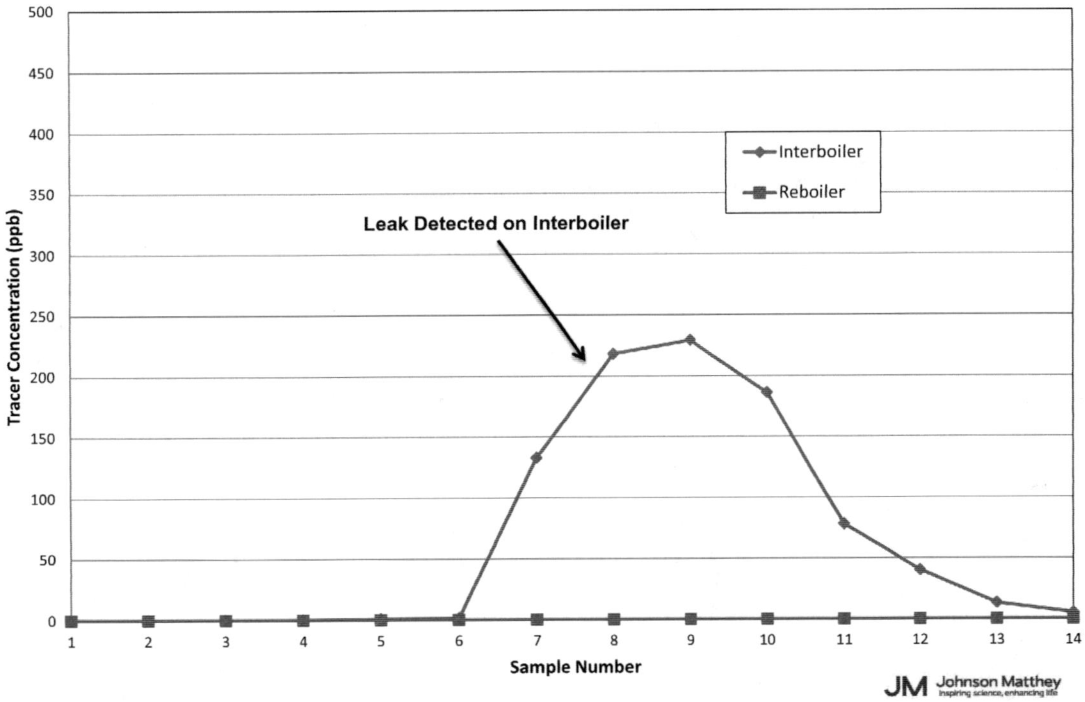

Tracerco
Providing Insight Onsite

JM Johnson Matthey

Identifying CW Leaking Into Amine

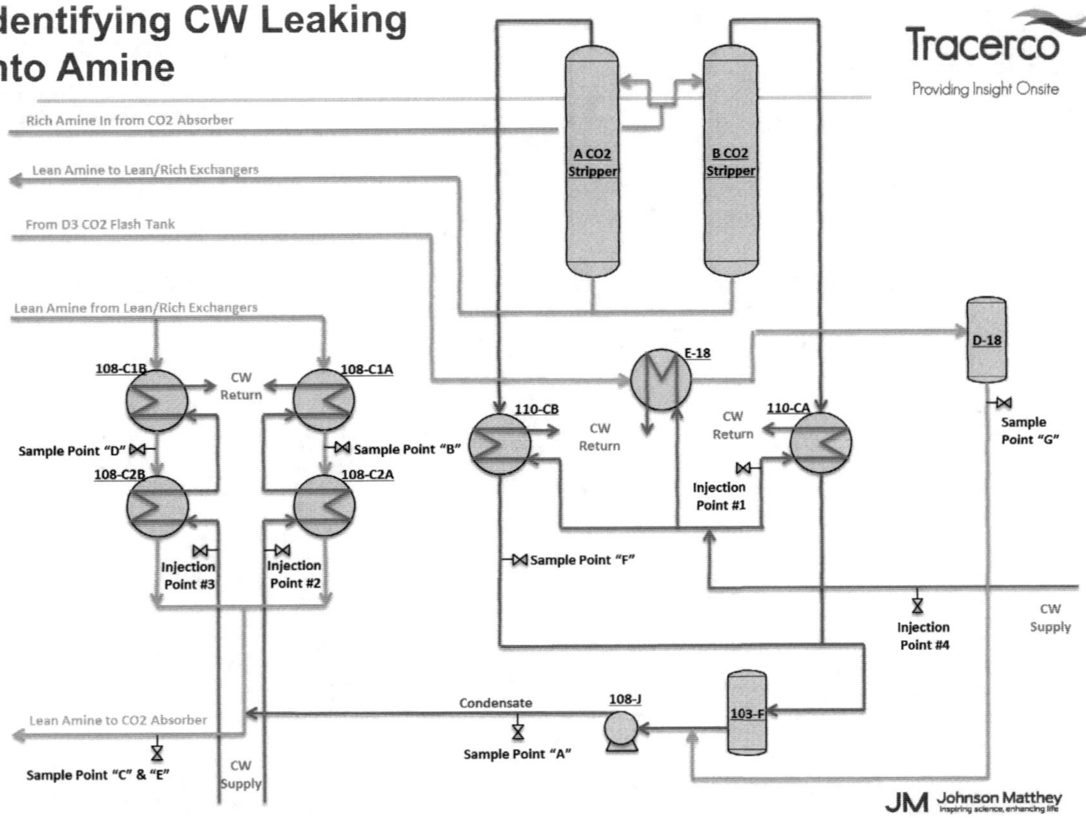

Tracerco
Providing Insight Onsite

JM Johnson Matthey

Identifying CW Leaking Into Amine

Tracerco
Providing Insight Onsite

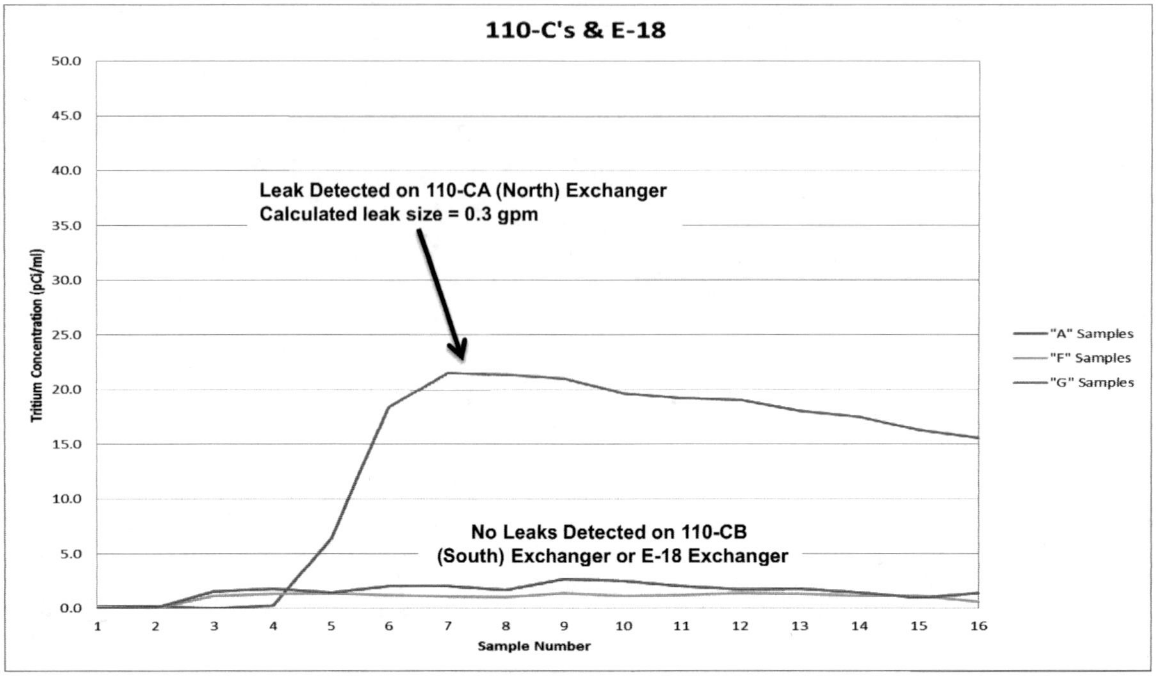

JM Johnson Matthey
Inspiring science, enhancing life

Identifying CW Leaking Into Amine

Tracerco
Providing Insight Onsite

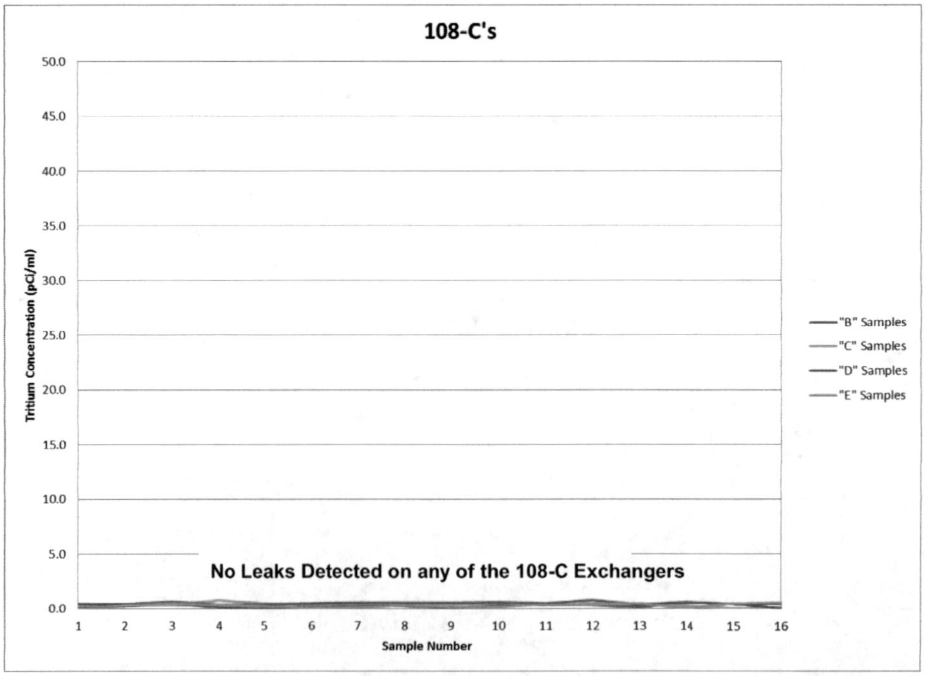

JM Johnson Matthey
Inspiring science, enhancing life

Tracerco

Providing Insight Onsite

Flow Study and Flare System Leak Studies

Johnson Matthey

Tracerco Diagnostics™ Flow study

Tracerco

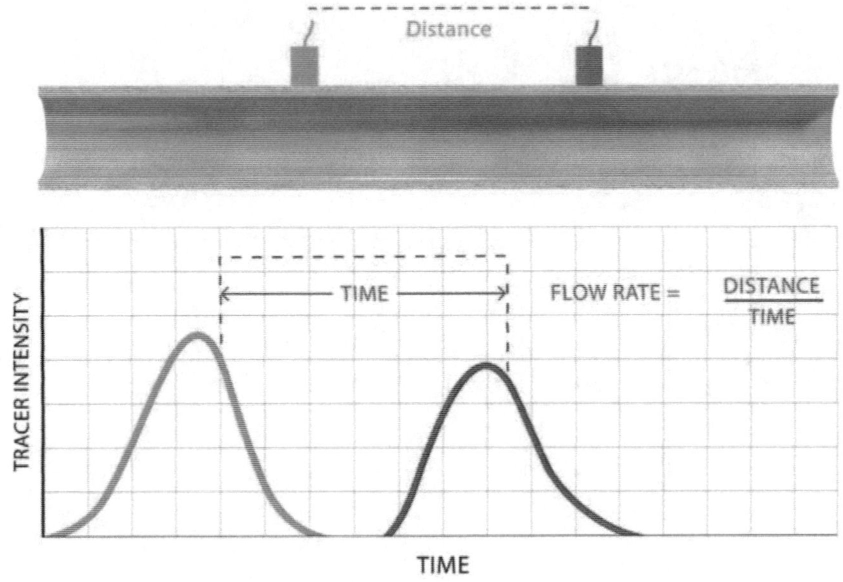

JM Johnson Matthey
Inspiring science, enhancing life

Tracerco Diagnostics™ Flare study

Tracerco
Providing Insight Onsite

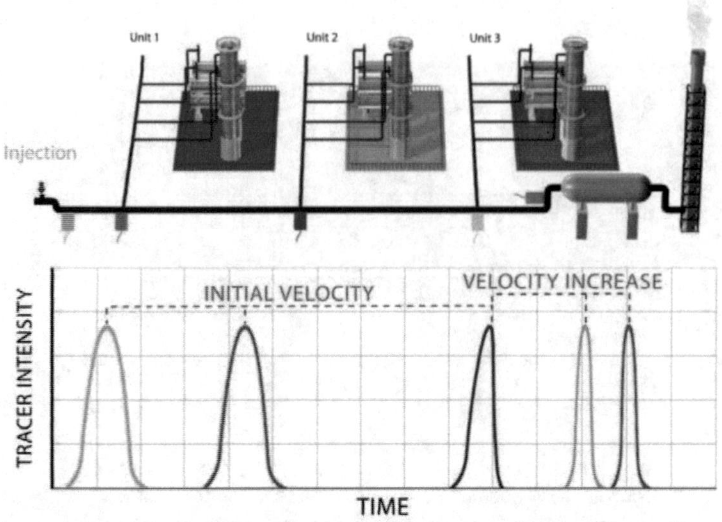

Tracer velocity increases after Unit 2 signifying additional vapor is entering the flare line here

JM Johnson Matthey
Inspiring science, enhancing life

Flare Flow Studies & PSV Leak Tests

Individual PSV Leak Test Example – Injection Upstream of PSV

Tracerco
Providing Insight Onsite

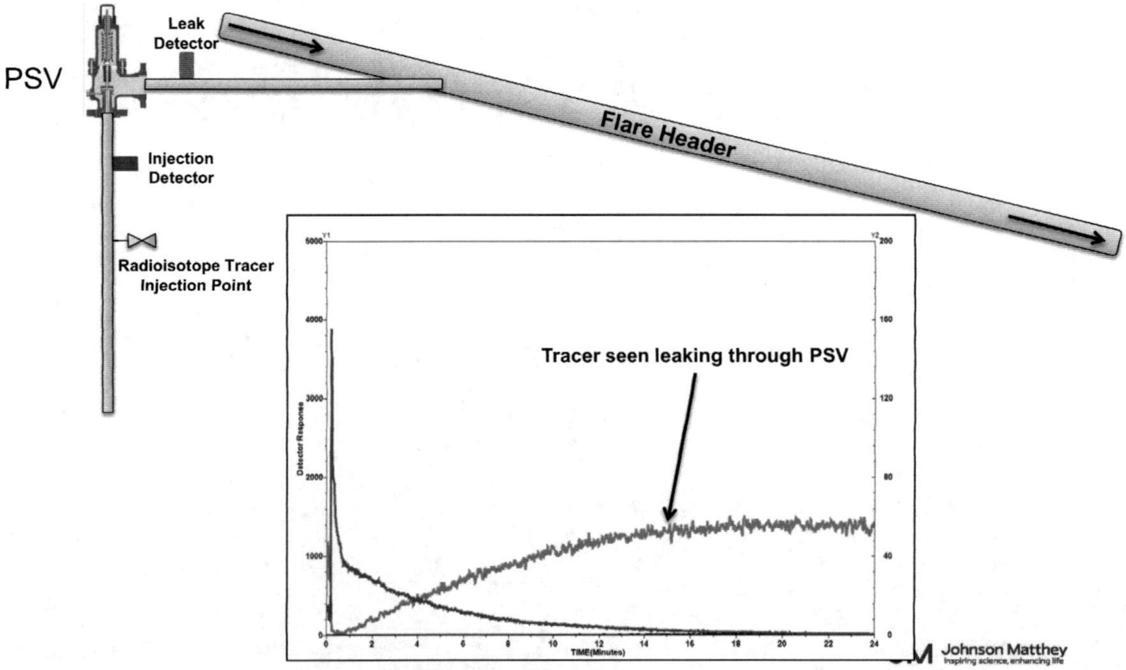

JM Johnson Matthey
Inspiring science, enhancing life

Flare Flow Studies & PSV Leak Tests

Individual PSV Leak Test Example – Injection Downstream of PSV

Tracerco
Providing Insight Onsite

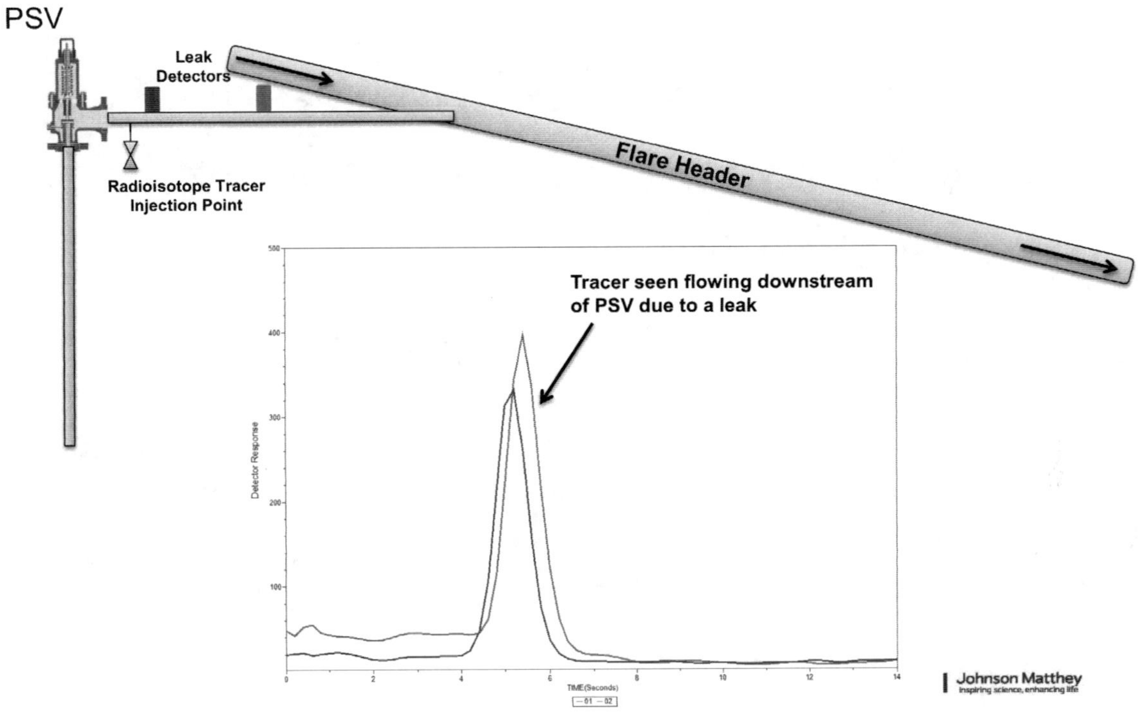

Johnson Matthey
Inspiring science, enhancing life

THANK YOU....

Tracerco
Providing Insight Onsite

Any Questions?

JM Johnson Matthey
Inspiring science, enhancing life

AFPM

An Action Plan to Improve FCC Unit Performance at the Marathon Galveston Bay Refinery

Marathon Petroleum Company LP

ALBEMARLE®

AM-17-48 Page 1 2017 Annual Meeting

AFPM

Authors

M. Skurka and D. Zalewski
Marathon Petroleum Company LP
Catlettsburg, KY

B. Dinkel and J. Becerra
Marathon Petroleum Company LP
Texas City, TX

N. Larsen
Marathon Petroleum Company LP
Findlay, OH

G. Yaluris and E. Rutan
Albemarle Corporation
Houston, TX

AM-17-48 Page 2 2017 Annual Meeting

AFPM

Outline

- Catalyst selection strategies
- MPC's catalyst selection philosophy
- MPC's catalyst testing results
- Albemarle's ACTION^{TM} catalyst technology
- Validating in unit performance

AM-17-48 Page 3 2017 Annual Meeting

AFPM

Catalyst Selection

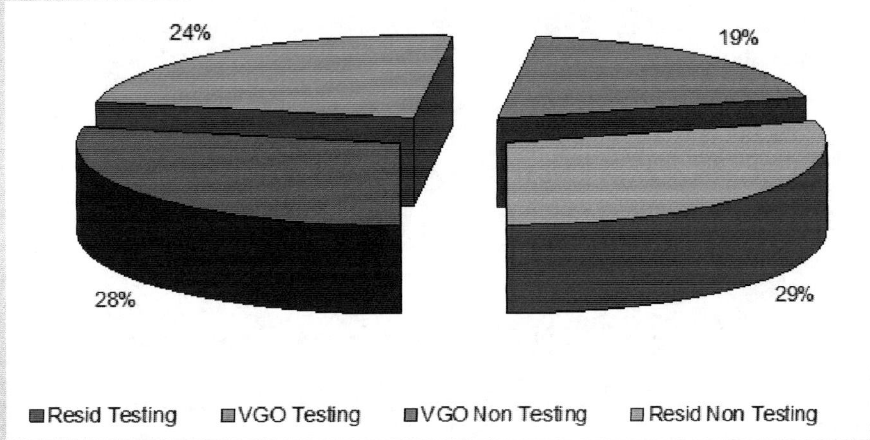

- About half of FCC unit operators select catalyst for their units by lab or pilot plant testing.

AM-17-48 Page 4 2017 Annual Meeting

Catalyst Selection Strategies

AFPM

- **Testing**
  - Catalyst selection by lab or pilot plant testing followed by in-unit trial.
  - Catalyst testing to narrow list of candidates to a few selected for in-unit testing.

- **Non-Testing**
  - In-unit trials of proposed catalysts to determine the long-term catalyst for the unit.
    - Usually one from each selected supplier
  - One catalyst selected for in-unit trial based on supplier proposals, references, technology reputation, past experience.

AFPM

Marathon Refining Analytical and Development (RAD)

GBR FCCU3 Catalyst Testing and e-cat Post Auditing

MPC Catalyst Testing Philosophy and Capability

AFPM

- Major catalyst decisions are based on competitive testing.
- Testing is customized to the commercial FCC application.
- RAD utilizes a Davison Circulating Riser (DCR) complemented by an ACE unit and internal analytical support.
- Catalyst suppliers are engaged in the process.
- Testing starts with benchmarking the test unit to the commercial base yields.

Circulating Riser Benchmarked to Commercial FCCU3 RMS Yields

AFPM

Circulating Riser Deltas from RMS at Constant Coke

	Base Unit e-cat	Lab Deactivated (no metals)	Lab Deactivated (with metals)
Conversion, wt%	0.3	0.4	-0.1
Dry Gas, wt%	0.2	0.1	0.2
Total LPG, wt%	0.3	-0.3	-0.3
Propylene, wt%	0.5	0.3	0.3
Butenes, wt%	0.3	0.2	0.2
Gasoline, wt%	-0.2	0.6	0.0
LCO, wt%	0.2	0.1	0.3
CSO, wt%	-0.6	-0.5	-0.2
Coke, wt%	0.0	0.0	0.0

FCCU3 Study Result and Follow-up Testing

AFPM

- Albemarle ACTION-536 catalyst was selected based on 2014 testing results without metals.
- Follow-up testing was done in 2015 with an updated commercial baseline.
  - ACTION-748 catalyst tested to track formulation change
  - Testing conducted with and without metals
- Intermediate unit e-cat samples were tested to track progress.
- Final post audit testing will be done with RMS materials obtained at the end of 2016.

Circulating Riser Post Audit with FCCU3 E-cat

AFPM

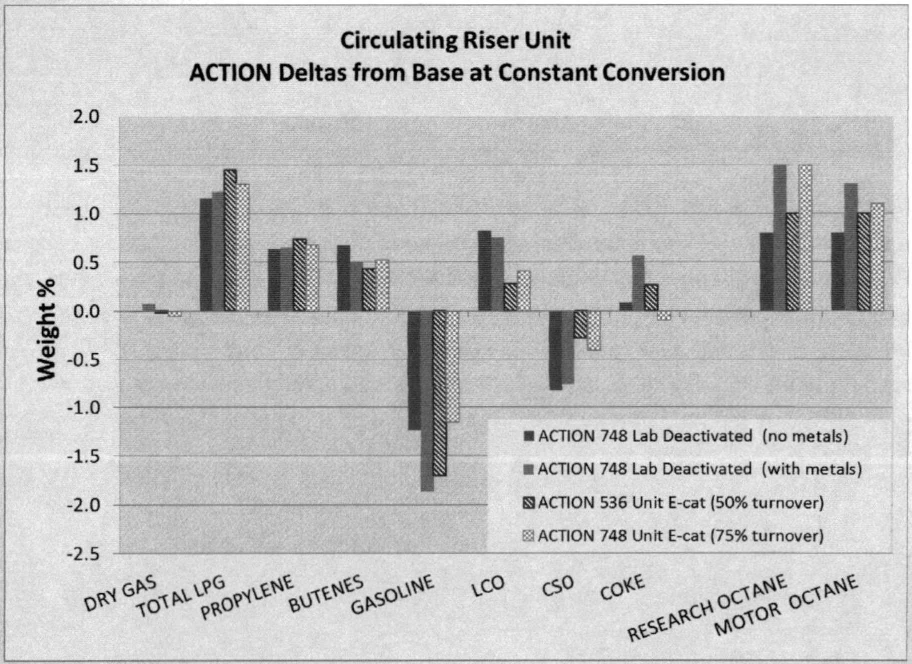

Albemarle's ACTION Catalyst Technology

AFPM

- **High Si/Al ratio zeolite technology (ADZ-100)**
  - Activity and stability
  - Tuned hydrogen transfer for maximum alky feed production (butylenes)
  - Shape selective for isomerization and octane increase
- **Matrix**
  - ADM-20 alumina matrix for high accessibility and metals resistance
    - o Maximum coke selective bottoms upgrading
    - o Enhanced strippability
    - o Decreased rate of secondary reactions
    - o Fe and Ca tolerance
- **Formulation flexibility**
  - Proven in low metals gas oil and hydrotreated feeds to heavy oil resid applications

FCCU LPG Olefins Yields

AFPM

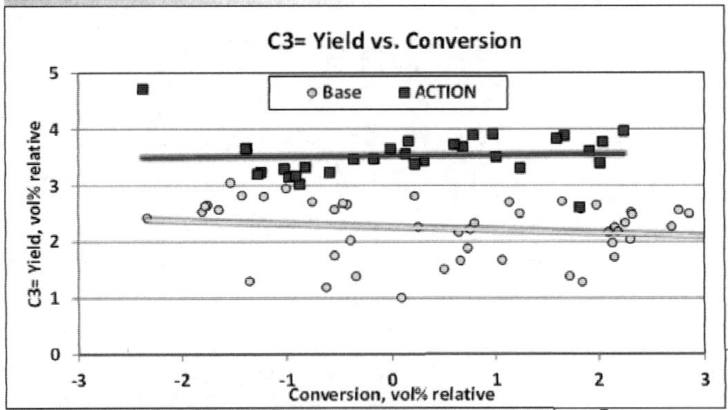

- Propylene up by 1.3 vol% and butylene up by 1.5 vol%
- Unlike ZSM-5 additives, ACTION technology increases butylenes more than propylene

- Data selected at relatively constant feed rate and riser temperature

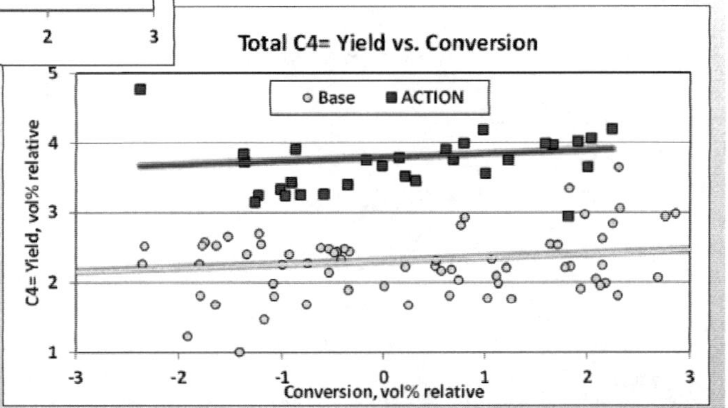

FCCU LPG Olefinicities

AFPM

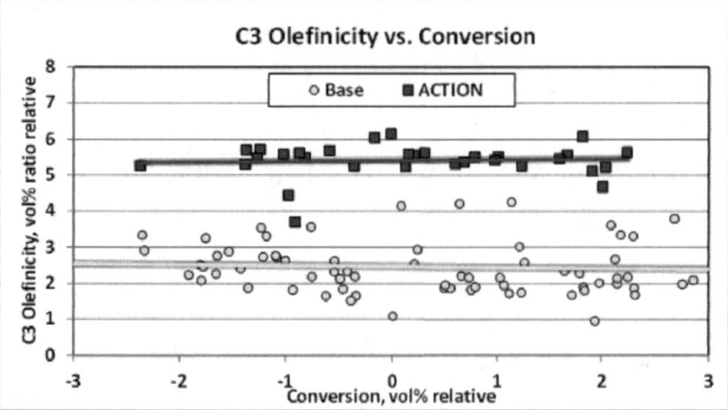

- C3 olefinicity up by 3 and C4 olefinicity up by 4.3
- Unlike ZSM-5 additives, ACTION technology increases butylenes more than propylene

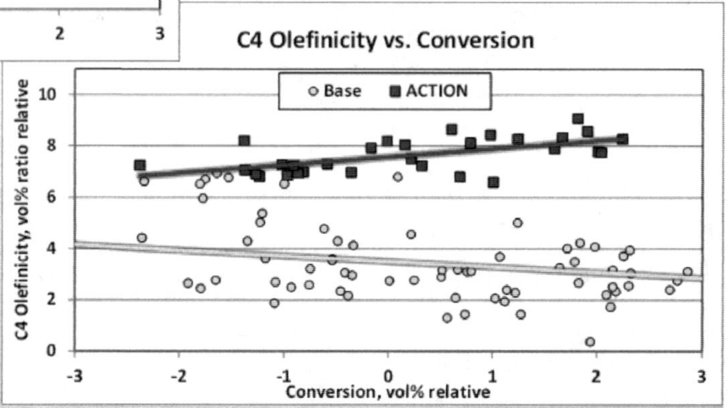

- Data selected at relatively constant feed rate and riser temperature

AM-17-48 Page 13 2017 Annual Meeting

E-cat Testing Confirms LPG Yields Improvement

AFPM

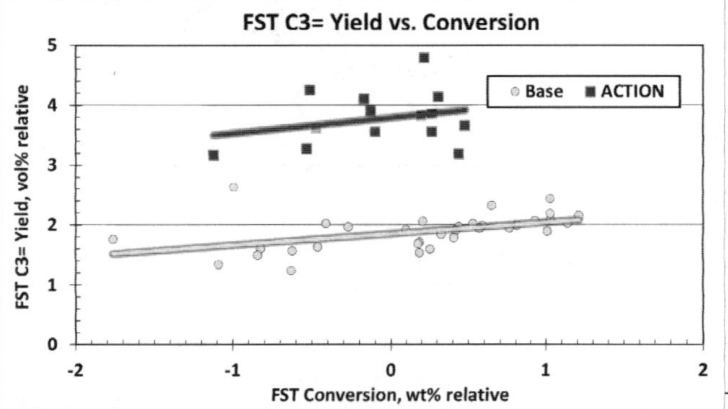

- Note that bench scale testing does not fully capture the in-unit butylene selectivity of ACTION catalysts
- Albemarle does not recommend the use of ACE to test this family of catalysts

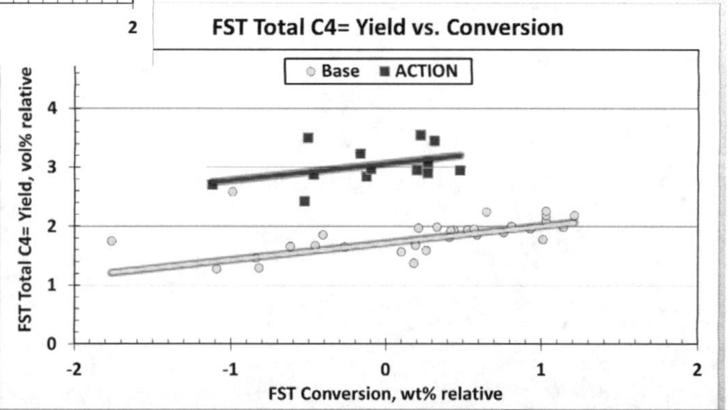

- Same periods for each catalyst as those used for unit data comparisons
- FST is a fluidized fixed bed unit Albemarle uses for e-cat testing

AM-17-48 Page 14 2017 Annual Meeting

Validating in-Unit Performance with the Use of a Kinetic Model

- MPC and Albemarle use FCC SIM
- Data periods with similar feed quality, feed rate, e-cat, riser T
- Measured data available for all product streams
- Good mass balance
- Catalyst additions after subtracting unit losses
- Cases calibrated and data confirmed to be error-free
- Averages of catalyst factors from each catalyst set of cases represent each catalyst
- FCC SIM predicts what would have happened if the ACTION catalyst had been in the unit during the base catalyst period and vice versa
- For both catalyst periods, the performance difference of the two catalysts is calculated at the same unit conditions and feed quality.

Validation of the FCC SIM Model Calibration

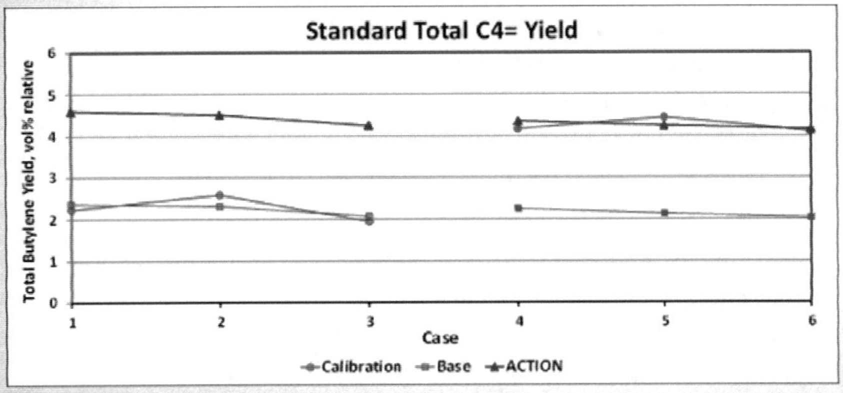

- For each catalyst period the model must be able to predict the actual unit data
- Calibration data same as base catalyst prediction during base period
- Calibration data same as ACTION prediction during ACTION period
- Catalyst performance difference should be the same in both periods

FCC SIM Comparison results

AFPM

FCC-SIM Predict Result ACTION vs Base During Base Period		
Yields	**Base Period**	**ACTION delta Yields, LV%**
C2- (wt%)	Base	-0.1
C3	Base	0.4
C3=	Base	**2.3**
iC4	Base	0.3
nC4	Base	0.1
C4=	Base	**2.4**
LPG	Base	5.4
TCN	Base	-0.7
LCO	Base	-2.0
DCO	Base	-0.6
Vol. Exp.	Base	**2.1**
Conversion	Base	**2.7**
Total Profit delta ($/bbl):		**1.66**

FCC-SIM Predict Result ACTION vs Base During ACTION Period		
Yields	**Base Period**	**ACTION delta Yields, LV%**
C2- (wt%)	Base	-0.1
C3	Base	0.4
C3=	Base	**2.2**
iC4	Base	0.3
nC4	Base	0.1
C4=	Base	**2.3**
LPG	Base	5.2
TCN	Base	-0.5
LCO	Base	-2.1
DCO	Base	-0.6
Vol Exp	Base	**2.0**
Conversion	Base	**2.7**
Total Profit delta ($/bbl):		**1.60**

- FCC SIM comparison is valid and shows margin improvement >$1.6/bbl
- Increase of 0.2 octane-barrels (not included in margin calculation)

Conclusions

AFPM

- MPC proprietary catalyst selection method.
  - Matching pilot plant yields with unit yields
  - Validating that expected unit performance improvements are achieved by using e-cat testing and rigorous unit data analysis tools
- ACTION catalyst selected for Galveston Bay FCCU.
- Post-audit analysis shows:
  - Increased alky feed production
  - Better selectivity improvement for C4 olefins than for propylene
  - $1.60/bbl margin improvement
  - Performance improvement 8% better than forecasted during the selection process
- By optimizing the unit catalyst, significant improvement in unit profitability can be achieved.

A New Experimental Method for a Fast and Reliable Determination of Saturates, Aromatics, Resins and Asphaltenes, SARA, in Crude Oils

AIChE
Southwest Process Technology Conference

Conference: Southwest Process Technology Conference

Year: 2017

Proceeding: 9th AIChE Southwest Process Technology Conference

Group: Southwest Process Technology Conference

Session: Meet The Industry Poster Reception

Time: Friday, October 13, 2017 - 3:00pm-6:00pm

Authors:

Rezaee, S., Rice University
Doherty, R. H., Rice University
Vargas, F., Rice University

Due to the complex nature of the crude oil, its compositional characterization is a very challenging task. One of the most conventional ways to characterize crude oil is SARA (Saturates, Aromatics, Resins, and Asphaltenes) analysis. This fractionation is based on the difference between the solubility and polarity of the four fractions: Saturates, Aromatics, Resins, and Asphaltenes[1].

Currently, three main methods of SARA analysis are used by the industry: Thin Layer Chromatography with Flame Ionization Detection (TLC-FID) or Iatroscan for SARA analysis, ASTM D2007 and High Pressure Liquid Chromatography (HPLC) for maltene (deasphalted crude oil) analysis. Reproducibility of the results has always been a challenge, in addition to the different limitations of each of these methods[2].

In this study, we propose a new methodology for maltene analysis that aims to overcome some of the limitations of the previous methods. After removing the asphaltene fractions, the remaining mixture, called maltenes, is fractionated in two steps; in the first step, maltenes are separated by selective adsorption into the clay and the alumina columns, and in the second step these components are desorbed and recovered.

In the adsorption step, two chromatographic columns, which are filled with clay and silica gel, are connected in series with the ability to recirculate the effluent if needed. Based on the polarity, the resins are selectively adsorbed onto the clay in the first column while the aromatics are adsorbed onto the silica gel of the second column. The effluent of the silica gel column contains the saturate fraction. The efficiency of the process is verified by UV spectroscopy.

Once the adsorption process is completed, the next step is desorption of resins and aromatics in two separate processes. The columns are washed with a relatively polar solvent. The solvent is recovered in a distillation column, and the residues contain the resins and aromatics, respectively. In the end, the solvent is evaporated by a hot plate, and all fractions are weighted.

In comparison to the ASTM method, our new technique requires between 90 to 95% less solvent and is about three to four times faster. The collected data are expected to be more reproducible because the volatile components are not separated and are measured as a part of saturates or aromatic fractions, and the process automation significantly reduces human error. Another advantage is that the equipment is fairly inexpensive and could be easily assembled in any laboratory.

References:

References:

1. Fan, T., Wang, J. & Buckley, J. S. Evaluating Crude Oils by SARA Analysis. in (Society of Petroleum Engineers, 2002).

2. Fan, T. & Buckley, J. S. Rapid and Accurate SARA Analysis of Medium Gravity Crude Oils. Energy Fuels 16, 1571â1575 (2002).

A New Proactive Scheduling Methodology for Front-end Crude Oil and Refinery Operations under Uncertainty of Shipping Delay

Southwest Process Technology Conference

Conference: Southwest Process Technology Conference

Year: 2017

Proceeding: 9th AIChE Southwest Process Technology Conference

Group: Southwest Process Technology Conference

Session: Meet The Industry Poster Reception

Time: Friday, October 13, 2017 - 3:00pm-6:00pm

Authors:

Xu, J., Lamar University
Xu, Q., Lamar University

The petroleum industry plays an important role in the economic development of the country. Advanced decision-supporting techniques such as planning and scheduling are needed to cope with challenges in crude oil supply chain management. Which have been studied and practiced for a long time, especially in the last two decades driven by increasingly intensive global competitions, more volatile feedstock and product markets, as well as stricter environmental regulations. In reality, the crude scheduling activities are highly vulnerable to disruptions brought by various uncertainties. For example, the shipping delay of crude vessels may disrupt the associated activities and cause the originally well-set schedule operations to become suboptimal or even infeasible.
In this research, a new large-scale continuous-time scheduling model has been developed for crude unloading, transferring, and processing (CUTP) systems to simulate and optimize the front-end and refinery crude-oil operations simultaneously. The scope of the scheduling problem includes crude oil unloading from vessels to storage tanks at onshore berths, transfer of crude from these tanks to charging tanks, charging crude distillation units and further processing of crude inside the refinery, which includes crude distillation, cracking, coking, reforming, hydrotreating, product blending and component recovery. The general objective of the proposed CUTP model is to maximize the total production profit under uncertainty of shipping delay; meanwhile, operation and product specifications, inventory limits, and production demands have to be satisfied.

To efficiently solve this large-scale mixed integer nonlinear programming problem without compromising problem complexity, a general mixed-integer nonlinear global optimization solver, ANTIGONE, has been employed. The profit maximization and the production flexibility maximization are generally two contradictory aspects that should be well balanced. The efficacy of the proposed scheduling model has been demonstrated by industrial-scale case studies. This work has three major contributions: (i) Simultaneous scheduling of front-end crude transfer and refinery processing with shipping delay uncertainty has been achieved. (ii) The shipping delay uncertainty has been taken into account to avoid refinery shut-down events by introducing an inventory-related time flexibility index. (iii) The relationship between the total profit and the minimum flexibility threshold is also given for a plant to manage its operational risks. It greatly increases the potential profitability and production flexibility of refineries.

Ag-Promoted Dehydroaromatization of Ethylene over ZSM-5 Catalysts

Southwest Process Technology Conference

Conference: Southwest Process Technology Conference

Year: 2017

Proceeding: 9th AIChE Southwest Process Technology Conference

Group: Southwest Process Technology Conference

Session: Meet The Industry Poster Reception

Time: Friday, October 13, 2017 - 3:00pm-6:00pm

Authors:

Zhou, Y., University of Houston
Hsieh, M. F., University of California, Santa Barbara
Thirumalai, H.
Grabow, L. C., University of Houston
Rimer, J., University of Houston

Industrial production of aromatics heavily relies on catalytic reforming in refineries.[1] Driven by the limited supply of fossil fuels and the recent shale gas revolution, it is highly desirable to develop methane- and/or light olefin-based processes to meet the growing demand for aromatics. Ethylene dehydroaromatization (DHA) over Ag-ZSM-5 is one of the promising technologies given the availability of ethylene feedstock. Ag-ZSM-5 was chosen for the ethylene DHA reaction, in part because Ag-ZSM-5 was reported to promote aromatic selectivity in the isobutane aromatization,[2] and to facilitate non-oxidative methane upgrading at moderate reaction temperatures.[3] Here, we will report the influence of Ag species in Ag-ZSM-5 on ethylene DHA without and with co-feeding methane. We observe that Ag-ZSM-5 greatly promotes aromatic selectivities by ~3-fold. In contrast, H-ZSM-5 tends to produce light olefins and other aliphatic hydrocarbons under otherwise identical conditions.[4] With methane co-feed, Ag- and H-ZSM-5 both exhibit very similar catalytic behavior compared to the conditions without methane. Interestingly, methane is activated over Ag-ZSM-5 (â¤ 5% conversion), but H-ZSM-5 is inactive for methane conversion.

To rationalize these observations, we use periodic and van der Waals corrected density functional theory (DFT) to investigate the adsorption and activation of methane and ethylene over H- and Ag-ZSM-5. The results reveal that ethylene reacts with a moderate activation barrier over H-ZSM-5, whereas the energy barrier for methane can be considered practically inert. In the case of Ag-ZSM-5, methane activation requires a lower barrier than ethylene; however, ethylene adsorbs strongly to Ag sites and blocks site accessibility for methane, consistent with experimental observations. Through systematic studies of Ag/H-ZSM-5 at different reaction conditions, we have been able to differentiate the role of Ag species in methane/ethylene activation and product selectivities. This work provides insights for designing metal-containing zeolite catalysts for non-oxidative natural gas upgrading.

[1] aH.-G. Franck, J. W. Stadelhofer, Industrial Aromatic Chemistry, Spring, 1988; bH.-J.Arpe, S. Hawkins, Industrial Organic Chemistry, 5 ed., 2010.

[2] Y. Ono, Catalysis Reviews 1992, 34, 179-226.

[3] Y. Ono, T. Baba, Physical chemistry chemical physics : PCCP 2015, 17, 15637-15654.

[4] M.-F. Hsieh, Y. Zhou, H. Thirumalai, L. C. Grabow, J. D. Rimer, ChemCatChem 2017, 9, 1675-1682.

CO and Hydrocarbon Oxidation on Pd/Ceria-Zirconia/Alumina Three-way Catalysts

Southwest Process Technology Conference

Conference: Southwest Process Technology Conference

Year: 2017

Proceeding: 9th AIChE Southwest Process Technology Conference

Group: Southwest Process Technology Conference

Session: Meet The Industry Poster Reception

Time: Friday, October 13, 2017 - 3:00pm-6:00pm

Authors:

Lang, W., University of Houston
Hubbard, C., Ford Motor Company
Laing, P., Ford Motor Company
Cheng, Y., Ford Motor Company
Harold, M., University of Houston

Three-way catalyst (TWC) light-off (LO) performance is critical in cost effectively meeting automotive emission standards. Improvements in catalyst LO should include lower precious group metal (PGM) loading, use of a lower cost PGM such as Pd, and catalyst modifications and components that minimize inhibition effects of exhaust species and enhance catalyst activity. The use of ceria-zirconia (CeO2-ZrO2) within the TWC provides the key benefit of oxygen storage, compensating for deviations from stoichiometric operating conditions by supplying oxygen for CO and hydrocarbon oxidation under rich transient conditions. Pd/Al2O3 and Pd/Ceria-Zirconia monolith catalysts were synthesized and catalyst performance experiments were conducted to study the role of Pd and support compositions and loadings on inhibition during oxidation of CO, acetylene (C2H2), and propylene (C3H6) under near-stoichiometric conditions (Î» = 1.01). Transient and steady-state bench scale reactor studies with simulated exhaust gas mixtures were conducted to extract oxidation kinetics, compare LO behavior, and develop a predictive model for understanding and optimizing catalyst performance.

CO, C2H2, and C3H6 were observed to be self-inhibiting. For the same concentrations of carbon (ppm C1 basis) in the feed mixture, C2H2 oxidation occurs at higher temperatures than C3H6 oxidation. Mutual inhibition effects were also examined for binary and ternary mixtures of CO, C2H2, and C3H6; C2H2 was observed to strongly inhibit CO and C3H6 oxidation in mixtures. For singular and mixture oxidation experiments, LO temperatures were lower using the Pd/Ceria-Zirconia catalyst compared to using Pd/Al2O3, demonstrating the promotional effects of using ceria. Steady-state kinetics measurements during individual species oxidation were conducted to find reaction orders and activation energies of oxidation. Ongoing work involves the use of a mechanistic-based kinetic model incorporated into a low-dimensional monolith reactor model.

The reactor model comprises coupled, transverse-averaged species and energy balances and aforementioned kinetics. The reactor model is used to verify and predict experimental LO behavior and trends during individual and co-oxidation using catalysts and feed mixtures of different formulations.

Anchor-Tenant Models for the Synthesis of Eco-Industrial Parks through Carbon-Hydrogen-Oxygen Symbiosis Networks

Southwest Process Technology Conference

Conference: Southwest Process Technology Conference

Year: 2017

Proceeding: 9th AIChE Southwest Process Technology Conference

Group: Southwest Process Technology Conference

Session: Meet The Industry Poster Reception

Time: Friday, October 13, 2017 - 3:00pm-6:00pm

Authors:

Topolski, K., Texas A&M University
Noureldin, M., Dow Chemical
Eljack, F. T., Qatar University
El-Halwagi, M., Texas A&M University

An eco-industrial park (EIP) is a network of entities who choose to interact by exchanging byproducts, wastes and/or energy to gain benefits greater than unintegrated operations. The concept is particularly attractive for developing industrial clusters and integrating processing facilities within an industrial city.

Recently, the concept of Carbon-Hydrogen-Oxygen Symbiosis Network (CHOSYN) has been introduced [1] for the integration of hydrocarbon processing facilities. The key idea of CHOSYN is to use atomic-level information to set targets for the integration opportunities and to employ a multi-scale systems approach to devising macroscopic strategies to attain the atomic-based benchmarks. CHOSYN takes advantage of the common ground of the chemical industry, utilizing streams containing carbon, oxygen and hydrogen atoms, to create synergism among multiple entities. Optimization approaches and shortcut methods have been developed to synthesize implementation alternatives of a mass integrated network that reduces raw material usage and waste disposal while offering significant economic benefits [1, 2].

In synthesizing new CHOSYNs or retrofitting an existing system, it is important to account for the various relationships among the EIP participants. The objective of this paper is to adopt the Anchor-Tenant model in the synthesis of CHOSYNs. Anchors are first invited as the key participants in the EIP. âTenantsâ are potential plants that could be developed and integrated with the existing âAnchor(s)â thus creating a genesis of an EIP. A multi-scale optimization approach is developed to identify and screen the tenants and to determine performance benchmarks for individual plants and for the whole EIP. Different cooperative and non-cooperative scenarios are examined. A case study is developed and solved to demonstrate the key theoretical concepts and the applicability of the new approach.

References

[1] Noureldin, M.M.B. and M.M. El-Halwagi, Synthesis of C-H-O Symbiosis Networks. AIChE Journal, 2015. 61(4), 1242-1262.

[2] El-Halwagi, M.M., âA Shortcut Approach to the Multi-Scale Atomic Targeting and Design of C-H-O Symbiosis Networksâ, Process Integration and Optimization for Sustainability (DOI: 10.1007/s41660-016-0001-y, 2017)

Chemical Kinetics Mechanism Modeling of Sour Gas Combustion

Southwest Process Technology Conference

Conference: Southwest Process Technology Conference

Year: 2017

Proceeding: 9th AIChE Southwest Process Technology Conference

Group: Southwest Process Technology Conference

Session: Meet The Industry Poster Reception

Time: Friday, October 13, 2017 - 3:00pm-6:00pm

Authors:

Gai, H., Lamar University

Sour gas combustion is very important for gas processing plants and refineries. Combustion behavior of sour gas is different from other hydrocarbon gases due to the lower heat capacity of sulfur containing components and different radical pool, etc. However, researches on how to combust sour gas stably, effectively, while producing minimal emission is still a big problem. Chemical kinetic modeling provides us an efficient way to assess to this approach. The reaction mechanism is a combination of exciting mechanism of oxidation of C1-C3 hydrocarbonsand H_2S. Optimization work is done to compensate the radical pool changes and interaction between C containing and S containing component. The optimized mechanism was validated against real experiment data to achieve a reliable overall mechanism.

Key Words: Sour gas, Kinetics mechanism, Combustion, Validation

Density Functional Theory (DFT) Based Machine & Deep Learning Strategies for Rational Screening and Directed Design of Catalytic Promoter Materials and Supported Nanoparticle Interfaces for Renewable Fuel Applications

Southwest Process Technology Conference

Conference: Southwest Process Technology Conference

Year: 2017

Proceeding: 9th AIChE Southwest Process Technology Conference

Group: Southwest Process Technology Conference

Session: Meet The Industry Poster Reception

Time: Friday, October 13, 2017 - 3:00pm-6:00pm

Authors:

Kasiraju, S., University of Houston
Grabow, L. C., University of Houston

The high oxygen content (35-40 wt. %) of bio-oil produced from fast pyrolysis of biomass reduces its heating value and stability, and limits its subsequent use as a transportation fuel. Bio-oil may be upgraded to renewable bio-fuels via catalytic hydrodeoxygenation (HDO) and MoO_3 is a promising catalyst candidate for this reaction with high reactivity and selectivity. [1] HDO requires the initial creation of an oxygen vacancy site where the feed molecule can adsorb. However, the oxygen vacancy formation on MoO_3(010) is slow (high activation barrier) and highly endothermic. We investigated transition metal promotion (Fe, Co, Ni, Cu and Zn) on MoO_3 and determined that the transition metal promotion facilitates oxygen vacancy formation. Therefore, we expanded our search for the promoters across all transition, alkali and alkaline earth metals based on machine learning techniques. With this approach, we aim to predict ab-initio/experimental reactivity trends, by using just the intrinsic physical properties of promoter materials used in the screening study.

Previous experimental work observed the presence of molybdenum carbide or an oxy-carbide phase along with the oxide with potential implications to the HDO mechanistic chemistry and overall activity.[2] To understand the individual role of the oxide and the carbide phase for HDO chemistry, molybdenum carbide supported on the oxide was modeled based on earlier HRTEM characterization.[3] A novel two-step approach was implemented using Deep Learning techniques such as artificial neutral networks (ANN) to account for the absence of detailed structural and phase composition information, and also to reduce the computational tediousness of sifting through all possible combinations of nano-particle support interfaces using ab-initio DFT calculations. Preliminary results suggest a drastic reduction in the total computational time required to predict the structure and thermodynamic feasibility of the interface with this approach. Oxygen vacancy formation, and H_2 spillover at the interface were also investigated.

The ability to rapidly screen promoters for MoO_3 and rationally design mixed oxy-carbide interfaces will ultimately lead the way to rationally designing novel HDO catalysts for the commercial upgrade of biomass to chemicals and fuels.

(1) Prasomsri, T.; Nimmanwudipong, T.; Román-Leshkov, Y. Energy Environ. Sci. 2013, 6 (6), 1732.
(2) Wang, H.; Liu, S.; Smith, K. J. Energy & Fuels 2016, 30 (7), 6039.
(3) Delporte, P.; Meunier, F.; Pham-Huu, C.; Vennegues, P.; Ledoux, M. J.; Guille, J. Catal. Today 1995, 23 (3), 251.

Deposition Behavior of the Asphaltene-Stabilized Emulsion and its Interfacial Rheology at Oil-Brine Interface

Southwest Process Technology Conference

Conference: Southwest Process Technology Conference

Year: 2017

Proceeding: 9th AIChE Southwest Process Technology Conference

Group: Southwest Process Technology Conference

Session: Meet The Industry Poster Reception

Time: Friday, October 13, 2017 - 3:00pm-6:00pm

Authors:

Lin, Y. J., Rice University
He, P., Rice University
Zhang, Z., Rice University
Kuci, S., Rice University
Biswal, S. L., Rice University

Asphaltenes are polyaromatic molecules in crude oil that are known to precipitate and deposit in pipelines, leading to significant flow assurance problems. Asphaltenes are surface-active and stabilize water-in-oil emulsions by flocculating and coagulating at interfaces. Stability of these emulsions is widely studied, but their role on the deposition behavior in the multiphase flow system is unclear. In this study, we elucidate the deposition behavior of asphaltene-stabilized water/brine-in-oil emulsions in microfluidics and further use the double-wall ring (DWR) rheometer to characterize the rheological responses of the asphaltene interface. The interface formed with 1mM to 1M of NaCl and $CaCl_2$ showed different rheological responses at oil-brine interface as well as the deposition rate in microfluidic devices. Higher salinity lowered the elasticity of the interface resulting in larger emulsions and eventually altered the blockage of the microchannel. Also, various oil-to-water ratios (water concentration: 0 wt% to 20 wt%) are studied. Lower the ratio gives rise to higher deposition rate. Finally, this multiphase deposition in microchannel is characterized by confocal laser scanning to determine the composition of the deposit. This combined imaging in microfluidics and interfacial rheology platform provides an effective method to understand deposition rate of the emulsions to the interfacial rheological properties.

Design of Diesel Oxidation Catalysts with Improved Activity at Reduced Temperature Using Computational Methods

Southwest Process Technology Conference

Conference: Southwest Process Technology Conference

Year: 2017

Proceeding: 9th AIChE Southwest Process Technology Conference

Group: Southwest Process Technology Conference

Session: Meet The Industry Poster Reception

Time: Friday, October 13, 2017 - 3:00pm-6:00pm

Authors:

Song, Y., University of Houston
Grabow, L. C., University of Houston

Low temperature combustion (LTC) diesel engines have higher fuel efficiency and produce less NOx and particulate matters (PM) compared to traditional diesel engines. However, this benefit comes at the cost of higher concentrations of CO and unburned hydrocarbons (HC) emissions.1 Meanwhile the low exhaust temperature results in a reduction in the activity of commonly used diesel oxidation catalysts (DOC). Thus, improved DOC with superior activity at reduced exhaust temperatures are needed.
In this work, we use a combination of density functional theory and descriptor-based microkinetic modeling to computationally screen a large number of bimetallic alloys. This approach enables us to quickly search for promising DOC under different reaction conditions.2 Based on the computational results, we propose that better low temperature DOC can be made by the alloys of coinage metals with Pt or Pd. Compared to the commonly used DOC which are Pt and Pd alloys, these coinage metal alloys will not only achieve higher activity at the reduced temperature, but also prevent the mutual inhibition between CO and NO oxidations. Besides, manufacturing cost of the catalytic converter is reduced by using coinage metals. These promising alloys have been synthesized and their activity has been experimentally tested using temperature programmed oxidation.

In addition, we build reactor models to optimize the DOC efficiency and catalysts loading. Since under real working conditions, the exhaust CO, NO and HC are oxidized along the DOC reactor and different reaction conditions require different catalyst compositions, the overall DOC performance can be improved by designing a DOC reactor with metal concentration gradients, where the local alloy formulation is optimized for the local temperature and exhaust gas compositions. To this end, we have built and optimized a preliminary reactor model when only CO is present. The optimal catalyst for each position is identified from microkinetic models. Our results show the reactor length is reduced by 70% by using the non-uniform metal distributed DOC, which implies the use of metal catalysts is significantly decreased. Our ongoing efforts aim towards our ultimate goal of optimizing the DOC composition in the presence of CO, NO and HC. With this optimized DOC, we are hoping to lead to significant advances to meet tailpipe emission regulations at the low exhaust temperatures of a LTC engine, which will enable the widespread use of LTC engines.

References:

1. Musculus, M. P. B., Miles, P. C., and Pickett, L. M. Prog. Energy Combust. Sci. 39, 246 (2013).
2. Nørskov, J. K., Bligaard, T., Rossmeisl, J., and Christensen, C. H. Nat. Chem. 1, 37 (2009).

Development and Testing of Amine-Type Scavengers for Elimination of H2S from Liquid Sour Crudes

Conference: Southwest Process Technology Conference

Year: 2017

Proceeding: 9th AIChE Southwest Process Technology Conference

Group: Southwest Process Technology Conference

Session: Meet The Industry Poster Reception

Time: Friday, October 13, 2017 - 3:00pm-6:00pm

Authors:

Piler, K., Lamar University

Sour crudes (i.e. > 0.5% sulfur) are of low quality and cause harm to equipment, the environment, and to humans. Chemical agents, known as scavengers, are used to scrub H_2S from crude oils, which may even be necessary to conform to legal transport laws prior to the crude entering a pipeline or railcar. Scavengers are either of the regenerative or non-regenerative type, and are used depending upon whether the pH or water content of the crude stream. The non-regenerative scavengers are usually triazines containing various sidechain constituents that make the scavenger more soluble depending on the type of crude (i.e. API, pH, water content, etc.). However, the conversion of triazines with H_2S lead to byproducts which foul downstream refinery equipment, particularly heat exchangers on overhead distillation streams.

In this project, typically used scavengers have been surveyed from the open literature to reveal those absorbents that allow the most effective scavenging of H_2S at wellhead applications, depending on crude oil quality. In addition, the use of ionic liquids as replacements for regenerative amines (i.e. monoethanolamine (MEA)) was also studied. Ionic liquids offer improved absorbent qualities while being easier and more economical to recover than MEA. The chemical processing, including recovery of scavenger or end use of products, will be presented for decision-making from the wellhead to the refinery. This work serves at the confluence of chemistry and chemical engineering for hydrogen sulfide removal and seeks how and why some scavengers perform better for some crudes and environmental conditions than others with the intent to develop designer scavengers and processing conditions for the crude oil industry.

Modeling Alkane Partitioning and Phase Behavior in Non-Permeable and Permeable Slit Pores

Southwest Process Technology Conference

Conference: Southwest Process Technology Conference

Year: 2017

Proceeding: 9th AIChE Southwest Process Technology Conference

Group: Southwest Process Technology Conference

Session: Meet The Industry Poster Reception

Time: Friday, October 13, 2017 - 3:00pm-6:00pm

Authors:

Liu, J., Rice University
Chapman, W. G., Rice University

In unconventional shale reservoirs, hydrocarbon adsorption and absorption both affect reservoir fluid distribution and phase behavior, which makes the prediction of fluid thermodynamic properties (pressure/volume/temperature) increasingly challenging. It has been recognized that the nano-confinement lowers the bubble point and critical point of fluid mixtures, and solvation of hydrocarbon in kerogen is also significant. For a hydrocarbon mixture, the pore fluid composition has a large deviation from extracted bulk fluid due to different surface affinities of different components. Although sorption can be described by some macroscopic models, density functional theory (DFT) can provide detailed information of the fluid structure in the pore-wall system.

In this presentation, we apply interfacial Statistical Associating Fluid Theory (iSAFT) to investigate the fluid adsorption in a non-permeable graphite pore and both adsorption and absorption in a permeable polymer pore. We show the shifted pure and binary phase diagrams in a non-permeable 3 nm pore and the pore size dependency of pure component critical properties, where the results all have a reasonable agreement with molecular simulation. The competitive adsorption of mixtures is predicted and suggests different dominating factors, enthalpic or entropic, determine the pore selectivity depending on the bulk conditions. For a permeable pore wall, we show the adsorbed density in the pore and the penetrated density in polymer wall, and meanwhile the sorption isotherms obtained from density distribution. We compare the enhanced sorption of fluid in a pore-wall system with bulk phase equilibrium solvation and discuss the phase equilibrium in different pore width and wall thickness.

Operating Procedure Analysis with Consequence Prediction by Dynamic Simulation

Southwest Process Technology Conference

Conference: Southwest Process Technology Conference

Year: 2017

Proceeding: 9th AIChE Southwest Process Technology Conference

Group: Southwest Process Technology Conference

Session: Meet The Industry Poster Reception

Time: Friday, October 13, 2017 - 3:00pm-6:00pm

Authors:

Fang, Y., Lamar University
Richmond, P. C., Lamar University

Faults due to human errors cost the petrochemical industry billions of dollars every year. In this work, we propose an operating procedure analysis (OPA) to estimate the probabilities of faults due to human error while following standard operating procedures (SOP) such as maintenance procedures and equipment switching procedures. These procedures occur frequently during the operation of continuous chemical process plants. The proposed OPA is similar to event tree analysis (ETA) with branches and events, but has been specifically designed for analyzing operating procedures. Transitions involving more frequent or more complicated operator procedures are likely at increased risk of faults.

In this work we demonstrated the OPA methodology for both a high frequency daily dryer switching scenario and a relatively low frequency monthly reflux pump switching scenario. Both equipment switching scenarios are commonly accomplished manually by operators. First, SOPs were developed and potential operator errors were determined for each procedural step. Then, dynamic simulation was applied to determine the consequences of each error. Finally, the frequency of consequences resulting from those scenarios was calculated. An Aspen HYSYS dynamic simulation was developed to determine the specific consequence of each fault.

We found that the probability for a typical ethylene plant to flare during a dryer switching procedure is more than 90% per year. These results emphasize the importance of careful analysis of manual operating procedures for safe and environmentally friendly operation. Due to the high probability of flaring, the human factor and SOPs should be evaluated for changes that could reduce the procedure abnormalities. One alternative is to emphasize operator training and safety education; another is to redesign the procedure to reduce the probability of mistakes. Simplifying procedures would make them less susceptible to operator errors and decrease the probability of faults.

Optimization of Flare Operation

Southwest Process Technology Conference

Conference: Southwest Process Technology Conference

Year: 2017

Proceeding: 9th AIChE Southwest Process Technology Conference

Group: Southwest Process Technology Conference

Session: Meet The Industry Poster Reception

Time: Friday, October 13, 2017 - 3:00pm-6:00pm

Authors:

Wang, A., Lamar University

Flare operation is a challenging task since the flowrate, net heating value, composition of the vent gas and weather conditions vary significantly. The industry is in urgent need of a viable approach to achieve smokeless flaring that can meet the combustion/destruction efficiency standard set by the agencies, while minimize stream and makeup fuel consumption.

In this work, a novel optimization approach was developed based on flare performance models. The controlled variables include flowrate of steam/air assist and flowrate of make-up fuel. For a flaring vent with a specific combustion zone net heating value and particular composition, the most economic solution to achieve smokeless flaring and high combustion/destruction efficiency objective can be identified.

Optimized Design of Shale Gas Processing and NGL Recovery Plant under Uncertain Feed Conditions

Southwest Process Technology Conference

Conference: Southwest Process Technology Conference

Year: 2017

Proceeding: 9th AIChE Southwest Process Technology Conference

Group: Southwest Process Technology Conference

Session: Meet The Industry Poster Reception

Time: Friday, October 13, 2017 - 3:00pm-6:00pm

Authors:

Reddy Asani, R., Texas A&M University

Shale gas is an increasingly booming resource and it has been predicted to increase from 1% in 2000 to 40% in 2035 of the total US domestic gas produced. Since shale gas is both industrially economical and environmentally clean compared to oil or coal as a resource, many studies are focused on developing technologies to monetize shale gas. However, one of the key challenges in utilizing shale gas is its fluctuating flow rate and compositional behavior. The flow rate of a well dwindles over a period of time and the composition of shale gas differs from well to well in the same shale play. This provides a challenge in designing a plant of optimum size for shale gas processing and NGL recovery. In this study, this uncertainty in shale gas feed flow rate and composition is addressed while designing a shale gas processing and NGL recovery plant. First, different shale gas flow rates are chosen over a period of well life based on the average shale gas declination curve of a shale play. Second, two different process flow sheets are developed â (i) using conventional technology and (ii) using novel technology. In the novel technology, the NGL recovery section of the conventional technology is modified to accommodate novel changes such as using a divided wall column or pre-fractionated sequence to separate methane, ethane and propane. Later these process flow sheets are simulated in Aspen plus for comparing economics of different plant sizes. Furthermore, heat integration and optimization of individual units of the process flow sheets are carried out using pinch and sensitivity analyses, respectively. Lastly, the economic analysis of a plant of optimum size with constant feed flow rate over its plant life is evaluated. In this, shale gas from different wells are collected in a header and adjusted such that the shale gas flow rate is constant to the plant. The technical and environmental impacts of various cases would also be discussed.

Optimizing ZSM-11 Catalysts for Methanol-to-Hydrocarbon Reactions

Southwest Process Technology Conference

Conference: Southwest Process Technology Conference

Year: 2017

Proceeding: 9th AIChE Southwest Process Technology Conference

Group: Southwest Process Technology Conference

Session: Meet The Industry Poster Reception

Time: Friday, October 13, 2017 - 3:00pm-6:00pm

Authors:

Shen, Y., University of Houston
Le, T. T., University of Houston
Rimer, J., University of Houston

ZSM-11 (MEL) is an emerging catalyst for a wide variety of applications. The MEL framework is very similar to ZSM-5 (MFI), which is one of the most widely used zeolite catalysts in industry. Recent studies have shown that ZSM-11 has better catalytic performance in several different reactions compared to ZSM-5 [1-3], possibly due to improved intra-crystalline diffusion within the less-tortuous straight channels of the MEL topology. However, there are relatively few systematic studies of ZSM-11 synthesis, specifically with respect to tailored physicochemical properties such as size, shape, and Si/Al ratio (SAR). Here, we will present a systematic study of ZSM-11 catalyst preparation and testing for C1 applications. Through the judicious selection of synthesis parameters, we successfully developed a facile protocol to generate a library of ZSM-11 catalysts with tunable size, shape and SAR. Further optimization of the synthesis process was focused on reducing the use of organic structure-directing agents (OSDAs), which significantly reduced the cost of synthesis. These methods for tailoring zeolite crystallization will be discussed along with catalytic studies using methanol-to-hydrocarbon (MTH) reactions. Collectively, our findings are part of a broader initiative to establish quantitative structure-performance relationships as a platform for optimizing zeolite catalysts.

References

[1] Kustova, M. Y.; Rasmussen, S. B.; Kustov, A. L.; Christensen, C. H., Applied Catalysis B- Environmental 2006, 67 (1-2), 60-67.

[2] Zhang, L.; Liu, H.; Li, X.; Xie, S.; Wang, Y.; Xin, W.; Liu, S.; Xu, L., Fuel Processing

Technology 2010, 91 (5), 449-455.

[3] Bleken, F.; Skistad, W.; Barbera, K.; Kustova, M.; Bordiga, S.; Beato, P.; Lillerud, K. P.; Svelle, S.; Olsbye, U., Physical Chemistry Chemical Physics 2011, 13 (7), 2539-2549.

Techno-Economic Analysis of Alternative Pathways for Isopropanol Production

Warissara Panjapakkul, Mahmoud M. El-Halwagi
Artie McFerrin Department of Chemical Engineering, Texas A&M University
ARTIE MCFERRIN DEPARTMENT OF CHEMICAL ENGINEERING TEXAS A&M UNIVERSITY
TEXAS A&M UNIVERSITY

Introduction

- Isopropanol has growing demands for households, cosmetic and personal care products, pharmaceuticals, veterinary, electronics, and automotive industry [1,2].
- Traditionally, isopropanol was manufactured largely from propylene.
- Shale gas boom caused a shift from naphtha cracking to ethane cracking which led to propylene shortages. Therefore, it is imperative to find alternative pathways to produce propylene derivatives for a long-term cost-effective and sustainable [3].

Objectives

To assess alternative pathways from different feedstocks for isopropanol production and to select recommended pathways through a techno-economic analysis.

Methods

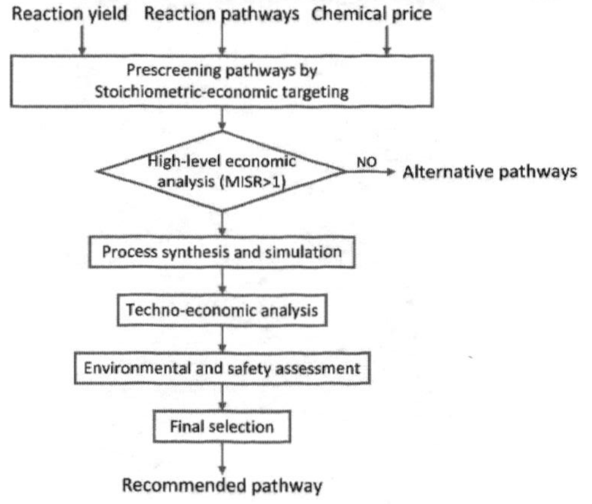

Results

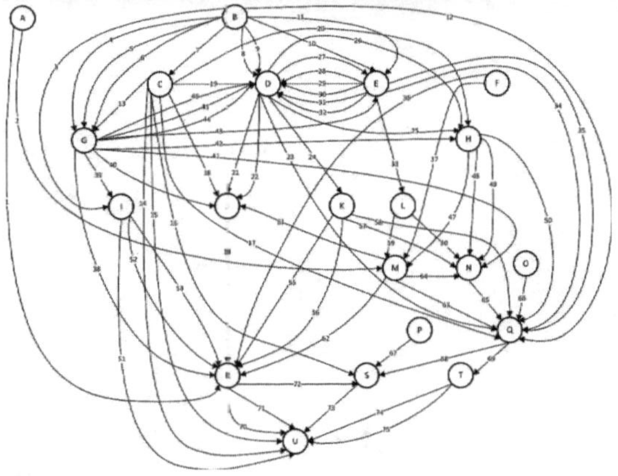

Chemical species:
A = Naphtha, B = Biomass, C= Sugar/Carbohydrate, D = Syngas, E = Methane, F = Ethane, G = Glycerol, H = Ethanol, I = Propane, J = Propanol, K = Methanol, L = Acetylene, M = Ethylene, N = Acetaldehyde, O = Butane, P = Benzene, Q = Acetic acid, R = Propylene, S = Acetone, T = Isopropyl acetate, U = Isopropanol

Figure 1. Generation of alternatives

Stoichio-nomic indicator:

Metric for Inspecting Sales and Reactants (MISR) is defined as [4]:

$$MISR = \frac{\sum_{p=1}^{N_{products}} F_p \times S_p}{\sum_{r=1}^{N_{products}} F_r \times C_r}$$

Where F_p is the annual production rate of product p, S_p is the selling price of product p, F_r is the annual feed rate of reactant r and C_r is the purchase cost of reactant r.

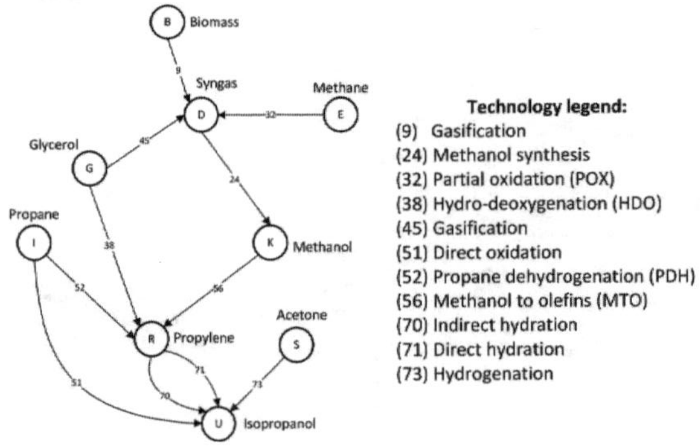

Figure 2. The synthesized pathways after the prescreening step

In this research, the number of synthesis is limited to two maximum steps per route.

Table 1. Possible pathways result from prescreening step

Nodes	Feed	Product	Route	MISR
G → R → U	Glycerol	Isopropanol	Hydro-deoxygenation, Hydration	4.67
I → R → U	Propane		Dehydrogenation (PDH), Hydration	3.54
K → R → U	Methanol		Methanol to olefins (MTO), Hydration	2.51
I → U	Propane		Direct oxidation	2.13
S → U	Acetone		Hydrogenation	1.74

Ongoing work

- Process synthesis and simulation of promising pathways
- Techno-economic analysis for pathways selection
- Environmental and safety assessment

References

1. *Global Isopropanol Market Research Report*. MicroMarketMonitor.
2. *Propanol (Isopropanol & N-Propanol) Market By Applications (Solvents, Chemical Intermediates, Pharmaceuticals & Others) & Geography - Trends And Forecasts To 2018*. 2013, marketsandmarkets.com.
3. *Bell, A.T., et al., The Changing Landscape of Hydrocarbon Feedstocks for Chemical Production: Implications for Catalysis: Proceedings of a Workshop. 2016, National Academies of Sciences, Engineering, and Medicine, Washington, DC (United States).*
4. *El-Halwagi, M.M., Sustainable design through process integration: fundamentals and applications to industrial pollution prevention, resource conservation, and profitability enhancement, Second Edition, 2017: IChemE, Elsevier.*

Synthesis and Characterization of Tin, Tin/Aluminum, and Tin/Boron Containing MFI Zeolites

Meysam Shahami, Ross Ransom, Daniel F. Shantz*
Department of Chemical and Biomolecular Engineering, Tulane University, LA, 70118, USA

Introduction

Zeolites are an intellectually fascinating and technologically important class of materials used in the chemical conversion and separation of small molecules. Their ability to achieve these unit operations is a result of the crystalline microporous structures which enabling highly size-selective separations [1, 2]. Coupling that with the ability to impart acidity to the zeolites through the introduction of trivalent elements such as aluminum and boron into the framework has led to zeolites finding a wide range of applications in the chemical and petrochemical industry [3].
We report here a study of the synthesis and characterization of tin and mixed tin containing MFI materials made in alkaline media. We believe this coupling of a material possessing both Brönsted and Lewis acidity may be of interest for numerous chemistries.

Synthesis of Tin, Tin/Aluminum, Tin/Boron Containing MFI Zeolite

Zeolites were hydrothermally synthesized across a wide composition space. Tin-containing silicalite-1 was synthesized as well as mixed Sn, Al and Sn, B phases. The Si/Sn molar ratio was either 200 or 100. For mixed samples the Si/Sn was fixed at 100 and the Si/X (X: Al, B) was varied. The gel composition was 1 SiO_2: X $SnCl_4.5H_2O$: Y $Al(OH)_3$: Z $B(OH)_3$: a TPAOH: b H_2O, with typical values of X = 0.005 – 0.01, Y = 0.01 – 0.02, Z = 0.01 – 0.02, a = 0.20, and b = 30. Hereinafter, the calcined samples at 550 °C will be denoted as Sn-X-MFI (c, d), where c and d are the molar ratio of Si/Sn and Si/X (X: Al, B).

Characterization

X-ray diffraction patterns of all synthesized samples after calcination are shown in Figure 1. No amorphous or crystalline impurities including SnO_2 are observed. Table 1 summarizes the inductively coupled plasma (ICP) results of the various samples made in this work.

Sample Name	Theoretical wt%		ICP wt%		Uptake (%)	
	Sn	X	Sn	X	Sn	X
Sn-MFI (200)	0.98	-	0.98	-	100	-
Sn-MFI (100)	1.94	-	1.84	-	95.0	-
Sn-Al-MFI (100, 100)	1.93	0.44	1.63	0.32	84.5	73.9
Sn-Al-MFI (100, 50)	1.92	0.87	1.13	0.65	58.8	74.8
Sn-B-MFI (100, 100)	1.93	0.18	1.6	0.05	82.7	30.0
Sn-B-MFI (100, 50)	1.93	0.35	1.69	0.09	87.5	26.5

Table 1. Theoretical wt%, actual wt% and percent uptake of heteroatoms for all samples. (X: Al or B).

Figure 2 shows UV-Vis spectra of the calcined samples as well as SnO_2.
☞ Signal at ~ 200 nm: Tetrahedral coordinated Sn
☞ Weak Shoulder at 232 nm: Penta-coordinated Sn

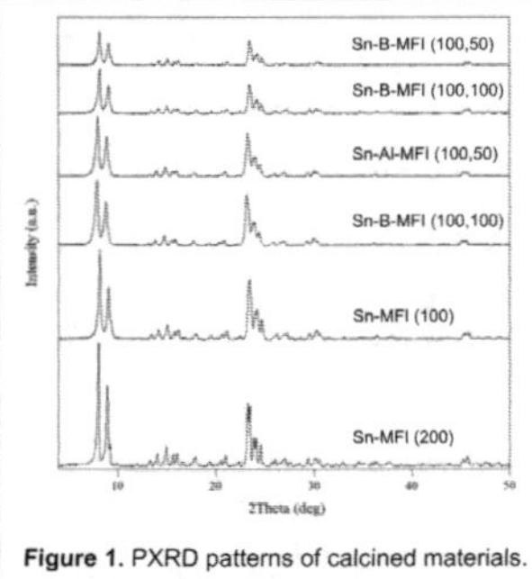

Figure 1. PXRD patterns of calcined materials.

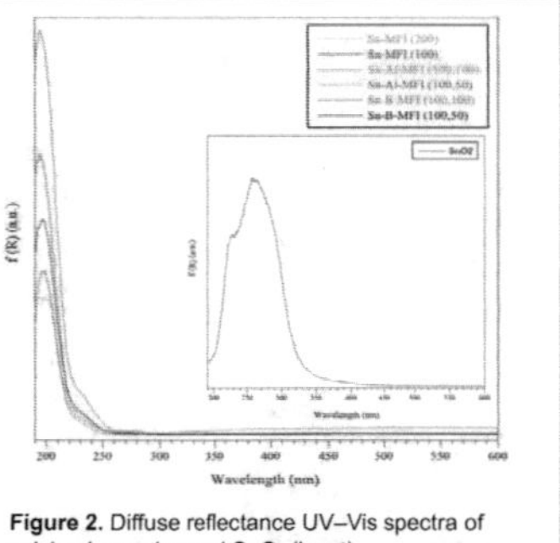

Figure 2. Diffuse reflectance UV–Vis spectra of calcined samples and SnO_2 (inset).

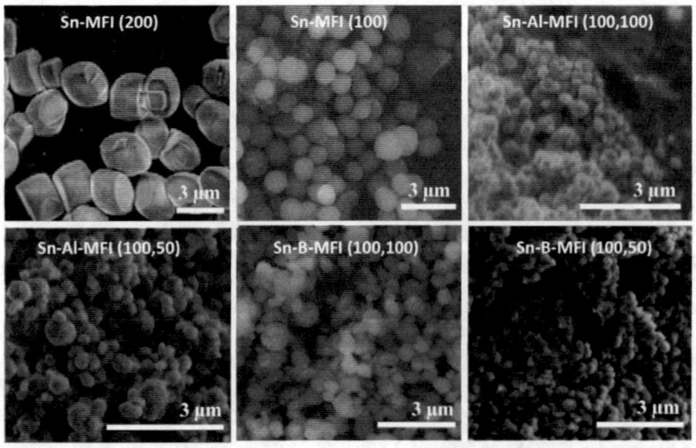

Figure 3. FE-SEM images of calcined Sn-X-MFI samples, (X: Al or B). The scale bar in all images is 3 µm.

FE-SEM images show that addition of more tin into the gel not only changes the morphology but also the size of the crystals. The samples made using boron and aluminum with tin also in general lead to spherical crystals but of smaller size.

In-situ IR spectroscopy of d-acetonitrile adsorbed on the surface of these materials shows Sn has been incorporated tetrahedrally in all of these samples due to presence of a single peak at 2307 cm^{-1}. Moreover, there are not any bands at higher frequencies indicating that hydrolyzed Sn sites do not exist in these sample. The pronounced peak at around 2274 cm^{-1} has not been assigned in the literature. But we believe that this peaks originates from interactions of d-acetonitrile with the silanol groups or might be also due to presence of some internal defects.

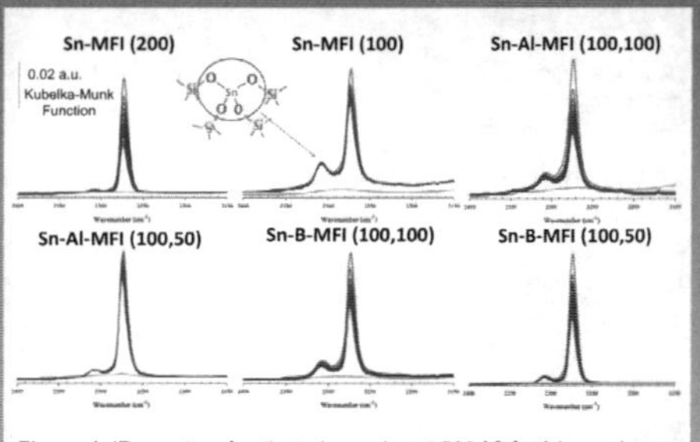

Figure 4. IR spectra of activated samples at 500 °C for 3 hours (green line, spectra obtained at 21 °C and 10^{-5} Torr), after introduction of d-acetonitrile at 21 °C (highest blue line), desorption of d-acetonitrile after 5 minutes (second blue line) up to 120 minutes (rest of blue lines) under the high vacuum (10^{-5} Torr).

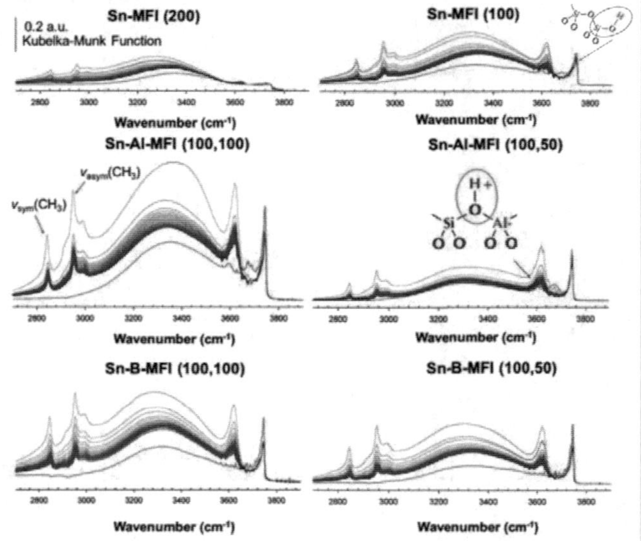

Figure 5. IR spectra of activated samples at 500 °C for 3 hours (red line, spectra obtained at 21 °C and 10^{-5} Torr), after introduction of methanol at 21 °C (highest blue line), desorption of methanol after 5 minutes (second blue line) up to 120 minutes (rest of blue lines) under the high vacuum (10^{-5} Torr).

Also, in-situ IR was performed to study how methanol interacts with the various types of sites in the zeolites. It has been shown that methanol has weak interactions with silanol groups that are not associated with acid sites in Sn-MFI (200). Presence of Brönsted acid sites have been confirmed in Sn-Al-MFI samples. Additionally, methanol does not only interact with silanol groups but also it directly interacts with the boron centers.

Conclusions

A series of Sn, Sn-Al, and Sn-B containing MFI zeolites were successfully synthesized and well characterized. PXRD and FE-SEM showed that introduction of more heteroatoms into the initial gel leads to decreasing the crystal sizes and changing the morphology. Based on the UV-Vis spectra, tin was substituted tetrahedrally into the framework. IR spectra of chemisorbed d-acetonitrile on the surface of these materials also corroborate the UV-Vis results. In-situ IR of adsorbed methanol indicates that the nature of number of acid sites in these materials changes in going from pure Sn-MFI (weakly acidic) to Sn-Al-MFI (mildly acidic).

References

1. Barrer, R.M., Hydrothermal Chemistry of Zeolites. 1982, London: Academic Press Inc.
2. Čejka, J., Introduction to Zeolite Science and Practice. Vol. 168. 2007: Elsevier.
3. Kokotailo, G.T., et al., Structure of Synthetic Zeolite ZSM-5. Nature, 1978. **272**(5652): p. 437-438.

Acknowledgements

The authors acknowledge partial financial support from SABIC.

Tulane University
Chemical & Biomolecular Engineering

Palladium Dendron Encapsulated Nanoparticles Supported on MCM-41 and SBA-15

Southwest Process Technology Conference

Conference: Southwest Process Technology Conference

Year: 2017

Proceeding: 9th AIChE Southwest Process Technology Conference

Group: Southwest Process Technology Conference

Session: Meet The Industry Poster Reception

Time: Friday, October 13, 2017 - 3:00pm-6:00pm

Authors:

Lou, Y., Tulane University
Shantz, D. F., Tulane University

Dendrimer-encapsulated nanoparticles (DENs) have attracted interest since they were first introduced in 1998 [1]. The nanoparticles formed have been investigated for numerous applications including catalysis, such as Heck coupling, allyl alcohol hydrogenation, and electrocatalysis [2]. In this work, ordered mesoporous silica (OMS) was used as supports for dendron growth. In the previous work, melamine-based dendrimers were successfully grafted to OMS supports through the bridge of organosilanes [3]. Here the dendrimers were used as templates to form nanoparticles, with palladium explore initially. Multiple methods were used to validate both the structural integrity of the dendrimers and the nature of the metal nanoparticles formed. The metal complexation and reduction are monitored by UV-Vis spectroscopy and shows that the palladium loading can be varied from 1 wt.% to 3 wt.%. The STEM images show the formation of uniformly sized palladium particles approximately 2 nanometers in size within the OMS pores. In contrast to prior work with dendrimers in solution where G4 or higher generation moieties were needed here it is shown that highly uniform nanoparticles can be formed in small dendron fragments, i.e. G2. The accessibility of both the metal sites and dendron ligand sites were demonstrated by catalytic probe reactions and CO_2 uptake experiments. The TOF values for the Heck reaction vary from 160 to 239 mmol iodobenzene consumed/mmol Pd demonstrating that the encapsulated palladium particles are catalytically active. The initial rate for the Nitroaldol (Henry) reaction decreases from 14.9 to 7.3 mmol aldehyde consumed/g catalyst-h, but nonetheless shows that the amine sites are catalytically active in the palladium-containing samples. Carbon dioxide uptake measurements showed minimal perturbations to the CO_2 uptake after palladium particle formation. These hybrid materials have potential uses in reactions where both the organic and metal center can be used sequentially such as Cascade reactions.

[1] Zhao, M., Sun, L. and Crooks, R. M., J. Am. Chem. Soc. 120, 4877 (1998).

[2] Myers, V. S., Weir, M. G., Carino, E. V., Yancey, D. F., Pande, S. and Crooks, R. M., Chemical Science 2, 1632 (2011).

[3] Acosta, E. J., Carr, C. S., Simanek, E. E. and Shantz, D. F., Advanced Materials 16, 985 (2004).

Thermodynamic Interactions in Model Polydiene/Polyolefin Blends

AIChE
Southwest Process Technology Conference

Conference: Southwest Process Technology Conference

Year: 2017

Proceeding: 9th AIChE Southwest Process Technology Conference

Group: Southwest Process Technology Conference

Session: Meet The Industry Poster Reception

Time: Friday, October 13, 2017 - 3:00pm-6:00pm

Authors:

Qiu, J., University of Houston
Mongcopa, K. I. S., University of Houston
Han, R.
López-Barrón, C. R., University of Delaware
Robertson, M., University of Houston
Krishnamoorti, R., University of Houston

Polydiene/polyolefin blends have been used in commercialized elastomer manufacturing for a while and are model system to study thermodynamics. But the previous study mostly focused on polydiene/polydiene or polyolefin/polyolefin blends. There is little quantitative information available for polydiene/polyolefin blends. Most of the polymer blends are phase separated. Upon phase separation, the properties and morphologies are inter-related. To fully understand properties of a given system, itâs essential to predict its phase behavior. The phase behavior is related to the free energy of mixing which can be computed through Flory-Huggins interaction parameter Ï.Ï is used to describe the interactions between polymers and is of crucial importance in determining the processing conditions for polymer mixtures and block copolymers. We have chosen the 1,2-polybutadiene (1,2-PBD) based system, which provides multiple attractive features: 1,2-PBD is an amorphous polymer and doesnât crystalize. Therefore we have a wide temperature range over which can be characterized, and we employed anionic polymerization techniques, allowing facile control over 1,2 content and producing polymers of low polydispersity (<1.1). It is possible to achieve very high 1,2 content (>99%) and therefore we avoid the complication of a random copolymer system. The temperature dependence of Ï between 1,2-PBD and saturated 1,2-PBD was investigated. The synthesized 1,2-PBD was saturated with deuterium to provide contrast for small-angle neutron scattering (SANS). SANS technique enables isotope labeling to provide contrast without changing structure drastically. Two series of blends were prepared with different molecular weight. Values of Ï were extracted from fitting the Random Phase Approximation (RPA) to SANS data. Additionally, Ï was extracted by Zimm analysis, using the low-angle scattering intensity. Ï extracted by RPA and Zimm analysis were in good agreement. The temperature dependencies of Ï characterized for two separate blends of different molecular weight were consistent with one another. The measured Ï (1/T) behavior of our system were compared to literature values of other systems. At low temperatures, Ï of the 1,2- PBD system approaches that observed in a blend of polyisoprene (PI) and saturated PI of high 1,4-content. Ï of the 1,2-PBD system decreases more rapidly at high temperatures than that of the 1,4-PI system. At low temperatures, the Ï parameters for both the 1,2-PBD and 1,4-PI systems are significantly higher than that measured for a blend of 1,4-PI and saturated 1,2-PBD. The strong temperature dependence of our system indicates its highly tunable properties. Our measured Ï(1/T) behavior provided quantitative information for processing.

Preparation of PVMS-b-PMMA Block Copolymer with Potential for Oil-Water Separation

Southwest Process Technology Conference

Conference: Southwest Process Technology Conference

Year: 2017

Proceeding: 9th AIChE Southwest Process Technology Conference

Group: Southwest Process Technology Conference

Session: Meet The Industry Poster Reception

Time: Friday, October 13, 2017 - 3:00pm-6:00pm

Authors:

Lwoya, B. S., Tulane University
Uddin, M. F.
Chattejree, S.
Albert, J., Vector-Borne Infectious Diseases Research Center

Several years ago, the oil spill in the Gulf of Mexico resulted in great degradation of the environment. Additionally, it opened up discussion on effective methods of purifying oil and water in case of a similar incident occurrence. Membrane technology serves this purpose, and development of effective membranes includes careful material selection to achieve certain properties such as: (1) high-throughput of either oil/water, (2) prevention of fouling (3) great selectivity (4) mechanical strength. We herein report synthesis of poly(vinylmethylsiloxane-block-methyl methacrylate) (PVMS-b-PMMA) by a combination of anionic synthesis of PVMS, ATRP of PMMA, and then co-joining the end-group functionalized blocks with a click reaction. In using this combination of methods, we are able to effectively control both molecular weight and polydispersity in polymers that were otherwise challenging to synthesize by sequential block addition. Following thin film nanostructure alignment by solvent annealing, exposure to UV light cross-links the siloxane copolymer block, which imparts mechanical integrity, and degrades the sacrificial PMMA block in a single step to produce a nanoporous film. Morphology of the block copolymer was studied by both atomic force microscopy and scanning electron microscopy. The hydrophobicity and nanoporosity of these films are desirable characteristics for their application as nanoporous membranes for oil/water separation.

Process Intensification Using Reactive Distillation: Conversion of Crude Glycerol to Propanediol

Southwest Process Technology Conference

Conference: Southwest Process Technology Conference

Year: 2017

Proceeding: 9th AIChE Southwest Process Technology Conference

Group: Southwest Process Technology Conference

Session: Meet The Industry Poster Reception

Time: Friday, October 13, 2017 - 3:00pm-6:00pm

Authors:

Agbroko, O., Lamar University
Benson, T. J., Lamar University

Combining reaction engineering and distillation into a single unit operation presents some interesting challenges, yet, reduces the operational footprint as well as the operating and capital costs of manufacturing. Operational and design challenges include mismatched temperatures between the separating components and the reacting components, unfavorable volatilities, and slow reaction rates generally due to poor reactant/catalyst/product mass transfer rates. Reactive distillation is a likely pathway towards process intensification to minimize overall energy consumption, particularly since distillation accounts for 20 % of total energy used within the U.S. manufacturing sector. Increased reflux ratios typically yield higher product purities but at the cost of reboiler heat duties.

In this work, AspenPlus was used to study the feasibility of utilizing a Reactive Distillation column for the catalytic hydrogenolysis of glycerol using isopropanol as a donor molecule for hydrogen. In short, isopropanol was converted to acetic acid, yielding a hydrogen atom that would be available for the conversion of glycerol to 1,2-propanediol and 1,3-propanediol. To perform the simulation, a series of preliminary tests were performed using a CSTR to generate initial parameters for the reaction and a shortcut column to generate initial distillation column operating conditions. Finally, the column design was implemented with the RADFRAC model employing a partial condenser. Important operating parameters such as column pressure, Distillate to feed ratio, reflux ratio, feed temperature, feed flowrate, feed stage, reaction zone, changes in feed composition molar ratio and liquid holdup where analyzed for optimum values to see how the system responds to changes in these variables. Using the Non-Random Two Liquid activity coefficient model, the number of trays, conversion, and product selectivities were estimated. It was found that the column was energy efficient with a reflux ratio value of 0.42 , condenser temperature of 62 ^{o}C with a boilup ratio of 2.7 which corresponds to a reboiler temperature of 209 ^{o}C for a glycerol conversion of 96 %. While this work is focused at providing economic avenues for the biodiesel industry, it can also be extended to other industrially-relevant processes (i.e. mild hydrodesulfurization of crude oils, manufacturing of surfactants and detergents, and lubricants).

Shale Gas Monetization: Designing Separation Units for Feedstock Variability

Southwest Process Technology Conference

Conference: Southwest Process Technology Conference

Year: 2017

Proceeding: 9th AIChE Southwest Process Technology Conference

Group: Southwest Process Technology Conference

Session: Meet The Industry Poster Reception

Time: Friday, October 13, 2017 - 3:00pm-6:00pm

Authors:

Bohac, E., Texas A&M University

Shale gas is abundant in the U.S. and this presents many opportunities in the chemical, petrochemical, and fuel industries. As with conventional natural gas, shale gas is primarily methane, but also may contain significant compositions of higher molecular weight hydrocarbons, and other impurities. This variability presents a technical challenge to processing the gas. This research is aimed at examining the economic tradeoff between treating and not treating gas of differing compositions.

In this work, the composition of the raw shale gas is based on literature data for wells in the Barnett Shale play located northwest of Dallas. Process simulators such as ProMax and ASPEN Plus were used to construct base-case scenarios, and to carry out process simulations for differing compositions with any necessary process modifications. A techno-economic analysis was done to estimate the associated costs and profitability of all the scenarios. The goal was to see if the additional revenue generated by treating gas of differing compositions is justified by the necessary process modifications.

Rational Design of Block-Copolymer Additives for Polymer/Small-molecule Emulsions using Experimental and Computational Approaches

Southwest Process Technology Conference

Conference: Southwest Process Technology Conference

Year: 2017

Proceeding: 9th AIChE Southwest Process Technology Conference

Group: Southwest Process Technology Conference

Session: Meet The Industry Poster Reception

Time: Friday, October 13, 2017 - 3:00pm-6:00pm

Authors:

Mok, J., Rice University
Verduzco, R., Rice University

Polymer Solar Cells (PSCs) have emerged as a promising technology for harvesting renewable energy from sunlight. Despite tremendous advances, the performance of PSCs is still poor compared to traditional silicon-based solar cells. Many studies have shown PSCs performance strongly depends on the morphology and the long-term thermal stability of polymer/small-molecule emulsions. Here, I implement computational and experimental methods to design and synthesize block-copolymers (BCPs) additives to intelligently control polymer emulsions. First, I use surface energy measurements along with Monte Carlo simulations to screen potential chemical building blocks to be used in the formulation of the block-copolymer. Then, I fabricate the block-copolymers based on the simulation results. The addition block-copolymers improved the performance of the solar cells in more than 100% due to significant changes in phase behavior and the thermal stability of polymer emulsion. These changes were predicted by computer simulations and confirmed by analytical characterization such as electron microscopy and X-ray diffraction. In summary, this works demonstrates that BCPs can be utilized as additives to control the morphology and improve thermal stability of Polymer Solar Cells.

Smart Manufacturing in Distillation

Southwest Process Technology Conference

Conference: Southwest Process Technology Conference

Year: 2017

Proceeding: 9th AIChE Southwest Process Technology Conference

Group: Southwest Process Technology Conference

Session: Meet The Industry Poster Reception

Time: Friday, October 13, 2017 - 3:00pm-6:00pm

Authors:

Efetobore, D., Prairie View A&M University

Smart manufacturing is the next revolution of manufacturing in the chemical process industry which will provide the platform for the next generation of production in chemical manufacturing process. The chemical plant of the future is where self-aware process equipment work in concert, wirelessly communicating their status with each other to produce the highest quality product in the least amount of time with near zero incidents, emissions, and waste. The plant is intimately connected with upstream and downstream processes such that it can respond in real time to market shifts, customer demand, global economics, and socio economic activities. From the economic perspective of revenue loss due to unscheduled plant shutdown and maintenance, the goal of smart manufacturing is to create knowledge-embedded chemical manufacturing operations that would enhance a facilityâs ability to predict problems and act proactively to prevent problems before they occur (Humphrey, 2007). Also, process improvement, productivity enhancements, conservation in mass and energy resources, reductions in the operating and capital costs of chemical processes are some of the unique advantages of the smart manufacturing application (Dunn and El-Halwagi, 2003). The idea is to transform the existing paradigm of todayâs chemical manufacturing process operations from reactive to proactive, response to prevention, compliance to performance oriented, tactical to strategic, and local to global. Although the cost of the technology infrastructure is capital intensive, the return on investment will outweigh the cost. This paper provides an overview of some of the developments, major driving forces and hurdles in distillation, being the most widely used unit operations technique in the chemical process industry to explore smart distillation plant so as to elucidate smart manufacturing applications in the chemical process industry. The fundamental aspects of this approach along with their incorporation in the overall methodology will be discussed. The key role of optimization in smart distillation plant which is crucial to profitability is discussed. An understanding of the common operating problems in distillation plant along with timely corrective actions achieved through the advanced process control will enable operating personnel take appropriate decision. There is still knowledge gap in the area of achieving real time material and energy balance determination in smart distillation plant. Future research directions are recommended and extensive lists of references are provided.

Tight Oil Drilling Problems and Solutions

Southwest Process Technology Conference

Conference: Southwest Process Technology Conference

Year: 2017

Proceeding: 9th AIChE Southwest Process Technology Conference

Group: Southwest Process Technology Conference

Session: Meet The Industry Poster Reception

Time: Friday, October 13, 2017 - 3:00pm-6:00pm

Authors:

Fernando, L., Prairie View A&M University

Crude oil, also known as petroleum or fossil fuel is found in some rock formation deep below the earthâs surface. Tight oil is a unconventional oil that is found in reservoirs with very low permeability. The oil contained within these reservoir rocks will not flow to the wellbore at economical rates without assistance from technologically advanced drilling and completion process (Canadian Society for Unconventional resources). For tight oil reservoirs, matrix permeability plays a key role in determining drainage area for each well and the optimal well spacing. In this work, steady-state method to measure liquid permeability on intact core plugs from tight oil bearing rock under reservoir stress condition will be presented.

Techniques for quantitative analysis of ultra-low permeability reservoirs completed with hydraulic fracturing and horizontal wells are in their infancy. Complications that have shown the development of these techniques include complex reservoir porosity such as dual porosity effect, multi-layer behavior, multiphase flow and non-static absolute permeability and complex flux behavior associated with the simultaneous flow of multiple hydraulic fracturing into horizontal wells. Drilling problems to access the wellbore and solutions will be discussed.

Investigating the application of enhanced oil recovery method is essential. Carbon dioxide (CO_2) injection as a huff-n-puff process is a preferred approach to improve oil recovery in tight reservoirs. The effect of molecular diffusion will be investigated. A series of sensitive studies to quantify the impacts of reservoirs properties such as permeability and fracture properties (fracture half-length, fracture conductivity, the number of fractures, and operation parameters) will be performed. The operation parameters such as CO_2 injection rate, injection time, soaking time, the number of cycle CO_2 diffusivity on CO_2 huff-n-puff process for enhanced oil recovery in the Bakken Formation will be discussed.

Time-resolved in situ Studies of Zeolite Crystal Growth

Conference: Southwest Process Technology Conference

Year: 2017

Proceeding: 9th AIChE Southwest Process Technology Conference

Group: Southwest Process Technology Conference

Session: Meet The Industry Poster Reception

Time: Friday, October 13, 2017 - 3:00pm-6:00pm

Authors:

Choudhary, M. K., Iowa State University
Kumar, M., University of Houston
Rimer, J., University of Houston

Zeolites are widely used in commercial processes spanning from ion exchange in detergents to catalysis in the (petro)chemical industry. Understanding zeolite growth mechanisms at the molecular level aids the a priori selection of synthesis parameters to tailor the physicochemical properties of zeolites. Despite tremendous effort to elucidate the mechanisms of nucleation and crystal growth, these pathways in zeolite synthesis are not well understood. This is due in large part to the inherent complexity of zeolite crystallization and the synthesis conditions (i.e., high pH, high temperature, etc.) that render in situ characterization challenging. Recently, our group developed a system to carry out solvothermal in situ atomic force microscopy (AFM) wherein we can observe zeolite surfaces at near molecular resolution under realistic growth conditions.[1] Here, we will present in situ AFM measurements of silicalite-1 (MFI) crystallization using zeolite growth modifiers (ZGMs) to modulate the shape of zeolite crystals. ZGMs are molecules or macromolecules that selectively bind to zeolite crystal surfaces and mediate the anisotropic rate(s) of growth to achieve desired crystal size and morphology. We have reported the efficacy of ZGMs in silicalite-1 bulk crystallization experiments.[2] Here, we will discuss in situ AFM studies of ZGM effects on silicalite-1 crystals wherein we observe differences in the relative rates of growth by two distinct pathways: classical processes involving molecule addition and nonclassical pathways involving the attachment of amorphous nanoparticle precursors. We show that these pathways can be influenced by the presence of ZGMs.

We will also present in situ AFM results of industrially relevant aluminum-containing zeolite LTA where we observe distinct growth regimes as a function of supersaturation. At high supersaturation, we observe the three-dimensional assembly and structural evolution of gel-like islands on LTA crystals leading to a mechanism that differs from kinetic roughening in classical theory. We also report a distinct switch in the growth mode at low supersaturation marked by two-dimensional nucleation of single layers with step heights corresponding to the composite building units of the crystal structure.

(1) Lupulescu, A. I.; Rimer, J. D.; Science 344 (2014) 729-732.

(2) Lupulescu, A. I.; Rimer, J. D.; Angew. Chemie Int. Ed. 51 (2012) 3345â3349.

Understanding Pyrolysis Chemistry to Improve Biomass-to-Chemical Process

Southwest Process Technology Conference

Conference: Southwest Process Technology Conference

Year: 2017

Proceeding: 9th AIChE Southwest Process Technology Conference

Group: Southwest Process Technology Conference

Session: Meet The Industry Poster Reception

Time: Friday, October 13, 2017 - 3:00pm-6:00pm

Authors:

Chen, L., Rice University

The selective production of platform chemicals from thermal conversion of biomass-derived carbohydrates is challenging. Anhydrosugars, such as levoglucosan, are recognized as promising platform chemicals to produce fuel and chemicals (e.g., biologically active compounds, surfactants and polymers). Most of their chemical applications require the use of nearly pure crystalline compounds. However, anhydrosugars are very difficult to synthesize in large quantities and without side products, due to the nonselective nature of carbohydrate dehydration during pyrolysis. This leads to high production costs and currently limits the availability and use of anhydrosugar industrially.

Researchers have studied anhydrosugar production since early 1900, but until now, have not provided any great improvements in this field because no one understands why the selectivity was low for anhydrosugar production on the molecular level.

To address this problem, I developed two new methods to enhance anhydrosugar selectivity: (1) âex-situ ring-lockingâ strategy: protecting the carbohydrate compounds before pyrolysis by alkoxy or aryloxy substitution at the anomeric carbon prior to the thermal treatment; or (2) âin-situ ring-lockingâ strategy: we eliminated the initial ex-situ ring-locking step by pretreating the carbohydrate substrate with an acid catalyst and an alkali salt. Both approaches can effectively suppress the formation of undesirable side products, and therefore significantly enhance the yield of the target product such as levoglucosan.

Furthermore, we received a grant from National Science Foundation Innovation-Corps (NSF I-Corps) to explore the commercialization potential for this patent-pending technology. We discovered a $300M market for our product after conducting 100+ customer interviews across the pharmaceutical, biochemical and the oil field chemical industries.

Since we have demonstrated our chemistry in a micropyrolyzer system, and validated the commercial potential of our product, we moved to next stage to validate the technology feasibility in larger scale. We studied the interplay of reaction and transport within biomass particle during fast pyrolysis. First, we developed a particle level model comprising of kinetics, energy and mass conservation equations. Through characteristic timescale analysis by using the non-dimenisonal numbers like Biot, Pyrolysis, Peclet, Damkohler and Lewis number, we categorize the operating conditions into three different regimes based on the dominant phenomena: chemical regime, mass transfer regime and thermal regime. We compared our experiments results with this map to identify the dominant phenomena for levoglucosanâs yield under different operating conditions.

Viscosity Modeling of Live Oil and Viscosity Prediction of Gas Injection Blends

AIChE
Southwest Process Technology Conference

Conference: Southwest Process Technology Conference

Year: 2017

Proceeding: 9th AIChE Southwest Process Technology Conference

Group: Southwest Process Technology Conference

Session: Meet The Industry Poster Reception

Time: Friday, October 13, 2017 - 3:00pm-6:00pm

Authors:

Zhang, J., Rice University
Abutaqiya, M. I. L., Rice University
Vargas, F., Rice University

Viscosity reduction is a key benefit of gas injection for enhanced oil recovery. This study aimed to predict the viscosity of the reservoir fluid and injected gas blends based on Peng Robinson equation of state and friction theory. Carbon dioxide gas injection, lean gas injection and rich gas injection results are reported in this work. This study found that SARA based characterization method gave more accurate modeling results compared to that from Chi-Square distribution characterization method for live oil modeling. The capability to accurately predict viscosity of the gas injection blends was also demonstrated.

AUTHOR INDEX

AUTHOR INDEX

AIChE
120 Wall Street, FL 23
New York, NY 10005-4020

ISBN 978-1-5108-5674-5

9 781510 856745

CURRAN ASSOCIATES INC.
proceedings.com